TOTAL QUALITY ASSURANCE
SECOND EDITION

by
WILBUR A. GOULD, Ph.D.
Food Industries Consultant
Emeritus Professor of Food Processing & Technology,
Department of Horticulture,
Ohio State University,
Ohio Agricultural Research & Development Center,
Former Director, Food Industries Center,
The Ohio State University and
Executive Director Mid-America Food Processors Association,
Worthington, Ohio

and

RONALD W. GOULD
Statistical Process Control Manager
General Mills, Inc.,
Toledo, Ohio

CTI Publications, Inc.
2 Oakway Road, Timonium, Maryland 21093-4247 USA
Voice 410-308-2080   FAX 410-308-2079

All rights reserved. No part of this book may be reproduced or altered or utilized in any form or by any means, graphic, electronic or mechanical, including photocopying, recording or by any information storage and retrieval system, without permission in writing from the copyright owner. Inquiries should be addressed to:

CTI Publications, Inc.
2 Oakway Road, Timonium, Maryland 21093-4247 USA
Voice 410-308-2080   FAX 410-308-2079

©COPYRIGHT 1993 by CTI Publications, Inc.
Baltimore, Maryland

Printed in the United States of America

ISBN Numbers are as follows:
0-930027-20-5

Library of Congress Cataloging – in – Publication Data

Gould, Wilbur A., 1920–
 Total quality assurance for the food industries / by Wilbur A. Gould and Ronald W. Gould. --2nd Edition.
  p. cm.
 Includes index.
 ISBN 0-930027-20-5 (hard)
 1. Food industry and trade – Quality control.
I. Gould, Ronald W. II. Title.
TP372.6.G68   1993   92-37494
664'.07--dc20   CIP

While the recommendendations in this publication are based on scientific studies and wide industry experience, references to basic principles, operating procedures and methods, or types of instruments and equipment are not to be construed as a guarantee that they are sufficient to prevent damage, spoilage, loss, accidents or injuries, resulting from use of this information. Furthermore, the study and use of this publication by any person or company is not to be considered as assurance that a person or company is proficient in the operations and procedures discussed in this publication. The use of the statements, recommendations, or suggestions contained, herein, is not be considered as creating any responsibility for damage, spoilage, loss., accident or injury, resulting from such use.

CTI Publications, Inc.
2 Oakway Road, Timonium, Maryland 21093-4247 USA
Voice 410-308-2080   FAX 410-308-2079

ABOUT THE AUTHOR

Wilbur A. Gould was reared on a farm in Northern New Hampshire. He received his Batchelor of Science degree from the University of New Hampshire in Horticulture-Plant Breeding. He started his graduate work at Michigan State University prior to service in the U.S. Navy during World War II. After military service, he completed his Master of Science and Ph. D. degrees at The Ohio State University.

Dr. Gould retired from The Ohio State University after 39 years on the faculty as Professor of Food Processing and Technology. He taught 9 courses during his tenure and advised over 900 undergraduate students, 131 Master of Science Students and 76 Doctoral students. His major research interests were in Vegetable Processing and Technology and Snack Food Manufacture and Quality Assurance. He has authored some 83 referred journal research publications, over 200 Food Trade articles, and 10 books.

Dr. Gould is a Member of Phi Kappa Phi, Phi Sigma, Phi Tau Sigma, Sigma Xi, Gamma Sigma Delta (Award of Merit in 1984), Alpha Gamma Rho, Institute of Food Technologists (Fellow in 1982), and American Society of Horticultural Science (Distinguished Graduate Teaching Award in 1985).

The following are some of the recognitions that Dr. Gould has received: The Ohio State University Distinguished Leadership to Students Award in 1963 and a Certificate of Recognition Award in 1986; Ohio Food Processors H.D. Brown Person of Year Award in 1971; Ohio Food Processors Association Tomato Achievement Award in 1985; Ozark Food Processors Association Outstanding Professional Leadership Award in 1978; 49er's Service Award in 1979; Food Processing Machinery and Supplies Association Leadership and Service Award in 1988; Ohio Agricultural Hall of Fame in 1989 and an Honorary Life Membership in Potato Association of America in 1990.

Dr. Gould presently serves as Executive Director of Mid-America Food Processors Association, Food Technology Consultant to the Snack Food Association, Secretary-Treasurer of The Guard Society, and Consultant to the Food Industries.

Dr. Gould's philosophy is to tell it as he sees it, be short and get right to the point.

This copy of
**Total Quality Assurance
For The Food Industries**
belongs to:

Other Titles From CTI Publications

FOOD PRODUCTION/MANAGEMENT editorially serves those in the Canning, Glasspacking, Freezing and Aseptic Packaged Food Industries.
 Editorial topics cover the range of Basic Management Policies, from the growing of the Raw Products through Processing, Production and Distribution for the following products: fruits; vegetables; dried and dehydrated fruit (including vegetables and soup mixes); juices, preserves; pickles and pickled products; sauces and salad dressings; catsup and tomato products; soups; cured fish and seafood, baby foods; seasonings and other specialty items. (Monthly Magazine). ISSN: 0191-6181

A COMPLETE COURSE IN CANNING, 12th edition, are technical reference and textbooks for Students of Food Technology; Food Plant Managers; Products Research and Development Specialists; Food Equipment Manufacturers and Salesmen; Brokers; and Food Industry Suppliers. The three books total 1,300 pages. ISBN: 0-930027-00-0.

CURRENT GOOD MANUFACTURING PRACTICES/FOOD PLANT SANITATION covers all Current Food Manufacturing practices as prescribed by the United States Department of Agriculture, Food and Drug Administration, as it applies to food processing and manufacturing. A total of 21 chapters, covering all phases of sanitation. ISBN: 0-930027-15-9

GLOSSARY FOR THE FOOD INDUSTRIES is a definitive list of food abbreviations, terms, terminologies and acronyms. Also included are 20 handy reference tables and charts for the food industry. ISBN: 0-930027-16-7.

RESEARCH & DEVELOPMENT GUIDELINES FOR THE FOOD INDUSTRIES is a compilation of all Research and Development principles and objectives. Easily understood by the student or the professional this text is a practical "How To Do It and Why To Do It" reference. ISBN: 0-930027-17-5.

TOMATO PRODUCTION, PROCESSING & TECHNOLOGY, 3rd edition, is a book needed by all tomato and tomato products packers, growers, or anyone involved or interested in packing, processing, and production of tomatoes and tomato products. ISBN: 0-930027-18-3.

TOTAL QUALITY MANAGEMENT FOR THE FOOD INDUSTRIES is a complete interactive instruction book, easily followed, yet technically complete for the advanced Food Manager. TQM is the answer to guide a food firm, its people, its quality of products, and improve its productivity. It's the right step to achieve excellence and the development of satisfied customers, as well as build your bottom line. ISBN: 0-930027-20-5.

For a brochure or further information on the above publications please contact:

CTI Publications, Inc.
2 Oakway Road, Timonium, Maryland 21093-4247 USA
Voice 410-308-2080   FAX 410-308-2079

THIS BOOK IS DEDICATED TO JESSIE FOR HER SYMPATHETIC UNDERSTANDING, CONSTANT LOVE AND WHOLEHEARTED SUPPORT.

PREFACE
First Edition

The quality of the foods we eat is of prime importance to each of us. The food industry has an obligation to produce uniform quality, nutritious, and safe foods. The responsibility rests with management and the food plant worker to produce, process and pack safe, wholesome, and nutritious quality foods. The quality assurance technologists role is to assure management and the ultimate consumer that the quality of the food meets the firm's expectation. Only through a total quality assurance program can a food firm expect to grow and provide a satisfactory return on the investment for the stockholders. Total quality assurance in a food firm works for the consumer, the food plant worker, and the firm's management. It is a necessary part of the successful operation of a food firm today.

This text is divided into two major parts. The first part sets forth the basic principles of total quality assurance for management and the ultimate employee. These principles are necessary for the successful operation of a food firm in these times. The second part of the text describes the various attributes and characteristics of food product quality and quality evaluation methods. Examples are given for the evaluation of a wide array of food products. The methods and procedures described in this text have been applied to most situations for control, evaluation and auditing of the quality of foods.

The authors are deeply indebted to former students now employed in various segments of the food industries for their suggestions and help in the formating of this text. Further, we are indebted to our colleagues for their constructive help and cooperation. We particularly thank Art Judge, and Nancy and Randy Gerstmyer for their support, interest and wholehearted cooperation.

June 1988

Ronald W. and Wilbur A. Gould

TOTAL QUALITY ASSURANCE

Chapter 21. Maturity/Character and Total Solids/Moisture (Specific Gravity) 295
Chapter 22. Texture-Tenderness, Crispness, Firmness............ 313
Chapter 23. Rheology—Viscosity—Consistency 325
Chapter 24. Defects—Imperfections or Appearance 345
Chapter 25. Drosophila and Insect Control..................... 357
Chapter 26. Enzyme Activity 369
Chapter 27. Alcohol in Foods and Beverages.................... 379
Chapter 28. Fats and Oils...................................... 385
Chapter 29. Total Acidity and pH.............................. 397
Chapter 30. Mold–Counting Methods and Principles 411
Chapter 31. Water Activity..................................... 437
Appendix I — Temperature Conversion Tables................. 443
Appendix II — Weights and Measures......................... 447
Suggested Basic Reference Texts and Journals.................. 449
Subject Index 451
Index to Tests, Tables, Figures and QC Data Forms............. 458

TOTAL QUALITY ASSURANCE FOR THE FOOD INDUSTRIES
Second Edition

Table Of Contents

PART I — FUNDAMENTALS
Chapter   1. Introduction to Total Quality Assurance............   3
Chapter   2. Modern Concepts of Quality Assurance   9
Chapter   3. Quality Control and Use of CEDAC 35
Chapter   4. Productivity and Total Quality Assurance 41
Chapter   5. Sampling for Product Evaluation and Line Control... 47
Chapter   6. Statistical Quality and Process Control............. 61
Chapter   7. Variable — Raw Materials and Ingredients 87
Chapter   8. Variable — Machines.............................. 91
Chapter   9. Variable — Methods 93
Chapter 10. Variable — People Power.......................... 95
Chapter 11. Food Plant Sanitation 101
Chapter 12. Fundamentals of Research and Development 113
Chapter 13. Communicating Quality and the Cost of Quality 121
Chapter 14. Specifications and Quality Standards 127

PART II PRODUCT EVALUATION
Chapter 15. Packaging & Container Integrity Evaluation.......... 137
Chapter 16. Sugar, Salt and Seasonings 161
Chapter 17. Flavor.. 189
Chapter 18. Odor... 215
Chapter 19. Physical Evaluation of Color 235
Chapter 20. Size, Shape, Symmetry, and Style 279

PREFACE
Second Edition

Changes in the food industries are constantly taking place, particularly in the field of Quality Assurance. Many new practices, procedures and concepts are coming to the forefront. Some of these are finding their rightful place for the control of quality on the production line. On-line sensors are finding wide application for line control and in the long run will eliminate much of the laborious laboratory work.

Perhaps the most important change since the first publication of the fore runner to this book was first published back in 1977 is the modern emphasis on training the line operator-employee to control his unit operation as part of the process. By empowering the employee to control his operation to given specifications or parameters, quality of food products should be more uniform and consistent in quality. The employee must be held accountable and he should be made part of a team to constantly improve the process. Only then will food firms move forward with assured confidence that their products are meeting the expectations of the customer all the time.

The authors are deeply indebted to the many readers of this text for their suggestions and interest in this field of quality assurance. The methods and procedures in this edition have been up-graded where necessary and new chapters have been prepared to help the reader gain more insight into this field.

Quality assurance in the food industries is an area of work that offers many opportunities for great achievement for the food employee and his or her firm. With the use of computers and optimization of production lines, the future looks very bright for those food employees who understand process and product control and food product evaluation.

We are, also, deeply indebted to all the personnel at CTI Publications for their constant encouragement, interest, and cooperation. We particularly thank Art Judge II and Randall Gerstmyer for their many considerations and dedicated help.

Ronald W. and Wilbur A. Gould

Part I —

Fundamentals of Quality Assurance and Process Control

CHAPTER 1

Introduction — Total Quality Assurance in the Food Industries

Total quality assurance in the food industries is a requirement for consumer enthusiasm and acceptance. Total quality assurance means processs and products are acceptable and in conformance to requirements. Total quality assurance must start with top management supporting the concept of quality and their taking time to explain to all personnel the need for manufacture and control of product quality. Total quality assurance must provide better job instructions for all employees as some people working in a food system may not know good practices from bad practices. Most importantly for a firm to manage quality of production, management must take the time to train all personnel in the concept of statistical techniques and the application of statistical practices to the production line to help solve problems of reproducing quality products.

One way to obtain an insight into the food industry is to look at the organization plan of a typical food firm as shown in Figure 1.1. A modern food plant is organized around the M's and the first M of the food industry

FIGURE 1.1 — Organizational Plan For A Food Firm

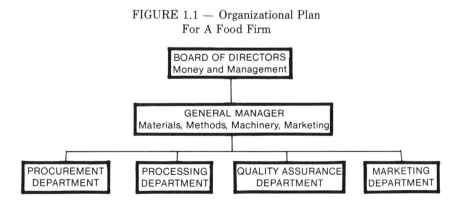

is Management. Management is responsible to the Board of Directors to see that the firm returns a profit on the invested capital. To return a profit, management must fully utilize the resources of Materials, Machines, Manpower, Methods, Money and the Managers of the respective departments in the plant. Perhaps the most important M is Manpower-womanpower and their achievement in producing a quality product at a profit.

Generally, workers in a food plant are a mirror image of the management. They will do their jobs right if given the proper tools, environment, knowledge, and direction or message. The environment is the most critical of the causes or lack of employees doing their job right. The environment is often called the "common cause" of production problems. "Common causes" are considered the faults of the system and can only be removed by management. Generally they represent up to 85% of the faults of the system. Some typical "common causes" are:
1. Poor supervision
2. Lack of instructions
3. Varying quality of incoming materials
4. Machines out of control
5. Uncomfortable working conditions (humidity, noise, confusion, too hot or too cold, poor ventilation, poor light, dirty, etc.)
6. Lack of statistical information or performance data
7. Poor design of process.

All of these are faults of the system and can only be corrected by management.

Total quality assurance implies the establishment of goals of quality improvement and the analysis of the costs associated with non-conformance of products and processes to established quality levels. The evidence of non-conformance must be conveyed to all personnel through newsletters, video-tapes, personal contacts, statistical charts, and in open forums.

The first requirement in establishing a total quality assurance system is for the Manager of the Quality Assurance program to thoroughly convince the management about the system so that there is excitement and enthusiasm for the program. The enthusiasm and the excitement must go all the way up and down the line and to each individual employee in the firm.

The function or job of the plant manager is to give consistency of purpose and continuity to the organization. He is solely responsible to see that the firm has a future. Workers work in the system, but the manager works on the system. He sees to it that it produces the highest quality product at the lowest possible cost. No one else is responsible for the system as a whole for its continued improvement. A successful manager relies on his staff to help make the many decisions as each staff member has a specialized knowledge and interest. Further, each one can focus on their area of expertise. All on the staff must be committed, visibly involved and they must project a strong leadership attitude. They must be truthful with each employee, be consistent with their facts, and be confident when dealing with others. For success, they must use "we" instead of "I". Most importantly, a good manager must remember where he came from. The 15 key traits to look for in a Manager are as follows:

1. Positive attitude
2. Initiative-willingness to dig in and get started
3. Ambitious-always broadening view, developing new skills, and willingness to take a risk
4. Self confident-a competitor, one who gets the job done
5. Courage and willingness to train a successor
6. Flexible-not set in his ways
7. Resilient-ability to bounce back
8. Stamina and mental attitude to cope with endless streams of stress
9. Ability to judge people and how to develop people
10. Goal setter-long range plans including budgets and deadlines
11. Collaborator
12. Imaginative
13. Creative
14. Objective
15. Stable with great self discipline

In the real world the only people who make the economic detailed observations of the system on a day-by-day basis are the workers. Thus, they should be part of the decision process as they are the ones who observe the system and the ones that best know its behaviour.

The performance of the worker is directly related to how the system operates. Problems within the system are usually first observed by the workers. If they complain about poor maintenance and have the statistical data and facts to back up their complaints; good managers do not consider them troublemakers; but, welcome their contributions to the success of the system. The manager, if he is working on the system, takes proper action based on the observations of the workers. Learning to share power and responsibilities will make the workplace more effective and will make managers more competent and productive. Workers want to be involved in the decision process. They will develop great pride, enthusiasm, and become more effective if given a voice in the operation of the system.

QUALITY CIRCLES

During recent years many American firms have adopted the quality Circle or similar Improvement systems for working with employees. Some firms use the term "task force committee" or "work teams" in lieu of Quality Circles. The terms refer to the work environment and have the common objective of resolving problems in a given situation. This management concept originated in the United States several years ago, but never really attracted much attention until after the automobile crisis when U.S. management started to question the great success the Japanese were having with employee motivation. The Japanese were utilizing the Quality Circle concept in many of their businesses.

A Quality Circle is a voluntary group of workers who have shared areas of responsibility. They meet together weekly to discuss, analyse and propose solutions to quality problems. They are taught group communication process, quality strategies, and measurement and problem-solving techniques. They are encouraged to draw on the resources of the company's management and technical personnel to help them solve problems and they generate and evaluate their own feed back. In this way, they are also responsible for communicating. The supervisor may become the leader of the circle. The leader is trained to work as a group member, but not as a "boss" during the circle meetings.

The Quality Circle may be defined as a small group of employees doing similar work who voluntarily meet for an hour each week to discuss their quality problems, investigate causes, recommend solutions and take corrective action. It is primarily a normal work crew, that is, a group of people who work together to produce a product on a given line.

The Circle leader goes through training in leadership skills, adult learning techniques, motivation, and communicating techniques. The leader is trained in the use of various measurement techniques and quality strategies, including Cause and Effect Diagrams and Cards (CEDAC), Pareto Diagrams, Histograms, Run Diagrams, Average (X Bar) and Range (R) charts. More advanced circles move on into sampling systems, data collection, scatter diagrams, charting techniques and Design of Experiments (DOE) and statistical interpretation of same. The purpose of all these techniques is to improve productivity for the firm.

Some of the topics that circles get involved in include: overview of product quality, line controls, sanitation, food regulations, waste, absenteeism, product rejection, accidents, poor work flow, excessive inventories, inefficiencies, spoilage, etc.

The whole concept of the circle provides an opportunity for the workers to develop their skills. It allows workers to have fuller participation in the operation of the firm and it provides a vehicle for allowing the worker to have a sense of dignity.

THE SUPERVISOR

The key to the success of the Quality Circle and the motivation of each employee is the supervisor. A good supervisor needs constant training in human relations. This training includes incentive standards, discipline, how to settle grievances and how to train others. Further, he needs training in cost analysis, how to manage, motivate and communicate, and leadership principles. A good supervisor sets a good example, he displays enthusiasm, he is always neat and clean, he is job oriented and interested in his people. He is a good listener, always respectful, tactful, and courteous. Most of all he is sound in his judgement.

The supervisor is between top management and the worker. He is in daily contact with each employee under his supervision. To the employee, the supervisor is management. Therefore, the morale of the supervisor affects the morale of the worker.

The general philosophy and thinking today by most successfully managed firms is that a shift has taken place from the quality assurance technologist and his or her laboratory to the line employee and their exercising their responsibility for producing quality products efficiently. The technologist is still an integral part of the company plan to assure product quality, but the accountability has shifted because the line employee now understands what the company's standard of quality is all about. They know their job depends on efficient production of a quality product and they know the laboratory will be auditing and evaluating the job they are doing.

SUMMARY

In summary, an effective food plant operation includes the obtaining of good people and the giving of adequate training to help them do what is expected of them. The right person is put in the right job. Through proper communications they can see their role and know how they are contributing to the success of the firm. Management, including the supervisor, must help them succeed and hold all personnel accountable for their performance. Those who perform up and beyond expectation must be rewarded accordingly. Most importantly, every effective organization has a good manager. The manager must plan carefully and he must work through his supervisors and they in turn with their workers. A good manager gives direction to the system, he coordinates all the activities and he controls the system to produce quality products efficiently. Management is the key to any firm's future.

CHAPTER 2

Modern Concepts of Quality Assurance

The relationship of the Quality Assurance Department to the other operating Departments in a typical food firm is shown in Figure 1.1. The important point to note in this Chart is the lines of communication among the Departments as well as to Management. Only by working together can a firm succeed. The key personnel are the quality assurance personnel as they must work across all departmental lines in doing their job for the continued growth of the firm. Quality Assurance personnel must take the lead in communicating to the respective departments as well the training of all production personnel for product control and assurance.

WHAT IS QUALITY ASSURANCE?

Quality assurance is the modern term for describing the control, evaluation and audit of a food processing system. Its primary function is to provide confidence for management and the ultimate customer, that is, in most cases the consumer. A firm is in the business of producing a product– intended for sale to a customer– from which the firm hopes to make a profit. The key word is the customer. The customer is the one a firm must satisfy and it is the customer who ultimately establishes the level of quality the firm must manufacture. The customer is management's guide to quality and this is what the firm builds its specifications and label requirements around. Only by having a planned program can a firm continue to succeed in supplying the customer with the desired products.

A large part of a quality assurance program is built around quality control. Quality control means to regulate to some standard. It is usually associated with the production line, that is, specific processes and operations. As used today, it is the tool for the production worker to help him operate the line in conformance with the predetermined parameters for any given quality level.

Quality evaluation is also a part of any quality assurance operation. It is the modern term used to describe or appraise the worth of the product. It generally means taking a measurement of the product in a laboratory. It is used to include the evaluation of all incoming materials, products in process, and or finished products.

A third part of a quality assurance program is to audit or verify or examine the products or even processes over time. It is a term used frequently with firms having multiple plants. It should be a part of any quality assurance program to verify products in the warehouse, in distribution, and/or competitors products in the market place.

A typical organization plan indicating the three primary concerns is shown in Figure 2.1 Some firms assign the quality assurance department the additional functions of product development, plant sanitation, waste disposal, and research on processes, equipment, ingredients etc. These are all specialized areas and require expertise well versed in these areas for success.

1. What is quality?
2. Why have a Quality Assurance program?
3. What are the fundamentals that should be considered for a successful quality assurance program?
4. What are the factors affecting quality?

It should be pointed out that many food firms today have attained their enviable position by the control of the many products they process. They do not process a product that is lacking in uniformity or one that will "just get by", but rather a product that will continue to build their business. Perhaps more important than their success is the fact that management knows at all times what kind of quality is being packed. Thus, a quality assurance program serves the management of a food firm by keeping them fully informed or assured of the quality and the condition of the products being packed as well as keeping management and the firm in line with the industry trends.

WHAT IS QUALITY

Quality makes a product what it is. Quality is conformance to requirements or specifications. Quality as defined in the USDA Marketing Workshop Report (1951): "It is the combination of attributes or characteristics of a product that have significance in determining the degree of acceptability of the product to a user". Industry defines quality as a measure of purity, strength, flavor, color, size, maturity, workmanship, and condition, and or any other distinctive attribute or characteristic of the product. Thus, the term quality, without being defined in terms of some standard, means very little. On the other hand, the trade generally uses the term to mean the finest product attainable. Food processors have learned from years of experience that high quality products never fail to sell. This is because the consumer recognizes brands that maintain their quality at the standard set for that particular product. Repeat sales are, thus, the outgrowth of quality assurance pratices.

Standards for Quality

What are the standards for quality evaluation? There are different ways of arriving at the many standards for product quality. However, the four common methods are:

Legal Standards–Legal standards are those commonly established by the World bodies, federal, state or municipal agencies and generally are mandatory. These mandatory standards are set up by law or through regulations and represent the Federal Food, Drug and Cosmetic Act minimum standards of quality, the various state minimum standards of quality or the municipal minimum standards of quality. They are generally concerned with freedom from adulteration. This may involve insects, molds, yeasts, and pesticides, maximum limits of additives permitted, or establishing specific conditions in processing so that foods are not contaminated with extraneous materials. Examples of all legal standards that we are concerned with are available from the various agencies involved.

Company or Voluntary Label Standards–The company or voluntary label standards represent the standards established by the various segments of the food industry. The voluntary standards generally represent a consumer image and may become a trademark or symbol of product quality. Generally speaking these voluntary standards are used by private firms or supermarkets and they tend to vary depending upon the particular requirements for any given label.

Industry Standards–The industry standards are those whereby an organized group attempts to establish given limits of quality for any given commodity. Normally these have become effective by pressure from marketing organizations or by specific commodity groups where legal standards are not involved. Examples of these are standards for cling peaches, peanut butter, and some of the frozen food standards.

Consumer or Grade Standards–The consumer standards represent the consumers' requirements of a product and generally are based on experience in use by the industry for consumers. Consumers are not very effective as a group, but individually they represent the day-to-day demands for any given product. The U.S. Department of Agriculture Standards for Grades represent the best standards in this area. Other examples are Military Standards and Veterans Administration Standards.

12   TOTAL QUALITY ASSURANCE

FIGURE 2.1

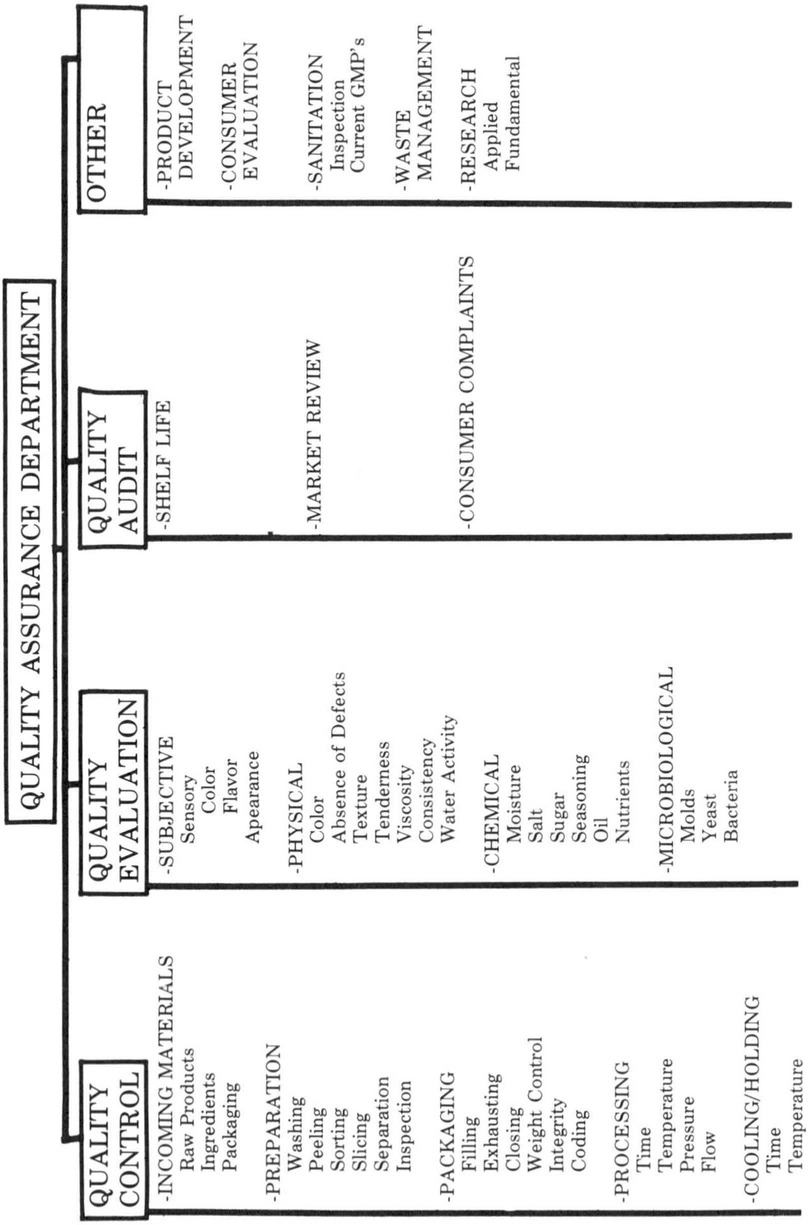

QUALITY ASSURANCE DEPARTMENT

QUALITY CONTROL

- INCOMING MATERIALS
  Raw Products
  Ingredients
  Packaging

- PREPARATION
  Washing
  Peeling
  Sorting
  Slicing
  Separation
  Inspection

- PACKAGING
  Filling
  Exhausting
  Closing
  Weight Control
  Integrity
  Coding

- PROCESSING
  Time
  Temperature
  Pressure
  Flow

- COOLING/HOLDING
  Time
  Temperature

QUALITY EVALUATION

- SUBJECTIVE
  Sensory
  Color
  Flavor
  Appearance

- PHYSICAL
  Color
  Absence of Defects
  Texture
  Tenderness
  Viscosity
  Consistency
  Water Activity

- CHEMICAL
  Moisture
  Salt
  Sugar
  Seasoning
  Oil
  Nutrients

- MICROBIOLOGICAL
  Molds
  Yeast
  Bacteria

QUALITY AUDIT

- SHELF LIFE

- MARKET REVIEW

- CONSUMER COMPLAINTS

OTHER

- PRODUCT DEVELOPMENT
- CONSUMER EVALUATION
- SANITATION
  Inspection
  Current GMP's
- WASTE MANAGEMENT
- RESEARCH
  Applied
  Fundamental

WHAT ARE THE METHODS FOR DETERMINING QUALITY?

Subjective Methods–Subjective methods of evaluating quality are based on the opinions of the investigators. It is usually a physiological reaction which is a result of past training, experience of the individual, influence of personal preference, and powers of perception. They are subjective because the individual is required to give his opinion as to qualitative and quantitative values of the characteristic or characteristics under study. These methods usually involve the various sense organs, and so they may be referred to as sensory methods. Examples of sense perceptions are flavor, odor, color, or touch.

Objective Methods–Objective methods of evaluating quality are based on observations from which the attitudes of the investigators are entirely excluded. They are based on recognized standard scientific tests and are applicable to any sample of the product or products without regard to its previous history or ultimate use. They represent the modern idea in quality control because the human element has been excluded. The methods can be divided into three general groups.

Physical Methods of Measurement–This is perhaps the quickest method and the type requiring the least training of the three. They are concerned with such attributes of product quality as size, texture, color, consistency, imperfections, or they may be concerned with process variables such as headspace, fill, drained weight, or vacuum. Usually instruments can be found or adapted for physical evaluation of product quality. See Table 1.1 for some common physical tests of food products.

Chemical Methods of Measurements–Standard food analysis methods are generally used for quantitative chemical evaluation of nutritive values and quality levels in most cases. However, these types of chemical analysis are too long and tedious, to say nothing of the expense involved in their determinations. As a result the industry and allied interested parties have developed methods that are termed "quick tests", such as those for enzyme, moisture, fiber, pH, or acidity. In many cases these tests can be closely correlated with the longer procedures and accurate values are determined. See Table 2.2 for common tests.

Microscopic Method–Microscopic methods have excellent application in a quality control program but usually require considerable training to properly interpret the results. They can be divided into two general categories:

  (1) Adulteration and Contamination. Examinations will indicate the presence of bacteria, yeast mold, insect fragments, insect excreta or foreign materials. Each test is specific and the technologist

must have the proper background to differentiate the various types of adulteration and contamination that might be present in the products.

(2). *Differentiation between Cell Types, Tissue Types, and Microorganisms of Various Stored Foods.* Examples of their applications are found in tissue testing for deficiency of fertilizer materials, stored food in the tissues of plant materials, micro-organisms causing spoilage and/or desirable fermentation changes.

TABLE 2.1
Common Physical Tests Of Food Products

Physical Factor	Test	Description
Color	Color difference meter	Measured differences in tristimulus values
	Munsell color system	Based on color discs or standards
	Spectrophotometry (Agtrons)	Measures light reflectance at different wave lengths
Viscosity Consistency Rheology	GOSUC or Ostwald viscosimeter	Flow through a capillary tube
	Rotating spindle	A rotating cylinder is immersed in the fluid and stress measured
	Falling weight	Measures the time required for a weight to fall through a tube containing the sample
Texture Tenderness	Finger feel	Gross test for firmness, softness and yielding quality
	Mouth feel	Gross test for chewiness, fibrousness, grittiness
	Texture values Tenderometer, Texture Meters (GOSUT) (Christel), and Shear Press	Gives indication of texture, tenderness, and shear values
Container	Weight, volume, vacuum	Gross, net, drained weight and fill of container
	Seal integrity	Seal evaluation
Symmetry, Size and Shape	Weight, volume, length, width and diameter	Uniformity and classification
Defects	Count or size measurement	Blemishes, extraneous matter, inedible parts

TABLE 2.2
Common Chemical Tests Of Food Products

Chemical Factor	Test	Description

Moisture
Solids
Specific
  Gravity
	Drying
	
	Hydrometer
	
	Titration
		Measures weight loss
		  due to evaporation
		Concentration of
		  dissolved solids
		Reaction of water with
		  specific chemicals

Fat—Oil	Ether extraction	Dried, ground material
		  is extracted by petroleum
		  ether

Protein	Kjeldahl—Macro and Micro	Determines total nitrogen

Carbohydrate	Molisch's general
	  test
		Color reaction with a
		  naphthol

Fiber	Sodium hydroxide
	  extraction residue
		Measures organic residues
		  including cellulose
		  and lignin

Ash and
  mineral
	Burn at 550°C	Determines total ash by
		  weight of residue after
		  incineration
		Specific minerals may
		  then be determined by
		  specific chemical
		  reactions

Enzymes	Catalase,
	  peroxidase
		Chemical reactions with
		  H_2O_2 and/or indicators

Vitamins	Specific chemical
	  or bioassay tests
	  required for each
	  vitamin
		Vitamin analyses should
		  be performed following
		  analytical procedures

pH, acidity	Hydrogen ion con-
	  centration pH
	  meter or titration
		Measures acidity or alkalinity
		  on a sample or on a continuous
		  basis

Chlorine	Chemical titration	Measurement of chlorine residual

WHY HAVE A QUALITY ASSURANCE PROGRAM?

Quality Assurance may be thought of as the scientific control of production. The primary objective of a quality assurance program is to obtain adequate information on all of the factors or characteristics of a product affecting the quality of the product. The intelligent interpretation of this information by the quality technologist provides management with and index of the entire operation. This means that the quality assurance technologist's information constantly serves as a guide for management regarding the exact quality that may be packed from a given quality of raw stocks, or it may provide management with the necessary information needed in the processing of a product to pack a given quality. Thus, the quality assurance technologist serves as the "nerve center" for management and each of the separate departments.

Quality assurance will also open the door to research. Charles Kettering said: "Research is a high hat word that scares a lot of people, but it needn't as it is rather simple. Essentially, it is nothing but a state of mind— a friendly, welcoming attitude toward change. This change may involve people, facilities, materials, and equipment." C. Kettering also said: "Research is something that if you don't do it 'till you have to, it's too late." In other words it is an investment in the future. Research, therefore, must establish the requirements. In fact, quality assurance is the application of ideas and techniques dervied from research and product development.

Some of the reasons for a quality assurance program are:
(1) Control over raw materials through setting of specifications
(2) Improvement of product quality
(3) Improvement of processing methods with resulting savings in cost of production and greater profits
(4) Standardization of the finished product according to label specifications
(5) Increased order and better housekeeping of a sanitary plant
(6) Greater consumer confidence in the uniformly high quality of the product

The basic responsibilities of a Quality Assurance Department are recording and reporting results from:
(1) Line inspection and control of
   (a) Supplies, materials and raw products
   (b) Operating procedures
   (c) Finished products
(2) Physical evaluation of raw and processed products
(3) Chemical evaluation of raw and processed products
(4) Microbiological evaluation of raw and processed products

(5) Warehousing conditions for shelf-life–time, temperature, and handling
(6) Sanitation control of products, processes and storage
(7) Waste disposal control
(8) Compliance with Federal, State and municipal requirements and standards
(9) Specification compliance during marketing and distribution for consumer confidence and assuring the integrity of the product and firm
(10) Self certification with government regulations

Additional responsibilities may include: training, problem solving, development of test and operational procedures, occupational safety and health regulations, and special research and development projects. The most important responsibility to remember is that the quality assurance technologist is a team player- a leader in producing a quality product efficiently.

WHAT ARE THE BASIC FUNDAMENTALS FOR A SUCCESSFUL QUALITY ASSURANCE PROGRAM?

There are six basic funadmantals that must be carefully considered and clearly worked out for the success of any quality assurance program. These are as follows.
(1) Organization of the quality assurance department
(2) The personnel
(3) Sampling (see Chapter 5)
(4) Standards and specifications (Chapter 14)
(5) Measurement
  (a) Laboratory
  (b) Equipment
  (c) Procedures
  (d) Reports
(6) Interpretation (see Chapter 6)

Organization

The organization of a quality assurance program is the first fundamental that must be carefully considered. The program must be desired by top management. The quality assurance department should be directly responsible to top management, not the raw products department, the factory operations or even under sales. Thus, the quality assurance technologist reports directly to management. Obviously, it is necessary for the quality assurance technologist to provide each of the other departments with specific information on the quality at the receiving platform, or on the line or even in the warehouse; but he is not responsible to these groups as such. Management must make the decision between quality and quantity, not one of the several departments of the company. The quality assurance technologist should, however, have the authority from management to cooperate with production to maintain production operations so that the product being packed at all times maintains the desired standard or standards. Thus, it should be quite obvious that the careful organization of the quality assurance department is most important. See Fig. 1.1 and Table 2.3

Personnel

The personnel in the quality assurance department will vary with the products being packed, the size of the operation, and the amount of control desired by management. The quality assurance technologist must have certain qualifications to fulfill the responsibilities necessary for a successful quality assurance program. Some of the more important of these are:

(1) Being truthful in reports, in decisions and above all in analysis
(2) Having sales ability
(3) Being able to speak the language of the industry and write intelligently
(4) Being an excellent cooperator–a team player
(5) Always being alert and responsive to necessary changes
(6) Being well mannered and always neat in appearance
(7) Always being on the job
(8) Being adequately trained
(9) giving instruction to production employees as to
 (a) What is to be done
 (b) How it is to be done
 (c) Why it must be done

At the 30th Annual Meeting of the Institute of Food Technologists the following definition and professional code was adopted.

A Professional Food Technologist is an individual, educated or experienced at least to the equivalent of a Bachelors Degree in the Sciences applicable to food and who demonstrates skill in the application of this knowledge in the chain of producing wholesome food.

The Professional Food Technologist Code–As a Food Technologist I believe that sound professional relationships are built upon integrity, dignity and mutual respect.

TABLE 2.3
Organization For Quality Assurance

I. Relationship chart
R = Responsible
C = Contribute
I = Informed

	General Manager	Controller	Marketing	Materials Procurement	Manufacturing	Quality Assurance
1. Determine needs of customer	I		R		I	I
2. Est. specifications level for busines	C	I	I	R	C	C
3. Establish manufacturing specifications	C				R	I
4. Produce products to specifications				C	R	C
5. Determine process capabilities					C	R
6. Incoming Materials evaluation	I			C	I	R
7. Plan quality control system	C	I	I	I	I	R
8. Feed back Quality Information	I	I	I	I	I	R
9. Gather complaint data	I		R	I	I	I
10. Evaluate complaint data	I		I			R
11. Compile quality costs	I	R			I	I
12. Evaluate quality costs	I	C				R
13. Quality audit	I	C			I	R
14. Product evaluation	I	I	I		I	R

I recognize my responsibilities to the public, to students, colleagues, associates and to my employer to be as follows:

I shall strive for highest performance, for raising professional standards, improving services, and providing a working climate which encourages sound, considered judgement and achieves conditions which attract to the profession only persons of high qualifications and integrity.

In the practice of my profession I shall avoid unethical exploitation of my professional status, and refrain from untrue, misleading, or exaggerated representations.

I shall be an exponent of the scientific method.

I shall give just recognition to the work of others, without distortion or discrimination.

I shall assume a personal responsibility for improved understanding of food technology, avoiding premature, false or exaggerated statements.

I shall discourage enterprises or practices inimical to the public interest or welfare.

I shall apply for, accept, or assign positions and responsibilities on the basis of professional and/or academic preparation. I shall represent my professional qualifications honestly and shall not knowingly withhold pertinent information.

I shall adhere to the terms of my employment and shall use the time granted me for the purpose intended. I shall keep in confidence such proprietary information as is obtained in the course of my professional services, unless disclosure is properly authorized or required by duly constituted legal authorities.

I shall strive to protect my employer from unjust financial, legal or personal injury, and from unjust embarrassment or disparagement.

I shall not use any position for private gain to the possible detriment of my employer. I shall not accept gifts, favors, or gratuities that might impair my professional judgement, nor shall I use coercive means, proffer gifts, or promise special treatment or favor in order to influence professional decisions or obtain special advantages.

Measurement

The facilities for a quality control program will vary with the size of the operation, the number of products being packed and the different qualities being packed.

The Laboratory–The following information should serve as a background for the formulation of a plan, layout design, and operation of a quality evaluation (QE) laboratory.

Uses of the QE Laboratory–The QE laboratory can serve as the nerve center for the control and evaluation of quality for all food plant operations. This means that the QE program starts with the purchase of seed and follows all the way through to the ultimate consumption of the finished product. Some of the many functions, duties, and line checks that can be made in a QE food laboratory are as follows.

(1) Determination of percentage germination and purity of seed
(2) Soil and tissue analysis
(3) Collection and summarization of weather data for use in scheduling of raw products, packaging materials, or labor
(4) Identification of crop diseases and insects
(5) Determination of raw product quality and other incoming materials
(6) Evaluation and continuous monitoring of processing variables affecting quality
(7) Determination of the efficiency of each processing operation as related to finished product quality
(8) Periodic and continuous monitoring of water supply, equipment, plant sanitation, and the waste disposal system
(9) Evaluation of the finished product quality and assurance of the storage life of the finished product
(10) Development of new products and improvement of present processing, production and quality evaluation methods.

The QE Laboratory–The location and design of the QE laboratory are factors that must be considered for each plant. It should be more than just the old cutting room; the modern QE laboratory must be completely designed and equipped to do the control job expected of it by management. The following are some of the more important factors to be considered.

The ideal location is one that is close to, but apart from the preparation lines. It should be located, if possible, so that north light is available. The lab should be designed so that it is dustproof, soundproof, well ventilated, and easy to keep clean. The walls and cabinets should be painted with a nongloss or a mat white paint.

The size of the laboratory depends primarily on the number of products being packed and the daily volume of business. The minimum recommended size is a floor space 12 x 18 ft. A typical floor plan for a medium sized QE laboratory is shown in Fig. 2.2. This laboratory is 18 x 25 ft and is laid out to handle the basic work for a medium size processing plant handling three or four processed food items. Many other innovations and arrangements are possible, but this layout has been found to be quite satisfactory.

The Grading Table–The grading table should be approximately 8 ft long, 3 ft high, and 30 inches wide. The top of the table should be painted with a hard-finish nongloss or mat-colored white paint, or covered with stainless steel. This latter top will wear well and is easy to keep clean. The table should have either north light or a recommended overhead light. In the case of the latter, the MacBeth Examolites over the grading table are ideal. (For a table of the above size, six MacBeth Examolites would be adequate.) These lights are the closest match to North daylight available and they are a good substitution for daylight. Of course, the obvious advantage is that the light is uniform day-in and day-out and one can grade at midnight as well as at midday.

In addition to the top, the grading table should be provided for the storage of grading equipment, such as vacuum, pressure, and headspace gauges; hydrometers and salometers; trays, pans, and screens. Standard equipment for the grading table also includes a heavy duty table-mounted can opener, and 1 or 2 grading scales. The scales should read to at least ¼ oz and should have a tare beam for tare weighing empty cans, screens, etc.

The Analytical Bench–The analytical or chemical bench should be approximately 3 ft high, 3 ft wide and, as long as space allows, up to 12 ft. (In Fig. 2.2 the bench is 8 ft long.) The bench should be equipped with water, gas, electric, vacuum, and air-pressure outlets. The top of the bench should have an acid-and alkali-resistant top with a 4 in. square lead drain sink running down the center of the bench. There should be a sink at the end of the bench with a heavy duty disposal. Above the bench top and over the center sink, a double deck chemical and reagent shelf (10 in. wide) should be constructed for the storage of chemicals and reagents used daily. Drawer and closed cupboard space should be provided under the bench for the storage of special analytical equipment and supplies.

The Taste Panel Table–The taste panel table should be approximately 30 in.high, 2 ft wide, and 8 ft long. Folding partitions 18 in. high should be constructed for the top of the table so that the panel members can work individually when evaluating the samples. The table should be open underneath so that the panel members can sit down to work. Overhead lighting with varying colored lights is desired so as to light the samples properly for unbiased taste evaluations. Obviously, a separate room operated from a separate kitchen is to be desired, but the above arrangement has been found to be very satisfactory for the practical taste evaluation of line samples.

FIGURE 2.2 — Floor Plan For A Quality Evaluation And Analysis Laboratory

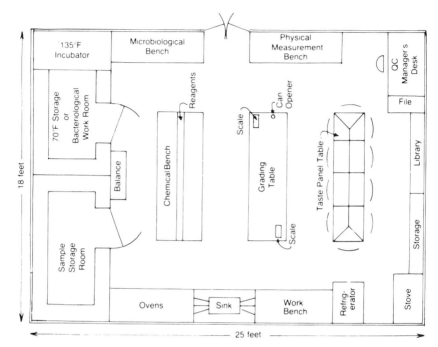

Microbiological and Physical Testing Table–The microbiological and physical testing tables should be approximately 36 in. high and 2 ft wide. Outlet facilities similar to those on the chemical bench are desired, but usually an electric line and a gas line will do. The bench top, drawer, and cupboard space should be constructed similarly to that suggested for the analytical bench. At one end of the microbiological bench, space should be provided for the analyst to sit down for microscopic evaluation. The space should be approximately 26 in. high and 30 in. wide with a 2 in. deep drawer.

The QA Manager's Corner–The quality assurance manager should have an adjacent office or one corner of the laboratory set aside for his desk, files, and library. The library and manuals should be available for use by his analysts at any time. Moreover space should be provided for the immediate and daily completion of reports.

Basic Equipment.—Laboratory supply houses and laboratory equipment manufacturing companies can supply the food processor with most of the following suggested basic equipment. In addition to

FIGURE 2.3 — Grading Table Arrangement Showing Location And Dimensions Of Macbeth Lights

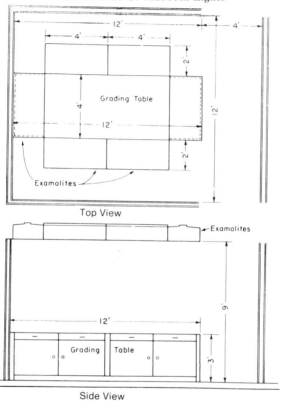

listing this equipment, brief remarks are given on efficient use, upkeep, and handling of the specific equipment. The grading equipment for fruits and vegetables should include the following.

Can Opener.—The can opener should be of the heavy duty type. It should cut the top of the can out and not cut the rim off, thus facilitating the measurement of the headspace of the can. For efficient operation and use, it should be oiled monthly and the teeth on the gripping wheel should be cleaned with a stiff wire brush periodically. Ideally, a heavy duty electric can opener should be provided.

Vacuum Gauge.—The vacuum gauge should be approximately 2½ in. in diameter and specially fitted for testing the vacuum in tin or glass containers having metal caps. Preferably the gauge should be a combination pressure and vacuum gauge. To keep the gauge in good working

FIGURE 2.4 — Determination Of Drained Weight Using Grading Scale And Screens

condition when taking a vacuum reading, the needle should be inserted as near to the outer edge of the can lid as possible, thus preventing the product from being drawn into the gauge. Of course, the vacuum or pressure reading should not be taken on cans that are swelled and/or badly dented. Periodically, the needle point of the gauge should be sharpened and the gauge tube should be cleaned with a piece of fine wire.

Headspace Gauge.—The headspace gauge should be calibrated in $1/32$ of an inch. The gauge should be kept dry and clean to prevent corrosion. Occasionally the gauge can be wiped with an oiled cloth, thus preventing rusting and facilitating operation.

Grading Scale.—The scale should have a capacity up to 12 lb, be equipped with a tare beam, and have an accuracy of $1/16$ oz. The grading scale should be kept on the grading table and preferably close to the can opener. Twice a year the scale should be dismantled and the knife edges cleaned lightly with a light emery cloth. After each use of the scale the weights should be returned to the weight rack. These weights and the scale pan should be kept clean to avoid inaccuracies in weighing.

Grading Screens.—Screens are used for many products to determine drained weight. The screens are of two diameters and of two meshes. For

No. 3 cylinder (404 × 700) size and smaller, an 8-in. diameter screen is used, and for cans larger than a No. 3 can, a 12-in. diameter screen is used. For products like R.S.P. cherries, pears, peaches, whole kernel corn, green and wax beans, a screen with 8 meshes to the inch (0.097-in. square openings) is used and for tomatoes, a screen with 2 meshes to the inch (0.446 in. square opening) is used. In addition to the above requirements, the diameter of the wire for the tomato screen should be 0.056 in. The grading screens should be washed immediately after use to prevent verdigris (green rust on copper). When purchasing screens (quantities of two or more) it is suggested that the manufacturing companies supply the screens of uniform weight, thus one tare weight would suffice for all screens when determining drained weights on two or more cans at one time.

Grading Trays.—White plastic trays should be used for grading food products. The trays should be of two depths, that is, ¾ in. deep for No. 3 cylinder (404 x 700) size can and smaller, and 2 in. deep for cans larger than No. 3 size. Trays approximately 10 in. wide by 15 in. long are satisfactory. The trays should be kept clean and stored in a convenient cabinet, preferably under the grading table.

Sizing Gauges.—Sizing gauges are a necessity for the grading of the commodities like peas, green and wax beans, lima beans and cherries. These gauges should be kept clean and stored in a drawer provided in the grading table. In addition to the standard sizing gauges for specific commodities, a vernier caliper, calibrated in $1/32$ of an inch, should be provided for sizing of commodities for which standard gauges have not been developed.

Brine and Sirup Cylinders.—Brine and sirup cylinders are essential for determining the grade for clearness and color of the liquor and the determination of the sugar content of the sirup. The cylinders should be of a clear, colorless glass, have a plain rim, and be approximately 12 in. in height and 2 in. in diameter. The cylinders should be kept clean and likewise stored in a separate compartment of the grading table.

Hydrometers and Salometers.—Hydrometers should be provided for the following different types of solutions and packing media: Salt concentrations (salometers reading in percent salt), sugar concentrations (Brix hydrometers reading in percent sugar in sirup), alcohol concentrations (hydrometers reading in percentage by volume), alkali concentrations (hydrometers reading in percentage by weight). Also, a wide range

MODERN CONCEPTS OF QUALITY ASSURANCE

of specific gravity hydrometers for making up reagents and checking the above hydrometers should be available. The range of any of the hydrometers is quite narrow; consequently, when ordering hydrometers consideration should be given to the solutions being evaluated. After any hydrometer has been used, it should be carefully washed, dried, and placed back in the hydrometer holder.

Thermometers.–Fahrenheit and/or Centigrade thermometers covering a wide temperature range should be provided. Also, piercing thermometers should be obtained for determining the center can temperature for acid products. The can center temperature testing machine should be obtained for determining the center can temperature for nonacid products. In some cases it may be desirable to obtain maximum and minimum thermometers and regular pocket thermometers for routine in-plant quality control work. All thermometers should be kept clean and in their proper holders. Thermometers and hydrometers should be kept in a drawer in the grading table provided for such instruments.

Miscellaneous Equipment.—Spoons, rubber tray scrapers, cylinder and tray brushes, marking pencils, rubber stamps, etc. are helpful and essential equipment and supplies for the efficient operation of the quality control laboratory.

Special Equipment–In addition to the above basic equipment for grading, in-plant control work and quality evaluation work, the quality assurance technologist should have special equipment for each commodity. These would include a colorimeter, pH meter, refractometer, and physical instruments for measuring consistency, viscosity, succulence, moisture texture, color, or other quality variables. The quality assurance technologist should also have basic quantitative equipment such as an analytical balance, beakers, burettes, condensers, crucibles, graduated cylinders, dessicators, filtering apparatus, volumetric flasks, funnels, pipettes and blenders.

For some commodities special apparatus will be needed such as equipment for fiber determination in snap beans, alcohol insoluble solids content of peas, sweet corn and lima beans, oil content of snack foods, pericarp content of sweet corn and peel oil of citrus products.

Finally, microscopic and bacteriological equipment is very essential for most products. Basic equipment includes a microscope (binocular type with low power, high power, and oil immersion lenses plus a mechanical stage), inoculation tubes, petri dishes, colony counter, dilution bottles, slides, autoclave, incubation oven, and other equipment. A few typical quality control procedures requiring micro- and bacteriological equipment are mold count, plate count, insect fragment count, water and sewage analysis and spoilage analysis.

Procedures–The specific procedures used to evaluate the quality of any product must be standardized. With the quick tests of quality, the quality assurance technologist must follow these procedures as given, as small deviations may cause large errors in the results. Specific details are given in this manual for most of the quality attributes. Additional information is to be found in the AOAC, USDA Standards for Grades, FDA Minimum Standards for Quality and Fill and in research articles and trade papers. The quality assurance technologist is expected to be thoroughly familiar with the methods of analysis for food, water, soil, pesticides, sewage, and the published literature on objective tests of quality. If methods are not available, he may have to develop techniques and procedures to meet his own specific set of conditions. At any rate, he should write out all procedures in complete detail- thus he will always be using the same basic procedure, until modified, year after year.

A general procedure for the evaluation of processed food for any given commodity should include obtaining the following detailed information, where applicable.

(1) Record the label information:
    (a) Name of the Product
        (1) Style
        (2) Type
    (b) Net Weight
    (c) Vendor
    (d) Other Information
        Artificial coloring, artificial flavoring, preservatives, packing medium including sirup concentration, nutritional information

(2) Record the following container information:
    (a) Code mark on container exactly as presented
    (b) External condition of container
    (c) Size of container

(3) Determine and record the GROSS WEIGHT of the container and its contents.

(4) On all thermal processed products, record the vacuum with the aid of a vacuum gauge. For carbonated beverages, record the pressure with the aid of a carbon dioxide gauge.

(5) Remove the lid on canned products by cutting it out from the bottom of the container (3 piece containers) or the uncoded end. (The coded end of the container may be used to evaluate the integrity of the doubleseam.) For glass-packed products the integrity of the seal should be evaluated for security.
(6) Pour the sirup or brine into a cylinder. If a sirup pack, determine the sirup strength with the aid of a hydrometer. If a brine pack item, note the clearness, color or general appearance of the brine. If a part of the grade, score in accordance with the applicable USDA Standard for Grade.
(7) Where applicable, empty the contents on a screen (use 8 mesh for all cut or small products and 2 mesh for halves or whole fruits or tomatoes). (Use 8 in. screen for 2½ size can or smaller and a 12 in. screen for cans larger than 2½ size.) Drain for 2 min and weigh the fruit on the screen without disturbing the contents thereof. Record the drained weight.
(8) Wash the can or package, dry and determine the NET WEIGHT by subtracting the weight of the empty container from the gross weight.
(9) Note and record the appearance of the interior of the container. Look for corrosion, discoloration or breakage of the enamel, if an enamel-lined container.
(10) Record the count or size of the commodity, where applicable.
(11) Determine the score points for each factor (attribute) of quality in accordance with the applicable standards for grades or quality for the particular commodity. Record the score points for each factor of quality.
(12) Ascertain the grade by totaling the scores for each factor of quality. Keep in mind the limiting rule (stopper) for any particular factor of quality, where applicable.
The U.S. Grade of Fruits are:
Fancy or Grade A
Choice or Grade B
Standard or Grade C
The U.S. Grades of Vegetables are:
Fancy or Grade A
Extra Standard or Grade B
Standard or Grade C

QC DATA FORM 2.1 — Examination Of Processed Foods.

Technoligist   Date

LABEL/BRAND					
Commodity					
Style					
Type					
Net Weight					
Vendor					
Code					
Condition of Container					
Container Size					
Gross Weight					
Vacuum					
Headspace					
Drained Weigth					
Net Weight					
Size/Count					
Brix					
ATTRIBUTES OF QUALITY	Score Points				
Total Score					
Grade					
REMARKS:					

Reports–The complete write-up of the results is just as important as the analysis of the samples. The quality assurance technologist (QAT) must complete report forms giving findings and recommendations. These should always be in writing with the original for management and copies to production, sales and the field operations when applicable. A copy, of course, should be retained for his files. The reports should give a complete history of his daily activities and recommendations where applicable. In all reports the QAT should keep in mind (1) what is he writing the report for, (2) for whom and to whom, and (3) type and amount of material to be presented. The details of the report will vary with the type of report. Minimum parts of the report should include:

(1) Title
(2) Purposes or objectives
(3) Scope of work
    (a) Accomplishments to date, including literature reviewed
    (b) Present efforts—what was done and what remains to be done
    (c) Statistical presentation of data—tabular and analysis
    (d) Newly developed techniques
(4) Discussion (if work far enough along)
(5) Summary of work by answering objectives
(6) Costs to date
(7) Future requirements
    (a) Laboratory procedures, supplies, apparatus, and space
    (b) Pilot plant equipment and use
    (c) Production runs
    (d) Personnel
    (e) Estimated costs to complete project
(8) Personnel involved in project work with project leader signing report.
(See Chapter 9).

WHAT ARE THE FACTORS AFFECTING QUALITY?

Quality of processed fruits and vegetables is affected by the following basic factors, either individually or in combinations.

Cultivar

The choice of the proper cultivar is perhaps the most important single factor when packing a quality fruit or vegetable product. Specific recommended cultivars for one area of the country or even within the State may not apply to another area. Thus, the cultivars of fruits and vegetables are specific for different growing areas, the intended use and the consumer's preferences. In the case of the growing area, specific recommendations are available upon request from the seed supply firms and the land grant universities. In the case of the intended use, little work has been done along this line until recently. Some examples are available upon request from the various agricultural experiment stations. As to the consumer's preference, very little has been done along this line, but considerable interest is being shown today in specific isolated areas.

Maturity

The maturity of any fruit or vegetable is perhaps more important than the specific cultivar in many cases. Any recommended fruit or vegetable cultivar for processing should mature uniformly, should stand relatively long in the field and should be resistant to insects and diseases. The exact stage of maturity for harvest depends upon the intended quality contemplated at the factory. The maturity changes quite rapidly after harvest unless handled properly and promptly. The fieldman may have the crop harvested at its optimum condition, but if it is not processed promptly the quality may drop down into the next lower grade in just a few hours.

Cultural Practices

Cultural practices from the standpoint of quality include such factors as organic matter, moisture, fertilizer, cultivation, and pest control methods. Any one of the above factors may be the limiting factor in producing a quality product. Perhaps the best example of a limiting quality factor is the use of insecticides that give good control of the pests, but they may impart an off-flavor to the processed product or leave a residue. Considerable work has been done in evaluating new insecticides from a flavor imparting aspect. However, there is room for much more work and evaluation of the actual amounts of residues left in the finished product. With the constant battle between insecticide residues and insect-contaminated products, the packer must always know the quality of the raw and finished products.

Harvesting and Handling

Harvesting and handling methods of fruits and vegetables for processing are factors that go hand in hand with maturity and other quality characteristics. The vegetable or fruit must be harvested at the desired stage of maturity and promptly delivered to the processing plant for immediate processing to preserve all the quality. With mechanical harvesters the processor, in most cases, has complete control of the harvesting of the crop. If he knows what quality he desires he should then get the crop harvested and handled according to specific schedules. Thus, he is able to gear the field operation to his factory operations and retain the maximum quality.

Processing

The actual processing methods are in-plant factors that may adversely affect quality. The processor knows from many years of experience that he cannot improve upon the raw product maturity and other quality characteristics by processing the crop. On the other hand, he can definitely lower the original raw product quality by using poor processing techniques. Consequently, the processor must exercise good in-plant quality control techniques to preserve the quality delivered to the plant. Some of the more important factors that must be carefully controlled are: efficiency of washing, trimming, cutting, inspecting and sorting; time and temperature of blanch or scald; fill weights; brine or sirup characteristics; closing machine vacuums; can seam formation; and processing (cooking and cooling) times and temperatures. Proper control of all these in-plant variables are necessary for quality retention.

Shelf-life

The storage temperature and time of processed products has been reported to have effects on retention of the quality of the finished products. However, normal storage temperatures are not detrimental to quality of most processed products up to a year of storage. On the other hand, extreme high temperatures or fluctuating temperatures may be very detrimental in a short period of time.

Use

The last factor that may affect the likes or dislikes of processed fruits and vegetables is the actual preparation and cooking of the processed products. A perfect product can be grown, processed and properly stored, but very easily ruined by poor preparation methods. Processors should be encouraged to inform the consumer of the best method of preparing their products. Informative, descriptive and grade labeling of processed products is a step in the right direction to improve consumer-processor relationships.

CHAPTER 3

Quality Control and the Use of CEDAC

In the manufacture of any food, there are many potential causes that can effect product quality. These causes can be broadly classified as follows: Materials, Machines, Men (workers), Methods, Environment, etc. These causes and their effects can be diagrammed as shown in Figure 3.1 (see Chapters 7, 8, 9, and 10). The diagram is often referred to as a "fish-bone" chart.

FIGURE 3.1 — Cause & Effect Diagram

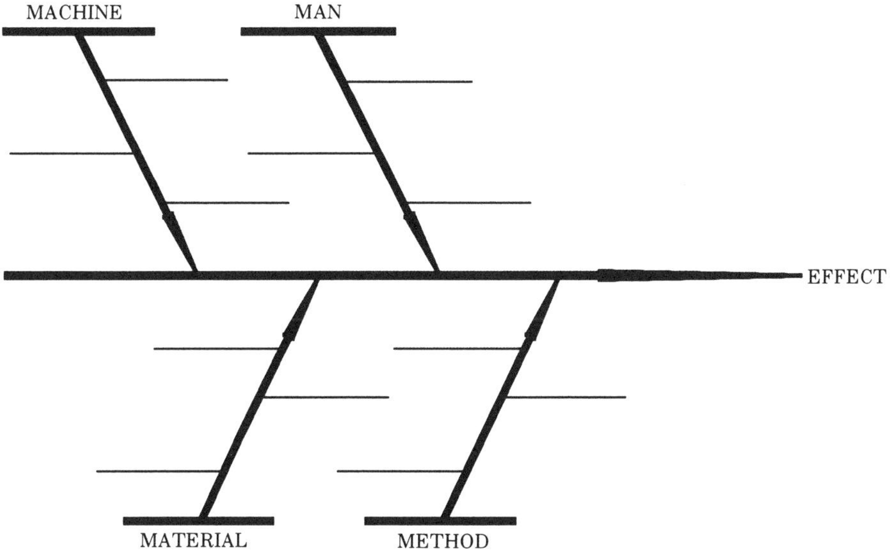

Cause and Effect diagrams are supplemented with Cards (CEDAC). The cards are used to explain the causes in short sentences, thus giving a more detailed explanation of the cause. As used in the real world, employees working with CEDAC write out their observation, their ideas, and their know how on a 3 x 5 card and place these on the diagram in the appropriate place. This information is then available for anyone to see, study, and add to with their own card. The new cards are added to help solve the problem. Thus, the main purpose of CEDAC is to classify the various causes using the information that is available and thought to have an effect on the cause to help arrive at a quantitative answer.

Process flow diagrams are valuable tools for analyzing a process to find potential hazards or causes and to enumerate the Critical Control Points (CCP). Hazard Analysis and Critical Control Points (HACCP) is a modern term to get at the Cause and Effect problems. A process hazard is a possible source of trouble along the processing line which might be defined as a failure to identify (1) critical material, (2) critical processing points, (3) adverse environmental conditions, and (4) human malpractices. Critical Control Points, that is, those points which if not properly controlled may result in unacceptable risks of adulteration, failure to maintain compliance with food standards, and failure to comply with net-weight declarations are identified and keyed to specifications provided in any unit operation, in packaging operating controls, and/or in storage requirements including shelf life.

Ryuji Fukuda states the following steps to prepare and apply CEDAC:
1. "Select a major quality problem that you want to improve and specify the target.
2. "Write out all the manufacturing conditions and technical know how needed to achieve good quality. From this, a selection of the needed information is made by the workers, foreman, production engineers, plant engineers, technologists, and all persons concerned. Put the selected information into diagram (fish-bone) form.
3. "Hang the diagram on a wall of the plant to show the many "Causes" and the "Effects" to all personnel concerned.
4. "If the quality cannot be kept within the control limits, there is room for improvement in the "causes". Thus, the root of the problem is sought by gathering and analyzing more "facts" and then further improvement of the technique or equipment is made.
5. "The content of the improvement is written on a new card, which is put over the corresponding old card. As a group of cards is gradually accumulated, they show not only the past record of the production process, but, also, the effects of each improvement."

Everyone participates in the process from points 2 through 5. With such a system, previously unknown information is brought up and necessary improvements are made so that the standard operation is formed step by step without any regression". These steps are more clearly pinpointed by studying Figures 3.2, 3.3 and 3.4. One will note that CEDAC can be called a Shewart Chart or X Bar Chart with limits and additional functions. Pictures of equipment, machines, products, and processes may be added along with the cards to help explain the causes and the effects.

FIGURE 3.2 — Example Of CEDAC For Potato Chip Manufacture

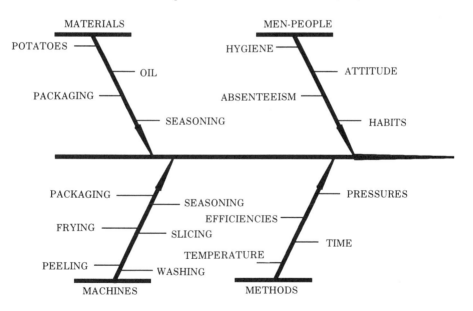

38   TOTAL QUALITY ASSURANCE

FIGURE 3.3 — CEDAC Showing Cards

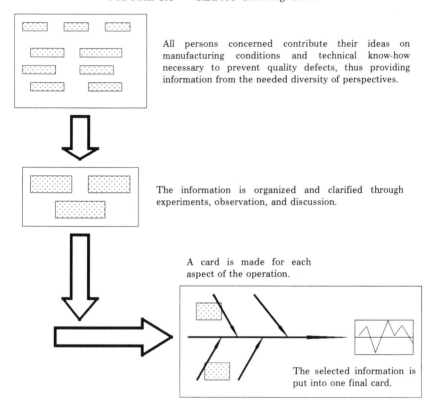

FIGURE 3.4. — CEDAC With Control Chart ($\overline{X}$ and LCL & LCL)

One short sentence describing the necessary conditions and technical know-how for controlling the characteristic quality identified at right.

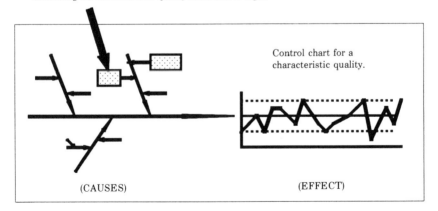

FIGURE 3.5 — Possible Factors Affecting
Flavor Of Potato Chips

POTATOES
   Variety (cultivar)
   Tuber maturity
     Reducing sugars
     Total Sugars
     Protein content
   Harvesting practices
   Handling Practices
   Holding times and temperatures
   Pesticides
   Diseased tubers
   Insect infested tubers

FRYING OILS
   Type of oil
     Chain Length
     Saturation
     Free Fatty Acid content
   Antioxidants
   Oil Frying Temperature and times
   Oil turnover
   Oil filtration
   Metal Contamination

PACKAGING MATERIALS
   M.V.T.R.
   Seal Integrity
   Compatability

SALT AND SEASONINGS
   Purity
   Amount
   Flavor pickup from applicator

ENVIRONMENT
   Relative Humidity
   Temperature
   Time-Shelf Life

CEDAC may be used (1) to analyze a process, (2) to solve problems, and (3) to make improvements. One way to use CEDAC includes the following steps:

  1. Form a team of persons concerned about a problem,

  2. Brainstorm and list all of the various problem areas (Causes),

  3. Select major causes by majority rule,

  4. Determine process capabilities and establish control charts, that is, X Bar and Upper and Lower Control Limits (UCL and LCL),

  5. Interpret the data and theorize as to the possible causes,

  6. Using the pareto principle, experiment to discover the actual causes,

  7. Implement the solution and install controls using X Bar and R Charts with limits for interpretation (Chapter 6).

Reference
Fukada, Ryuji Introduction to the CEDAC. Quality Progress November 1981.

CHAPTER 4

Productivity and Total Quality Assurance

Productivity is the key to the future of the food firm. And people are the key to productivity. People must be given more than their salary and benefits to be productive in this day and age. They want fulfillment and growth on the job. They want a voice in the decision processes in producing the products they are involved in. In most cases, they need help and they need to be taught how to think about their work assignments and activities. They need knowledge of what has to be done, how it is to be done, and why it is to be done a certain way. They need training in statistical methods, quality assurance concepts and control measurements, quality policies, and defect prevention. Thus, they need to be integrated into the system.

Stein describes the needs of people in a manual entitled Quality of Work Life (QWL) and he lists the following as elements that lead to QWL.

1. Automomy, that is reasonable freedom of action while on the job.
2. Recognition, that is, being known as an individual and being visible as a contributor by peers and supervisors.
3. Belonging and having shared goals and values.
4. Progress and development and sense of accomplishment.
5. Status or external rewards beyond pay, that is promotion, rank in position, and other visible benefits.
6. Decent working conditions.
7. Dignity, that is, treated with respect.
8. Involvement and being a part of the team.

These QWL's lead to motivation, greater productivity, and ultimately to a much more effective work force and organization. People want to grow and show progress. Without growth, people feel stuck in their tracks, they lose enthusiasm and they take on a 'don't care' attitude.

People want to be involved in problem solving. People want to show their commitment to their firm and to the product they produce. They want responsibility. They have pride in what they can do and they want to show it in their workmanship.

One of the ways of helping people to become involved in their work is to have them work in teams, that is team approach to the job as its done in many sports. Teams should be developed for groups of people working in the same area on similar tasks. They can share the same responsibilities for quality inspection of all materials, maintenance of the production schedules, control of the costs, and the solving of the technical problems associated in their area. They can handle the absenteeism problem, if it should be present. Further, they can be responsible for the work area and in many cases the maintenance of the equipment. The team helps to solve the problems of safety, they boost plant morale, and yes, they focus the individual employee directly on productivity. They help each other to work smarter, not harder. They all become more efficient.

People are an integral part of the food system and perhaps the most important part in the successful operation of any food firm. Management can make sure the system works if they (1) do the right job in the recruitment of individuals for the right job, (2) provide adequate and appropriate training, (3) provide proper supervision, and (4) take the time to follow up on a regular basis. High motivators or satisfiers for people are achievment, recognition, the work itself, responsibility, and advancement. The real test is to give a person the right job and the tools and the authority to do it, let him do it, and then reward him for doing it. Finally, give him credit if it is due.

People like to be treated as humans. The following are some commandments in human relations:

1. Speak to people and call them by name. There is nothing as nice as a cheerful 'hello' or a word of greeting.
2. Smile at people. It takes 72 muscles to frown and only 14 muscles to smile.
3. Be cordial and helpful.
4. Be interested in people and remember you can learn to like something about almost anybody. If you try.
5. Be generous with praise and be cautious with criticism.
6. Be considerate of the feelings of others.
7. Be thoughtful of the opinions of others. There are three sides to every controversy, that is, yours, the other fellow, and the right one.
8. Be alert to give service. Remember what counts most in life is what we do for others.

PRODUCTIVITY AND TOTAL QUALITY ASSURANCE

When a new person joins the firm, its most important to start them off right. The first prerequisite is the interview: are they the right person for the open position. If they are, the following should be carefully explained to them: The mission or purpose of the company, company rules and regulations, standard operating procedures, work schedules, work habits and ethics, absenteeism, dress code, and the expected standards of performance, and how to use statistical charts to show their records of achievment. The employee should be fully informed relative the firms history, company policies, goals, objectives of operation and their position on quality of finished products. All of the above should be in writing including the benefits program, opportunities for training, and salary schedules. The following is a suggested check off sheet to help each employee through the interview:

—Exact hours of work
—Where, when and how to use the time clock
—Lunch periods, breaks, and the lunch room
—Lockers and where to keep personal property
—Location of washrooms and rules on washing
—Smoking rules
—How supplies and tools are issued
—Safety rules and regulations
—Quality and production standards
—Location of Bulletin Boards
—Wearing of identification badges and uses of passes
—Removal of packages and properties from premises
—Use of telephone for emergencies or personal messages
—Absenteeism
—Overtime policies
—Housekeeping policies
—Personnel office
—Introduction of co-workers and job assignment

The key to employee relations, their motivation, and of course, their productivity is their supervisor. The ultimate aim of the supervisor is to help improve the performance of the worker and their specific operation, thus increasing production of the desired quality of the product. Supervisors should initiate work area dialogues with every employee to show that they care about them and want to help them do a better job. Supervisors should counsel each employee so that the employee knows that he or she has a friend. Supervisors must be visible by being around. Supervisors should provide open access to all resources needed by the employee to help them in their work. Supervisors should encourage employees to participate in fact finding groups or teams and promote the use of Quality Circles or other Quality Improvement practices.

There are many ways of describing the ideal supervisor. 1. In the selection of the supervisor one must look for an individual person who is consistent in work habits and performance. 2. How does the prospective person react in emergency situations, that is, is he quick to burn-up or is he stable in his actions. 3. Does the prospective supervisor have the ability to handle details and can he communicate what is expected from his superiors to his subordinates in a way that it is clearly understood. 4. Is the prospective supervisor a self starter and does he or she have stamina and perserverance. 5. Is he or she an innovator, that is, some one who looks for a better way—way, that is, quicker and less expensive to accomplish the end purpose. 6. Is the prospective supervisor loyal to the firm and does he meet time schedules. 7. Can this person do the job, can he or she inspire people, can he or she do the whole job. 8. Most importantly does the prospective individual want the promotion. Some firms do not believe in promotion from within as it may create problems. On the other hand, most authorities believe that promoting from within is a morale builder and a good outlook for other employees. Promotion is a sign of progress, not just for the individual being promoted but for others and the long range goals of the firm. In addition to the points above, the following are suggested supervisor leadership keys:

1. Supervisors should set good examples as employees will tend to emulate the leader. Good examples include appearance, attitudes, tactfulness, and being courteous.

2. Good supervisors give subordinates a sense of direction and goals to achieve. These must be reachable and they must be given ahead of time so that everyone knows and understands.

3. Supervisors keep employees informed, that is, work schedules, changes in the offing, or new plans.

4. Good supervisors do not give orders, but make requests and suggestions. They should always speak politely to employees.

5. Good supervisors emphasize skills, and rules. They should delegate as much as possible and enlist suggestions and exchange ideas with subordinates. A good supervisor gives credit where credit is due. Most importantly he praises in public and criticizes only in private.

6. Good supervisors seek out new ideas and display constant enthusiasm about the firm, their jobs, and the products they produce. Good supervisors make all employees feel that they belong and that they are a valued

member of the team. A good supervisor displays the fact that he is always job oriented.

There are times when problems do arise with people in any working environment and its very important that the supervisor and the personnel director have a policy of how to handle them. The following are suggested steps in handling personnel problems:

1. Get the facts, that is review the record, find out what rules and customs apply. Talk with the individuals concerned and gather opinions and feelings where applicable. Get the whole story.

2. Weigh all the facts and make a preliminary decision based on the facts and not hearsay. The decision should be based on practices and policies, the individual working group, and the possible bearing on his team or co-workers.

3. Take action and follow up the results. Watch for any changes in attitudes, work relationships, and output of product. Watch for recurring patterns.

4. Document your handling of the problem and your actions by filing a written report.

People are the most important part of any successful operation. The supervisor is the key representative to the employee and the supervisor must be well versed in human relations for the growth of the firm.

REFERENCES AND SUGGESTED READINGS

Dennissoff, B.A. Process Control Management. Quality Progress June 1980.
Dietrich, W.T. Why Managers Neglect Systematic Problem Solving. Quality Progress September 1974.
Kondo, Yoshio, Human Motivation: Basis for good Work. Quality Progress August 1977.
Schrock, E.W., How to Manufacture a Quality Product. Quality Progress August 1977.
Smith, M.R. Making Quality Control Management Effective. Quality Progress, December 1974.
Stein, Barry A. Quality of Work Like in Action-Managing for effectiveness. American Management Association 1983.
Tarver, Mae-Goodwin. Systematic Process Analysis: An SQC Tool. Quality Progress September 1974.
Walton, R.E. How to Counter Alienation in the Plant. Harvard Business Review November-December 1972.

CHAPTER 5

Sampling for Product Evaluation and Line Control

Acceptance sampling procedures are usually used to make decisions whether to accept and/or reject lots of product already produced as opposed to control charts used during manufacture.

The sample is the greatest limiting factor in the successful control of product quality. How many to use? Where to obtain? How many to evaluate? These are some of the questions that must be answered. Briefly stated, the sample must be representative of the lot of merchandise in question and it must be selected at random. The point to emphasize is that a poor, inadequate sample is worthless for inspection or evaluation of product quality if one is attempting to identify the true quality of a given lot. A typical model used is as follows:

Lot⟶Sample⟶Sample Plan⟶Decision⟶Accept Lot
⟶Reject Lot

The U. S. Department of Agriculture has adopted a statistical sample plan and regulations governing inspection and certification of processed fruits and vegetables and related products which should be followed when evaluating food product quality. The specific procedures, including tables for sample plans, definitions, statistical quality control and acceptance levels are reproduced to provide a handy guide for the quality control technologist. These are taken from Title 7 (Part 52) of the Code of Federal Regulations.

Regulations Governing Inspection and Certification
52.2 Definitions

Acceptance number: the maximum number of deviants permitted in a sample of a lot that meets a specific requirement.

Case: number of containers (cased or uncased) which, by the particular industry, are ordinarily packed in a shipping container.

Certificate of sampling: a statement, written or printed, issued pursuant to the regulations in this part, identifying officially drawn samples. It may include a description of condition of containers and the condition under which the processed product is stored.

Class: a grade or rank of quality.

Condition: the degree of soundness of the product which may affect its merchantability and includes, but is not limited to, those factors which are subject to change as a result of age, improper preparation and processing, improper packaging, improper storage or improper handling.

Deviant: a sample unit affected by one or more deviations or one that varies in a specifically defined manner from the requirements of a standard, specification, or other inspection document.

Deviation: any specifically defined variation from a particular requirement.

Inspection service; types of:
  (a) "Lot inspection" means the inspection and grading of specific lots of processed fruits and vegetables which are located in plant warehouses, commercial storage, railway cars, trucks, or any other conveyance or storage facility. Generally under "lot inspection" the inspector does not have knowledge of conditions and practices under which the product is packed and his grading is limited to examination of the finished processed product only.
  (b) "Continuous inspection" is the conduct of inspecting and grading services in an approved plant whereby one or more inspector(s) are present at all times when the plant is in operation to make in-process checks on the preparation, processing, packing, and warehousing of all products under contract and to assure compliance with sanitary requirements.
  (c) "Pack certification" is the conduct of inspection and grading services in an approved plant whereby one or more inspector(s) perform inspection and grading services on designated lots. The inspector(s) may make in-process checks on the preparation and processing of products under contract but is not required to be present at all times the plant is in operation.

Inspector or grader: any employee for the Department authorized by the Secretary or any other person licensed by the Secretary to investigate, sample, inspect and certify in accordance with the regulation in this part to any interested party the class, quality and condition of processed products covered in this part and to perform related duties in connection with the inspection service.

Lot: any number of containers of the same size and type which contain a processed product of the same type and style located in the same warehouse or conveyance, or which, under in-plant (in-process) inspection, results from consecutive production within a plant, and which is available for inspection service at any one time: *Provided,* That the number of containers comprising a lot may not exceed the maximum number specified for a sample size of 60 as outlined in the sampling plans in 52.38: *And further provided,* That (a) if the applicant requests a separate inspection certificate covering a specific portion of a lot, such portion must be separately marked or otherwise identified in such a manner as to permit sampling, inspection, and certification of such portion as a separate lot; and (b) under in-plant (in-process) inspection, the inspector is authorized to limit the number of containers of a processed product that may be included in a lot to the production of a single working shift when such production is not in compliance with specified requirements.

Person: any individual, partnership, association, business trust, corporation, any organized group of persons (whether incorporated or not), the United States (including, but not limited to, any corporate agencies thereof), any State, County, or Municipal government, any common carrier, and any authorized agent or any of the foregoing.

Plant: the premises, buildings, structures, and equipment (including, but not being limited to machines, utensils, vehicles, and fixtures located in or about the premises) used or employed in the preparation, processing, handling, transporting, and storage of fruits and vegetables, or the processed products thereof.

Quality: the inherent properties of any processed product which determine the relative degree of excellence of such product, and includes the effects of preparation and processing, and may or may not include the effects of packing media, or added ingredients.

Rejection number: the minimum number of deviants in a sample that will cause a lot to fail a specific requirement.

Sample: any number of sample units to be used for inspection.

Sample unit: a container and/or its entire contents, a portion of the contents of a container or other unit of commodity, or a composite mixture of a product to be used for inspection.

Sampling: the act of selecting samples of processed products for the purpose of inspection under the regulation in this part.

52.13 Basis of inspection and grade or compliance determination:

(a) Inspection service shall be performed on the basis of the appropriate United States standards for grades of processed products, Federal, Military, Veterans Administration or other government agency specifications, written contract specifications, or any written specification or instruction which is approved by the Administration.

(b) Unless otherwise approved by the Administrator, compliance with such grade standards, specifications, or instructions shall be determined by evaluating the product, or sample, in accordance with the

requirements of such standards, specifications, or instructions: *Provided,* that when inspection for quality is based on any United States grade standard which contains a scoring system the grade indicated by the average of the total scores of the sample units: *Provided further,* that:
(1) Such sample complies with the applicable standards of quality promulgated under the Federal Food, Drug and Cosmetic Act;
(2) Such sample complies with the product description;
(3) Such sample meets the indicated grade with respect to factors of quality which are not rated by score points; and
(4) With respect to those factors of quality which are rated by score points, each of the following requirements is met:
  (i) None of the sample units falls more than one grade below the indicated grade because of any quality factor to which a limiting rule applies;
  (ii) None of the sample units falls more than 4 score points below the minimum total score for the indicated grade; and,
  (iii) The number of sample units classed as deviants does not exceed the applicable acceptance number indicated in the sampling plans contained in 52.38. A "deviant," as used in this paragraph, means a sample unit that falls into the next grade below the indicated grade.
(5) If any of the provisions contained in subparagraphs (3) and (4) of this paragraph are not met, the grade is determined by considering such provisions in connection with succeedingly lower grades until the grade of a lot, if assignable, is established.

There are, at least, four reasons for sampling:
1. Lot size is too large to 100% inspect,
2. Destructive testing is involved,
3. Savings of time and money, and
4. Means of making an inference about the quality of materials received and or produced.

How samples are drawn by inspectors or licensed samplers:

An inspector or a licensed sampler shall select samples, upon request, from designated lots of processed products which are so placed as to permit thorough and proper sampling in accordance with the regulations in this part. Such person shall, unless otherwise directed by the Administrator, select sample units of such products at random, and from various locations in each lot in such manner and number, not inconsistent with the regulations in this part, as to secure a representative sample of the lot. Samples drawn for inspection shall be furnished by the applicant at no cost to the Department.

Accessibility for sampling:

Each applicant shall cause the processed products for which inspection is requested to be made accessible for proper sampling. Failure to make any lot accessible for proper sampling shall be sufficient cause for postponing inspection service until such time as such lot is made accessible for proper sampling.

How officially drawn samples are to be identified:
Officially known samples shall be marked by the inspector or licensed sampler so such samples can be properly identified for inspection.

How samples are to be shipped:
Unless otherwise directed by the Administrator, samples which are to be shipped to any office of inspection shall be forwarded to the office of inspection serving the area in which the processed products from which the samples were drawn is located. Such samples shall be shipped in a manner to avoid, if possible, any material change in the quality or condition of the sample or the processed product. All transportation charges in connection with such shipments of samples shall be at the expense of the applicant and wherever practicable, such charges shall be prepaid by him.

Sampling plans and procedures for determining lot compliance:
  (a) Except as otherwise provided for in this section in connection with in-plant inspection and unless otherwise approved by the Administrator, samples shall be selected from each lot in the exact number of sample units indicated for the lot size in the applicable sampling plans: *Provided,* that at the discretion of the inspection service the number of sample units indicated for any one of the larger sample sizes provided for in the appropriate plans.
  (b) Under the sampling plans with respect to any specified requirement:
    (1) If the number of deviants (as defined in connection with the specific requirements) in the nonprescribed sample does not exceed the acceptance number of the next smaller sample size, the lot meets the requirement.
    (2) If the number of deviants (as defined in connection with the specific requirement) in the sample exceeds the acceptance number prescribed for the sample size, the lot fails the requirement.
  (c) If in the conduct of on-line in-plant inspection of a product covered by a grade standard which does not contain sampling plans, the sample is examined before the lot size is known and the number of sample units exceeds the prescribed larger sample sizes, the lot may be deemed to meet or fail a specific requirement in accordance with the following procedure:
    (1) If the number of deviants (as defined in connection with the specific requirement) in the nonprescribed sample does not exceed the acceptance number of the next smaller sample size, the lot meets the requirements;
    (2) If the number of deviants (as defined in connection with the specific requirement) in the nonprescribed sample equals the acceptance number prescribed for the next larger sample size, additional sample units shall be selected to increase the sample to the next larger prescribed sample size;
    (3) If the number of deviants (as defined in connection with the specific requirement) in the nonprescribed sample equals the

acceptance number prescribed for the next larger sample size, the lot fails the requirement.
(d) In the conduct of on-line in-plant inspection, sampling may be performed on a time interval basis. The sampling frequency shall be specified in an applicable grade standard or other procedural instruction approved by the Administrator.
(e) In the event that the lot compliance determination provisions of a standard or specifications are based on the number of specified deviations instead of deviants the procedures set forth in this section may be applied by substituting the word "deviation" for the word "deviant" wherever it appears.
(f) Sampling plans referred to in this section are those contained in Tables 5.1 thru 5.4. For processed products not included in these tables, the minimum sample size shall be the exact number of sample units prescribed in the table, container group, and lot size that, as determined by the inspector, most closely resembles the product, type, container, size and amount of product to be sampled.
(g) (1) *Sampling plan for dried figs:* For each 10,000 lb (or fraction of 10,000 lb) of product—six sample units accumulated into one composite (at least 200 figs). A sample unit is approximately 35 figs. Each composite will be examined separately, and all must meet.
(2) *Sampling plan for dried fruits other than dates and figs:* For each 15,000 lb (or fraction of 15,000 lb) of product—six sample units accumulated into one composite. A sample unit is approximately 16 oz of product. Each composite will be examined separately and all must meet.
(h) Factors which influence the type of sampling plan to use:
1. Purpose for inspection
   A. Accept-reject lot
   B. Evaluate average quality—cannot reject lot in may cases—
      so determine the quality level for the purpose of estimating payment on the basis of a redetermined sliding scale.
2. Nature of material to be tested:
   A. Homegentiy— where an essentially homogenous material is to be sampled (true solution), one small sample is sufficient.
   B. Unit size.
   C. History of material and supplier.
   D. Cost of material– where raw material itself constitutes a substantial part of sampling cost— may be necessary to reduce amount to be sampled.
3. Nature of test/lab procedures:
   A. Importance of test– critical– major–minor defects.
   B. Destructive vs non-destructive testing.
   C. Labor and equipment consumption
4. Nature of lot/population:
   A. Size/number of units involved
   B. Sublots– how are they treated–separately or combined
   C. How loaded– can a random sample be obtained?

5. Level of assurance wanted:
   Usual basis is optimum point to get the amount of information for the dollar invested.
6. Characteristic to be sampled:
   A. Attribute—qualitative characteristics measured as either good or bad—applies to factors which are control measurements not possible or practical, (defects such as dents, counts, discolored areas, etc.).
   B. Variable— quantitative data— applies to factors measured as continuous scale used to establish a level of quantity (weight, length, etc.).

The regulations governing the sampling procedures for determination of the nutrient composition of foods are as follows from Title 21, Paragraph 1.17 Food: Nutrition labeling.

(e) Compliance with this section shall be determined as follows:
   (1) A collection of primary containers or units of the same size, type, and style produced under conditions as nearly uniform as possible, designated by a common container code or marking, or in the absence of any common container code or marking a day's production, constitutes a "lot."
   (2) The sample for nutrient analysis shall consist of a composite of 12 sub-samples (consumer units), taken one from each of 12 different randomly chosen shipping cases to be representative of a lot . . . .

The following are some techniques used for lot acceptance:

1. No testing (sampling) whatever—no knowledge is gained prior to production shipment of quality levels.

2. One hundred percent (100%) inspection and sorting—only can be used for non-destructive testing. The major disadvantage is inspector boredom and the time required.

3. Arbitrary sampling—This is an attempt to compromise between 100% checking and no testing.

   Spot sampling—stops defective material on those lots inspected.

   Constant percentage sampling—better than above, however, this type of plan places a burden of over inspection of larger lots—it does not define rules of accepting or rejecting lots.

4. Scientific sampling—this is again a compromise between 100% inspection and no testing. The advantages are: (a) distinction between critical and less critical characteristics, (b) risks of making wrong decision are known, (c) all materials of same classification are subjected to a similar discriminating power.

Before using any sampling plan, the user must define the type and severity of defect and tests to be used to measure quality.

Typical sampling plans are given in Tables 5.1 thru 5.4. These tables show the sample size based on size of lot and indicate the acceptance number for each lot size. The lot is accepted if the sample does not exceed the acceptance number (c). The lot is rejected if the sample exceeds the acceptance number.

There are sampling errors and/or risks involved, that is, two types of errors associated with a decision making:

	Lot within specification	Lot not in specification
Accept (No Change)	Accept OK Decision	Accept Type II error Beta- Consumer risk
Reject (Change)	Reject Type I error Alpha-Producers risk	Reject OK decision

Alpha– Producers risks- rejecting when you should accept, (doing something when you should not have to)

Beta– Consumers risk- accepting when you should reject, (not doing something when you should have).

Operational Characteristic Curves

Probably the most valuable aid in selecting any sampling plan is the operational characteristic curve. This permits one to evaluate the operation of the sampling plan under varying conditions of incoming materials and of the products produced. It is an excellent method of illustrating the risks inherent in all sampling plants.

The Operating Curve (OC) depends upon two (2) parameters, that is, the sample size "N" and the acceptance number "C".

The acceptable quality level (AQL) is the level of quality that you want the plan to accept a 'high' percentage of the time. This is usually set by company policy to balance the cost of:
        1. Cost of finding and correcting a defective unit,
        2. Loss incurred if a defective unit gets out, and
        3. The purchasing agreement.

To make an OC curve or sampling plan more to your liking, the acceptance number "C" can be increased by shifting to the right to help the producer get greater acceptance of the lot. On the other hand, if the consumer wants a tighter plan, the acceptance number can be increased and shifted to left. If one wants to change both as far as risks are concerned, increase "N" with "C" increased proportionately. The curve becomes steeper and the plan becomes more discriminating.

The index (AQL) used to define quality should reflect the needs of the consumer and the producer and not be chosen primarily for statistical convenience. The sampling risks should be known in quantitative forms (OC curve). The producer should have adequate protection against the rejection of good lots and the consumer should be protected against acceptance of bad lots. The plan should minimize the total cost of inspection of all products. This requires careful evaluation of the pros and cons of attributes and variable plans, and single, double, and multiple plans. The plan should have built in flexibility to reflect changes in lot sizes, quality of product submitted, and other pertinent facts.

The advantages of sampling plans include the following: Savings in cost, that is the economy of sampling a few vs 100% inspection. Sampling plans are a must if one uses destructive testing. Sampling plans permit less time to perform the evaluation, less handling of products, fewer inspectors required, an improved level of quality acceptance, and most importantly, improved vendor and vendee relationships. However, there are Alpha and Beta risks involved, it takes time to document the prescribed procedures and there is less information when compared to 100% inspection. In using sampling plans one must eliminate bias in measurements and judgements.

REFERENCES

Anon. Mil. Std. 105D Sampling Procedures and Tables for Inspection by Attributes. 29 April 1963.

Anon. Mil Std. 414 Sampling Procedures and Tables for Inspection by Variable for Percent Defective, 11 June 1957.

Duncan, Acheson J. Acceptance Sampling Plans. ASTM Standardization. News Sept. 1975.

TABLE 5.1
Canned Or Similarly Processed Fruits, Vegetables, And Products Containing Units Of Such Size And Character As To Be readily Separable Sampling Plans And Acceptance Levels

Container Size Group	Lot Size (Number of Containers)[1]							
Group 1: Any type container of a volume not exceeding that of a No. 303 size can.	3,000 or less	3,001–12,000	12,001–39,000	39,001–84,000	84,001–145,000	145,001–228,000	228,001–336,000	336,001–480,000
Group 2: Any type of container of a volume exceeding that of a No. 303 size can but not exceeding that of a No. 3 cylinder size can.	1,500 or less	1,501–6,000	6,001–19,500	19,501–42,000	42,001–72,500	72,501–114,000	114,001–168,000	168,001–240,000
Group 3: Any type of container of a volume exceeding that of a No. 3 cylinder size can, but not exceeding that of a No. 12 size can.	750 or less	751–3,000	3,001–9,750	9,751–21,000	21,001–36,250	36,251–57,000	57,001–84,000	84,001–120,000
Group 4: Any type of container of a volume exceeding that of a No. 12 size can.	Convert to equivalent number of 6-pound net weight containers and use group 3							
Lot inspection:								
Sample size (number of sample units)[2]	3	6	13	21	29	38	48	60
Acceptance number	0	1	2	3	4	5	6	7
On-line in-plant inspection:								
Sample size (number of sample units)[2]	3	6	6	13	21	29	38	48
Acceptance number	0	1	1	2	3	4	5	6

[1]Under on-line in-plant inspection, a 5 percent overrun in number of containers may be permitted by the inspector before going to the next larger sample size.

[2]When a standard sample unit size is not specified in the U.S. grade standards, the sample units for the various container size groups are as follows: Groups 1, 2, and 3—1 container and its entire contents. Group 4 approximately 2 pounds of product. When determined by the inspector that a 2-pound sample unit is inadequate, a larger sample unit may be substituted.

TABLE 5.2
Frozen Or Similarly Processed Fruits, Vegetables, And
Products Containing Units Of Such Size And Character
As To Be Readily Separable

Container Size Group	Lot Size (Number of Containers)[1]							
Group 1: Any type of container of 1 pound or less net weight	2,400 or less	2,401– 9,600	9,601– 31,200	31,201– 67,200	67,201– 116,000	116,001– 182,400	182,401– 268,800	268,801– 384,000
Group 2: Any type of container over 1 pound but not over 2½ pounds net weight	1,200 or less	1,201– 4,800	4,801– 15,600	15,601– 33,600	33,601– 58,000	58,001– 91,200	91,201– 134,400	134,401– 192,000
Group 3: Any type of container over 2½ pounds	Convert to equivalent number of 2½-pound containers and use group 2							
Lot Inspection								
Sample size (number of sample units)[2]	3	6	13	21	29	38	48	60
Acceptance number	0	1	2	3	4	5	6	7
On-line in-plant inspection:								
Sample size (number of sample units)[2]	3	6	13	21	29	38	48	60
Acceptance number	0	1	1	2	3	4	5	6

[1]Under on-line in-plant inspection, a 5% overrun in number of containers may be permitted by the inspector before going to the next larger sample size.
[2]When a standard sample unit size is not specified in the U.S. standards, the sample units for the various groups are as follows: Groups 1 and 2—1 container and its entire contents. Group 3 containers up to 10 pounds—approximately 3 pounds of product. When determined by the inspector that a 3-pound sample unit is inadequate, a larger sample unit of 1 or more containers and their entire contents may be substituted for 1 or more sample units of 3 pounds.

TABLE 5.3
Canned, Frozen, Or Otherwise Processed Fruits,
Vegetables, Related Products Of A Comminuted Fluid
Or Homogeneous State

Container Size Group	Lot Size (Number of Containers)[1]								
Group 1: Any type of container of 1 pound or less	4,500 or less	4,501– 18,000	18,001– 58,000	58,501– 126,000	126,001– 217,000	217,001– 342,000	342,001– 504,000	504,001– 720,000	
Group 2: Any type of container exceeding 1 pound but not exceeding 60 ounces	3,000 or less	3,001– 12,000	12,001– 39,000	39,001– 84,000	84,001– 145,000	145,001– 223,000	228,001– 336,000	336,001– 480,000	
Group 3: Any type of container exceeding 60 ounces but not exceeding 10 pounds	1,500 or less	1,501– 6,000	6,001– 19,500	19,501– 42,000	42,001– 72,500	72,501– 114,000	114,001– 168,000	168,001– 240,00	
Group 4: Any type of container exceeding 10 pounds	Convert to equivalent number of 6-pound containers and use group 3								
Lot inspection:									
Sample size (number of sample units)[2]	3	6	13	21	29	38	48	60	
Acceptance number	0	1	2	3	4	5	6	7	
On-line in-plant inspection									
Sample size (number of sample units)[2]	3	6	6	13	21	29	38	48	
Acceptance number	0	1	1	2	3	4	5	6	

[1]Under on-line in-plant inspection, a 5% overrun in number of containers may be permitted by the inspector before going to the next larger sample size.
[2]When a standard sample unit size is not specified in the U.S. grade standards, the sample units for the various container size groups are as follows: Groups 1, 2, and 3—1 container and its entire contents. A smaller sample unit may be substituted in group 3 at the inspectors' discretion. Group 4—approximately 16 ounces of product. When determined by the inspector that a 16-ounce sample unit is inadequate, a larger sample unit may be substituted.

TABLE 5.4
Dehydrated (Low Moisture) Fruits And Vegetables

Container Size Group	Lot Size (Number of Containers)[1]							
Group 1: Any type of container of 1 pound or less, net weight	1,800 or less	1,801–7,200	7,201–23,400	23,401–54,000	50,401–87,000	87,001–136,000	136,801–201,000	201,601–288,000
Group 2: Any type of container over 1 pound but not over 6 pounds net weight	600 or less	601–2,400	2,401–7,800	7,801–16,800	16,801–29,000	29,001–45,600	45,601–67,200	67,201–96,000
Group 3: Any type of container over 6 pounds	Convert to equivalent number of 5-pound containers and use group 2							
Lot inspection:								
Sample size (No. of sample units)[2]	3	6	13	21	29	38	48	60
Acceptance number	0	1	2	3	4	5	6	7
One-line in-plant inspection Sample size (No. of sample units)[2]	3	6	6	13	21	29	38	48

[1] Under on-line in-plant inspection, a 5% overrun in number of containers may be permitted by the inspector before going to the next larger sample size.
[2] When a standard sample unit size is not specified in the U.S. standards, the sample units for the various groups are as follows: Groups 1 and 2—1 container and its entire contents. Group 3 containers up to 10 pounds—approximately 3 pounds of product. When determined by the inspector that a 3-pound sample unit is inadequate, a larger sample unit of 1 or more containers and their entire contents may be substituted for 1 or more sample units of 3 pounds.

FIGURE 5.5 — Operating Curve

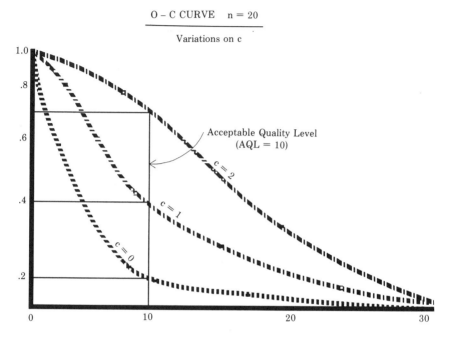

CHAPTER 6

Statistical Quality and Process Control

Statistical Quality Control (SQC) and Statistical Process Control (SPC) are most useful tools that can be of great value for the interpretation of processes and data. Many successful food firms today have established SQC and SPC procedures as part of their quality assurance program. They do not hire statisticians to do this but rely on well trained technical personnel to develop the statistical computer procedures and programs.

The interpretation of quality control data collected by the quality technologist is one of his more important functions in the successful operation of the plant. Management charges the technologist with the responsibilities of keeping the product within the specification limits of each grade for each product being processed.

In understanding a statistical quality control chart, one must first agree that variation is always present in the measured quality of manufactured products. This variation is composed of two components; that produced by "chance causes" and that produced by "assignable causes." Variation due to "chance causes" is inevitable. Variation due to "assignable causes" can usually be detected and corrected by appropriate methods.

SQC and SPC employs statistical principles and methods which have been developed to assess the magnitude of "chance cause variation" and to detect "assignable cause variation." SQC will indicate the limits beyond which these variations in the product should not go without correction. The manner in which the SQC-SPC program determines the variations is based on the laws of probability. Probability might be simply defined as the number of times an event occurs as to the total number possible. Thus, SQC-SPC is really sampling of the product, determining the quality variation of the sample, and relating the findings to the entire lot under consideration. In addition to the USDA Standards for Inspection by Variable and Determination of Fill Weights and the $\overline{X}$ and R charts, the following terms should be helpful to better understand the handling of SQC-SPC data.

(1) A *Histogram* is an easy and quick way to project distributions of variation. A histogram is defined as a vertical or horizontal bar

chart in which the adjacent columns have heights or lengths proportional to the number of observations as shown in Fig. 6.1.

FIGURE 6.1 — Histogram For Vacuums In Canned Soup

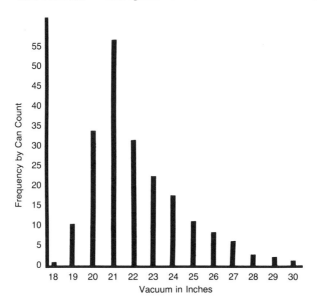

A frequency polygon is a line diagram where a line connects dots which represent the frequency of observations for a given dimension of value.

(2) *Measures of Central Tendency are:*
 (a) The *arithmetic mean* or *average* in statistics is referred to as $\overline{X}$, called X bar, and is defined as the quotient obtained by dividing the sum of a set (Σx) of readings or observations by the number (n) of observations or $\overline{X} = \Sigma x/n$ where Σ (S) means sum of. It usually does not describe any of the measurements from which it is computed.
 (b) The *median* is the reading or observation above and below which an equal number of observations fall.
 (c) The *mode* is the value that occurs most frequently; in the case of a Histogram it is the tallest or longest bar.

(3) *Measures of Dispersion*
  (a) The *standard deviation s,* called sigma, is the square root of the mean square of the deviations from the mean or $s = \sqrt{\Sigma(x - \bar{x})^2/n}$. It is a statistic used to express the amount of variation in a set of data.
  (b) *Range (R)* is defined as the difference between the largest and the smallest of a set of observations. It is one method of measuring the amount of variation.
(4) A *normal curve* is a distribution of individual values with the average, median, and mode the same. Further, the standard deviation divides the range of the set of data into six equal parts as follows:

Limits	% of Total Area
$\bar{X} \pm 1\ s$ (sigma)	68.26
$\bar{X} \pm 2\ s$ (sigma)	95.46
$\bar{X} \pm 3\ s$ (sigma)	99.73

See Fig. 6.2 for an illustration of this.

The sigma value is very useful when attempting to determine how large the variations are within the samples. In other words, the sigma

FIGURE 6.2 — Approximate Areas For A Typical Normal Curve

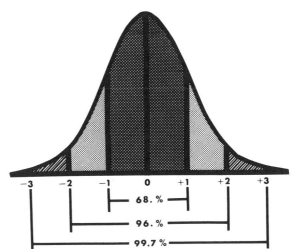

FIGURE 6.3 — Expected Weight Distribution Of Product In Control And Underweight

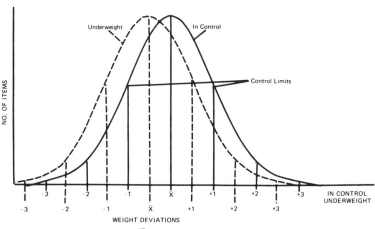

Note: $\overline{X}$ is the target weight.

value helps to determine what are the plus or minus limits from the average to include a portion of all of the values. One standard deviation indicates that 68% of the values will be within plus and minus one sigma from the average; two standard deviations indicate that 95% of the values will be within plus and minus two sigmas from the average; and three standard deviations indicate that 99% will be within plus and minus three sigma limits from the average. Thus, these values will inform management whether the product is within the specified tolerances.

Figure 6.3 illustrates the practical application of the expected weight distribution of items for a product or process in control, and for a product underweight or out of control.

SQC Data Form 6.1 for frequency distribution vacuum in a set of samples is typical of the data one may accumulate. Further, the calculations show these data (Fig. 6.1) skewed to the right indicating very high vacuums.

FIGURE 6.4 — Frequency Distribution For Vacuum In Canned Soup

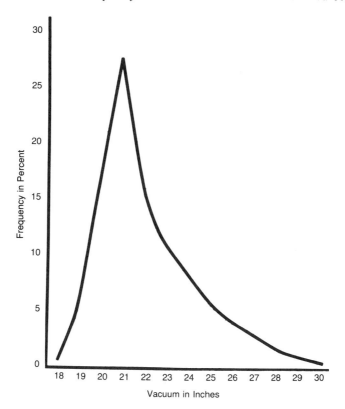

SQC DATA FORM 6.1
Frequency Distribution For Measured Variables

Class Interval (Value) (c)	Frequency	f	d	fd	fd²	%	Cumulative Frequency in % Ascending	Cumulative Frequency in % Descending
18	1	1	-4	-4	16	0.5	0.5	100.0
19	₩₩ ₩₩	10	-3	-30	90	5.0	5.5	99.5
20	₩₩ ₩₩ ₩₩ ₩₩ ₩₩ ₩₩ 111	33	-2	-66	132	16.5	22.0	94.5
21	₩₩ ₩₩ ₩₩ ₩₩ ₩₩ ₩₩ ₩₩ ₩₩ ₩₩ ₩₩ ₩₩	55	-1	-55	55	27.5	49.5	78.0
22	₩₩ ₩₩ ₩₩ ₩₩ ₩₩ ₩₩ 1	31	0	0	0	15.5	65.0	50.5
23	₩₩ ₩₩ ₩₩ ₩₩ 11	22	1	22	22	11.0	76.0	35.0
24	₩₩ ₩₩ ₩₩ 11	17	2	34	68	8.5	84.5	24.0
25	₩₩ ₩₩ 1	11	3	33	99	5.5	90.0	15.5
26	₩₩ 111	8	4	32	128	4.0	94.0	10.0
27	₩₩ 1	6	5	30	150	3.0	97.0	6.0
28	111	3	6	18	108	1.5	98.5	3.0
29	11	2	7	14	98	1.0	99.5	1.5
30	1	1	8	8	64	0.5	100.0	0.5
z=22	S (Total)	200		36	1030	100.0		

1. Standard Deviation (68%): $s = c\sqrt{\dfrac{Sfd^2}{f} - \left(\dfrac{Sfd}{f}\right)^2}$

$= 1\sqrt{\dfrac{1030}{200} - \left(\dfrac{36}{200}\right)^2} = \sqrt{5.15 - 0.0324} = \sqrt{5.1176} = 2.26$

2. Standard Deviation (95%): $2s = 4.52$
3. Standard Deviation (99%): $3s = 6.78$

$\overline{X}$ (Average) $= Z + \dfrac{c \times Sfd}{Sf} = 22 + \dfrac{1 \times 36}{200} = 22 + 0.18 = 22.18$

CV (Coefficient of Variability) $= \dfrac{s}{\overline{X}} \times 100 = \dfrac{2.26}{22.18} \times 100 = 0.102 \times 100 = 10.2\%$

STATISTICAL QUALITY AND PROCESS CONTROL   67

Figure 6.4 shows in percent the frequency distribution of these data, indicating 70% of the samples are 20-23 in. vacuum.

Figure 6.5 shows the ascending cumulative frequency distribution for these samples indicating that 10% of the samples have a vacuum less than 19 in.

Figure 6.6 shows the descending cumulative frequency data for these same samples indicating 10% have vacuums over 25 in. Thus, there are several methods of presenting statistical data.

FIGURE 6.5 — Ascending Cumulative Frequency Distribution
Of Vacuums For Canned Soup

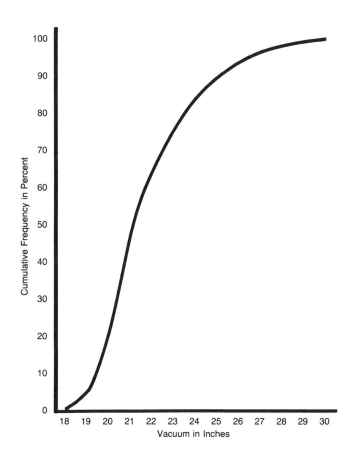

In addition to these SQC Data forms and Figures, see the USDA Standards for Inspection by Variables and Fill Weights (7 CFR 52.201 thru 52.210 and 52.221 thru 52.232).

FIGURE 6.6 — Descending Cumulative Frequency Distribution
Of Vacuums For Canned Soup

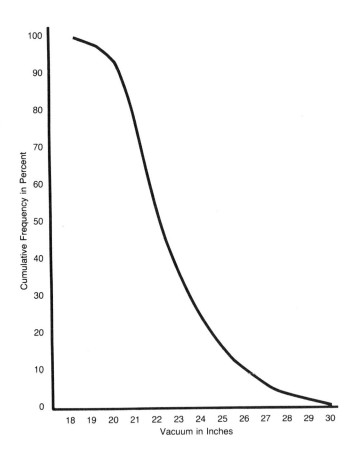

STATISTICAL QUALITY AND PROCESS CONTROL

TABLE 6.1
Definition Of Terms Used In Statistical Quality Control

Symbols	Read	Meaning
m	m	Number of subgroups in a sample
n	n	Total number of sample units or measurements in a sample $n = (m)(n_s)$
n_s	n sub s	Number of sample units or measurements in a subgroup
X	X	Value of an individual measurement for a variable
$\Sigma(S)$	sum of	Sum of a series of numbers (Σ means the sum of several measurements)
Mi	Mi	Median of all the individual measurements in a subgroup
$\overline{\text{Mi}}$	Mi double bar	Median of all the individual measurements or subgroup medians (Mi) in a sample
$\bar{X}$	X bar	Average of all individual measurements in a subgroup
$\bar{\bar{X}}$	X double bar	Arithmetic mean of all the individual measurements in a sample. When the average is calculated for each subgroup in a sample for conventional averages, $\bar{\bar{X}}$ is also the average of the subgroup averages
R	Range	A range of measurements, the difference between the highest and lowest measurement within a subgroup
$\bar{R}$	R bar	Average range of all the subgroup ranges
s	sigma	Standard deviation of the averages. The width of one zone in the normal distribution curve of individual items called a standard deviation. Plus and minus 3 sigmas from the average includes 99% of the normal curve
$s\bar{x}$	$s\ x$ bar	Standard deviation of the averages
*$\bar{X}'_{max}$	X bar prime max	A specified maximum lot average value
$\bar{X}'_{max}$ adjusted	X bar prime max adjusted	$\bar{X}'_{max}$ plus a sampling allowance
*$\bar{X}'_{min}$	X bar prime min	A specified minimum lot average value
$\bar{X}'_{min}$ adjusted	X bar prime min adjusted	$\bar{X}'_{min}$ minus a sampling allowance
*$\bar{R}'$	R bar prime	A specified average range value
*R_{max}	R max	A specified maximum range for a subgroup
$UCL_{\bar{x}}$	Upper control limit for averages	Upper and lower control limits for averages determine the pattern that sample averages should follow if a constant system of chance is operating. If the averages do not conform to this pattern, there is an assignable cause present
$LCL_{\bar{x}}$	Lower control limit for averages	
UCL_R	Upper control limit for ranges	Upper control limit for range sets the pattern within which sample ranges should fall. If the sample ranges do not conform to this pattern, there is an assignable cause present
*LRL	Lower reject limit for individual measurements	Lowest value an individual measurement may have without causing the production to be rejected for failure to meet prescribed requirements for individual measurements

TOTAL QUALITY ASSURANCE

TABLE 6.1 (Continued)

Symbols	Read	Meaning
*$LRL_{\bar{X}}$	Lower reject limit for subgroup averages or medians	Lowest value the average or median of a subgroup may have without causing the production to be rejected for failure to meet prescribed requirements for subgroup averages.
*LWL	Lower warning limit for individual measurements	This value serves as a warning point that the production may have reached a level where the chances of subsequently finding an individual measurement that will fall below the LRL have increased to a degree that the production may be in danger of rejection.
*$LWL_{\bar{X}}$	Lower warning limit for subgroup averages or medians	This value serves as a warning point that the quality of the production may have reached a level where the chances of subsequently finding a subgroup average or median that will fall below LRLx have increased to a degree that the production may be in danger of rejection.
URL	Upper reject limit for individual measurements	The highest value an individual measurement may have without causing the production to be rejected for failure to meet prescribed requirements for individual measurements.
URL_X	Upper reject limit for subgroup averages or medians	The highest value the average or median of a subgroup may have without causing the production to be rejected for failure to meet prescribed requirements for individual measurements.
UWL	Upper warning limit for individual measurements	This value serves as a warning point that the quality of the production may have reached a level where the chances of subsequently finding an individual measurement that will exceed the URL have increased to a degree that the production may be in danger of rejection.
UWL_x	Upper warning limit for subgroup averages or medians	This value serves as a warning point that the quality of production may have reached a level where the chances of subsequently finding a subgroup average or median that will exceed the URLx have increased to a degree that the production may be in danger of rejection.
Cp	Process Capability	This value serves as a guide as to how well the process is in control. It is calculated by dividing the specification width (Upper Specification-Lower Specification (6 sigma). Some users calculate the process capability index by $2A_r\bar{R}$ (See Table 6.2 for A_r) if the process is stable
CPK	Capability Index	This Index measures the improvement of the process as companies seek greater uniformity around the desired target. It is calculated by dividing the specification width by the process width. The greater the number the better the index.

*These limits and values are to be established and incorporated in the USDA Grade Standards for the various products. These values will be available upon request to: Chief Processed Products Standardization and Inspection Branch Fruit and Vegetable Division, AMS, U.S. Department of Agriculture, Washington, D.C. 20250

STATISTICAL QUALITY AND PROCESS CONTROL

Common causes are the basis for statistical process control to help the technologist or, more importantly, the operator in making a decision to adjust or not to adjust his or her process.

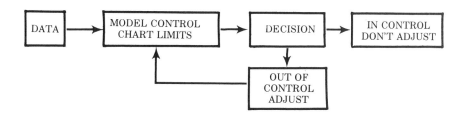

The control chart ($\overline{X}$ and R) takes time into consideration, whereas the distribution curve does not.

The control chart illustrates the three characteristics of data, that is, pattern, location, and variations.

If the points do not fall within the control limits, there is an assignable cause present. If the points do fall within the control limits, it means that a constant system of chance is operating. It is possible for a process to be statistically in control and not be at a satisfactory level. The $\overline{X}$ and R chart is used to study process capability and also to control a variable whose value can be expressed numerically. After a process has been brought under control at a satisfactory level, the control chart indicates significant changes. In control chart work, speed is the essence of effectiveness. If a *sample average* falls outside the control limits, the conclusion can be drawn that almost certainly the out-of-control *average point* does not come from the same system as the others; in operating terms it means that an assignable cause is prevalent and must be discovered and corrected.

The concept of the statistical control chart, very simply stated, is that if values which reflect the variation caused by "chance causes" present in a process are plotted on a time basis, the statistical limits can be determined within which such values will lie. Values falling outside these statistical limits will indicate the occurrence of significant changes in the "chance cause system," usually because of the presence of an assignable cause. On the $\overline{X}$ control chart these statistical limits are defined in the USDA Standards for Inspection, Part 52.201.

The range (R) chart is also a control chart. It indicates the difference between the highest and lowest value. Thus, it indicates the variance present in a set of samples. The R chart has an upper range limit (Rmax). The height of Rmax is determined by the "chance cause" variations and thus when a range exceeds this value it is usually due to an assignable cause.

It is necessary to use both the $\overline{X}$ and R chart in conjunction with each other because the $\overline{X}$ chart may indicate a consistent quality, but the range could vary from a minimum amount to a large extent showing the variability among the samples of product.

The following example may clarify any misconceptions: five samples were taken every hour for 10 hr off a given production line. The average, $\overline{X}$, was computed as well as the range. The average of the five samples was plotted on an $\overline{X}$ chart and similarly the range plotted on an R chart (SQC Form 6.2). A glance at the R chart indicates that there was a wide range for the first 2 hr, which narrowed down somewhat the third hour and widened the fourth hour and then gradually narrowed down to a low range the last three hours. The low range is indicative of greater uniformity. If the attribute of quality being evaluated has a low range which is within the UCL and LCL the operation is running under normal conditions.

SQC FORM 6.2
Statistical Quality Control Record And Data Form

Product_____ Size of Container_____ Code_____ Plant_____

Sample No.	\multicolumn{10}{c}{Frequency of Sample Sets by Day Shift for Line 4}	$\overline{\overline{X}}$	$\overline{R}$
	7am	8am	9am	10am	11am	12	1pm	2pm	3pm	4pm		
1	18.5	15.2	16.3	19.1	18.7	15.9	16.8	16.0	16.0	16.1		
2	17.0	15.3	14.8	18.4	18.3	15.2	15.8	16.1	16.2	16.0		
3	16.5	18.4	14.6	18.6	17.7	14.8	16.4	16.3	16.5	16.0		
4	16.8	15.0	15.1	16.1	16.2	14.1	15.8	16.0	16.1	16.1		
5	15.0	15.0	15.0	17.5	17.9	15.4	14.9	16.2	16.0	16.2		
Sum of X values	83.8	78.9	75.8	89.7	88.8	75.4	79.7	80.6	80.8	80.4		
$\overline{X}$	16.8	15.8	15.2	17.9	17.8	15.1	15.9	16.1	16.2	16.1	16.29	
R	3.5	3.4	1.7	3.0	2.5	1.8	1.9	0.3	0.5	0.2		1.88

Note 1: $UCL_{\overline{X}}$ (Upper control limit for average) = $\overline{\overline{X}} + A_2 \overline{R}$
$LCL_{\overline{X}}$ (Lower control limit for average) = $\overline{\overline{X}} - A_2 \overline{R}$
$UCL_{\overline{R}}$ (Upper control limit for range) = $D_4 \overline{R}$

Note 2: A_2 for 5 sample numbers in a set = 0.58.
D_4 for 5 sample numbers in a set = 2.11.
For these calculations see Table 6.2.

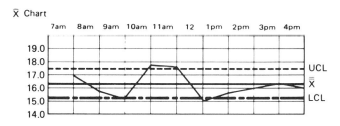

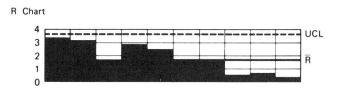

A flattening of the distribution attributable to material instability, operator carelessness, bearing wear in the machine, and innumerable other causes will be reflected by the appearance of a range of a sample above the URL. The process control chart thus provides certain and immediate information about the pattern of variation expected from the process, and affords prompt signals of trouble or of the absence of trouble.

When developing an $\overline{X}$ and R chart one should first determine the objectives in using these charts. In most cases this will be to provide a basis for taking corrective action, that is, when the process is out of control. The next decision is that one must select a method to measure these variables. The method and size of sample must next be determined. What sample size should be used for $\overline{X}$ and R chart technique? The answer to this question varies with the purpose of the chart and the characteristics of the process.

Samples should be selected so that they are representative of the time interval covered. All methods, measurements, and procedures should be followed strictly since an alteration in any one of these factors may cause a significant change in the data being collected.

When data are being recorded, any conditions that have changed since the last sample was taken should also be recorded. These include such items as changes in operators, product quality, and machine settings. The chart is plotted originally without the benefit of control limits until sufficient data have been collected so that the control limits computed will be reasonably reliable.

In setting up a SQC program, one should start with one given attribute of quality and learn all there is to know about this before moving on to other attributes. Take an example such as a filler on a given line. One should evaluate each "pocket" by drawing 5 samples with a 16 pocket filler; 80 samples would be taken (each, of course, carefully coded as to its pocket).

If it is a volume filler, determine the actual volume on product and plot it out on an $\overline{X}$ and R chart with the 16 pockets versus 5 samples for each pocket. If any of these are out of order, adjustments are made accordingly. Then resample as before to assure adjustments were made correctly. Then one would sample hourly, 5 samples at random and develop and $\overline{X}$ and R chart for each hour of operation. This chart should be prepared and mounted on a clipboard at the filler station for the operator to observe and make the proper changes during the process.

The $\overline{X}$ and R charts themselves will not correct a situation. They will only tell the operator where to look for the trouble. Once he has found the source of trouble, the operator must take appropriate action in order to control the particular unit operation within the established limits or specifications.

STEPS IN CONSTRUCTION OF VARIABLE CONTROL CHARTS (X-R)

1. Data should be taken in subgroups-units taken back to back over a short interval of time.
2. Subgroups should all be the same size.
3. Obtain the Average ($\overline{X}$)/subgroup.
4. Obtain the Range (R)/subgroup.
5. Obtain the Average Range ($\overline{R}$) for all subgroups.
6. Obtain the Average of the Averages or General Average ($\overline{\overline{X}}$).
7. Plot Average ($\overline{X}$) and Range(R) on the chart.
8. Calculate the Control Limits ($\overline{\overline{X}} \pm A_2\ \overline{R}$) around the target value.
9. Calculate Range Limits: ($D_3\overline{R}$) Lower Control Limit Range ($LCLR$) and ($D_4\overline{R}$), Upper Control Limit Range ($UCLR$).

Note: Factors A_2, D_3, and D_4 (Table 6.2) are calculated such that only one time in 1000 will a plotting fall outside of the control limits due to chance or natural variation or 999 times out of 1000 when a plot falls outside the calculated limits, it is due to assignable causes and must be corrected.

TABLE 6.2
Factors For Computing Variable Control Limits

Sample Size n	A_1	Factors for Averages A_2	Factors for Range D_3	Factors for Range D_4	Factor for Standard Deviation or Sigma d_2	Sample Size n
2	3.760	1.880	0.0	3.268	1.128	2
3	2.394	1.023	0.0	2.574	1.693	3
4	1.980	0.729	0.0	2.282	2.059	4
5	1.596	0.577	0.0	2.114	2.326	5
6	1.410	0.483	0.0	2.004	2.534	6
7	1.277	0.419	0.076	1.924	2.704	7
8	1.175	0.373	0.136	1.864	2.847	8
9	1.094	0.337	0.184	1.816	2.970	9
10	1.028	0.308	0.223	1.777	3.078	10
11	0.973	0.285	0.256	1.744	3.173	11
12	0.925	0.266	0.284	1.717	3.258	12
13	0.884	0.249	0.308	1.692	3.336	13
14	0.848	0.235	0.329	1.671	3.407	14
15	0.817	0.223	0.348	1.652	3.472	15

Formulas for computing control limits:

For Averages
$$UCL = \overline{\overline{X}} + A_2\overline{R}$$
$$LCL = \overline{\overline{X}} - A_2\overline{R}$$

For Range
$$UCL_R = D_4\overline{R}$$
$$LCL_R = D_3\overline{R}$$

Standard Deviation or Sigma (s)
$$s = \frac{\overline{R}}{d_2}$$

Note: $\overline{\overline{X}}$ = the average for a set of $\overline{X}$ values or the grand average
$\overline{R}$ = the average for a set of R values
All factors in Table are based on the normal distribution

Why are Control Limits Placed on Chart?

Control limits provide a basis for action, because they discriminate between causes of variability that can be discovered (assignable causes) and causes of variability that cannot be discovered (chance causes).

If a sample average falls outside the estimated control limits, the conclusion can be drawn that almost certainly the out-of-control average point does not come from the same system as the others; in operating terms this indicates a major source of extra variability between the samples; that is, an assignable cause so large that an investigation probably will uncover it, making possible corrective action.

Some SQC problems require the testing or measuring of the significance between two sample means. Howes (1954) devised a simplified form for carrying through a t^2 test using an F table and a table of reciprocals and the following formulas (see SQC form 6.3 & 6.4)

TOTAL QUALITY ASSURANCE

$$F = \frac{\dfrac{(\overline{X}_1 - \overline{X}_2)^2}{S'^2}}{\dfrac{1}{n_1} + \dfrac{1}{n_2}}$$

Step 1. Subtract the means and square the difference.
Step 2. Form the pooled sum of squares:

$$s'^2 = \frac{x_1^2 - x_2^2}{n_1 - n_2 - 2}$$

Step 3. Add the reciprocals of n_1 and n_2 ($1/n_1 + 1/n_2$)
Step 4. Multiply the pooled sum obtained in step 2 by the sum of the reciprocals obtained in step 3.
Step 5. Divide the squared difference between means obtained in step 1 by the product obtained in step 4.
Step 6. Compare with a standard F table for 1 and $n_1 + n_2 - 2$ degrees of freedom *(df)*. (Read 1 *df* at the top of the F table and $n_1 + n_2 - 2$ *df* at the left hand column of the F table.)

SQC DATA FORM 6.3
Statistical Analysis

Technologist ——————————————— Date ———

Code ————   Code ————

Sample	X_1	X_1^2	X_2	X_2^2
Sums				
Average				

STATISTICAL QUALITY AND PROCESS CONTROL 77

SQC FORM 6 4
Testing The Significance Between Two Sample Means

	Method I			Method II	
N_1	X_1	X_1^2	N_2	X_2	X_2^2
1	0.020	0.000400	1	0.010	0.000100
2	0.020	0.000400	2	0.022	0.000484
3	0.015	0.000225	3	0.029	0.000841
4	0.003	0.000009	4	0.009	0.000081
5	0.015	0.000225	5	0.036	0.001296
6	0.014	0.000196	6	0.023	0.000529
7	0.010	0.000100	7	0.010	0.000100
8	0.028	0.000784	8	0.025	0.000625
9	0.020	0.000400	9	0.021	0.000441
10	0.008	0.000064	10	0.034	0.001156
11	0.010	0.000100	11	0.010	0.000100
12	0.031	0.000961	12	0.010	0.000100

$X_1 = 0.194$ $X_1^2 = 0.003864$ $X_2 = 0.239$ $X_2^2 = 0.005853$
$\bar{X}_1 = 0.0161 = X_1/N_1$ $\bar{X}_2 = 0.0199 = X_2/N_2$

Notes:

1. $(\bar{X}_1 - \bar{X}_2)^2 = (0.0161 - 0.0199)^2 = 0.00001444$

2. $\Sigma X'^2_1 = \Sigma X_1^2 - (\Sigma X_1)^2/N = 0.003864 - (0.194)^2/12 =$
   $ 0.003864 - 0.003136 = 0.000728$
   $\Sigma X'^2_2 = \Sigma X_2^2 - (\Sigma X_2)^2/N = 0.005853 - (0.239)^2/12 =$
   $ 0.005853 - 0.004760 = 0.001093$

3. $(1/N_1 + 1/N_2) = (1/12 + 1/12) = 0.0833 + 0.0833 = 0.1666$

4. $S'^2 = \dfrac{\Sigma X_1^2 + \Sigma X_2^2}{N_1 + N_2 - 2} = \dfrac{0.009717}{22} = 0.00044168$

5. $(S'^2)(1/N_1 + 1/N_2) = (0.00044168)(0.1666) = 0.0000735839$

6. $F = (\bar{X}_1 - \bar{X}_2)^2 / ((S'^2)(1/N_1 + 1/N_2)) = 0.00001444 / 0.0000735839 = 0.19624$

The F table shows a 5% value of 4.30 for $N_1 + N_2 - 2 = 22$ degrees of freedom which is larger than 0.19624. Therefore, no significant difference between these means has been shown.

Analysis of Variance (ANOVA)

When a process has two or more factors affecting the variability, the interpretation of the relative contribution of each of these factors to the total variability can be made using a statistical technique known as "the analysis of variance." This statistical technique is based on the fact that

SQC FORM 6.5
Two-Way Analysis Of Variance

Summary of Data:

Column

Row	A	B	C	D	E	etc.	Sum	Mean
1								
2								
3								
etc.								
Sum			Sum of Columns				Grand Sum	

Analysis of Data

Source of Variance	Sum of Squares	Degrees of Freedom	Mean Square	F
Variate i-columns	C	(# of Columns -1)	$x=$ $\dfrac{(C)}{\text{\# of Col. -1}}$	x/z
Variate j-rows	D	(# of Rows -1)	$y=$ $\dfrac{(D)}{\text{\# of rows -1}}$	or y/z
Error	E	(# of columns -1) x (# of Rows -1)	$\dfrac{(E)}{(\text{Col. -1})(\text{Rows -1})} = z$	
Total	B			

A. (Grand Sum)^2 / total number of observations = c
B. (Square all observations and add) - c
C. [(Sum of observations in one column squared and add to same for other columns) / (# of rows)] - c
D. [(Sum of observations in one row squared and add to same for rows) / (# of columns)] - c
E. *(B) - [(C) + (D)]*

the total variance of the process is equal to the sum of the components' variances if the factors are acting independently. The advantage of this method is that where several factors are involved in a process or in the quality of a product, it is possible to rank their effects on the total variation in order of magnitude. Thus, the information is useful in the determination as to where to place efforts to reduce variability, for maximum improve-

SQC FORM 6.6
Two-Way Analysis Of Variance For Lbs. Yield Of Green Beans

	Year					
Cultivar	1	2	3	4	Sum	Average
GP 467	26	20	30	77	153	38.2
Early Gallaten	32	41	25	92	190	47.5
GP 68-115	50	55	45	59	209	52.2
GP 317	10	30	30	45	115	28.8
Wondergreen	35	40	36	85	196	49.0
Sum	153	186	166	358	863	
Average	30.6	37.2	33.2	71.6		

A. $863^2 / 20 = 37238.45$
B. $46001 - 37238.45 = 8762.55$
C. $213725/5 - 37238.45 = 5506.55$
D. $154831.00/4 - 37238.45 = 1469.30$
E. 1786.70
   See SQC Form 6.5 for equations.

Source	Sum of Squares	Degrees of Freedom	Mean Square	F
Year	5506.55	3	1835.52	12.33
Cultivar	1469.30	4	367.33	2.47
Error	1786.70	12	148.89	
Total	8762.55			

Book F values = 3.49 at 5% level of significance
              5.95 at 1% level of significance

LSD for Year:
  $= \sqrt{2/5 \times 148.89 \times 5.95} = 18.82$

Interpretation

Year	Average Yield
1	30.6
2	37.2
3	33.2
4	71.6

Thus the fourth year had a significantly higher yield than years 1, 2 and 3. There were no significant differences in the yield among the cultivars.

ment, with a minimum expenditure of time and effort. An example of presenting a two-way analysis of variance is given in SQC Form 6.5 and 6.6.

The interpretation of the F statistic is significant if it is equal to or greater than the value from the F-table at the level of significance selected. The level of significance commonly chosen is either 0.01 or 0.05. If F is significant then it can be stated that there exists a real difference within the variant. Significance at the 0.01 level means that the probability of the statement being incorrect is 1 in 100 and for the 0.05 level 1 in 20.

The value from the F-table is obtained by reading the degree of freedom of the variant in the rows or columns on the abscissa and the degrees of freedom of the error term on the ordinate. For example if the degrees of freedom of the row value is 9 and the degrees of freedom of the error term is 50, then the F value for 0.05 level is 2.07 and the F value for the 0.01 level is 2.78. (See Tables 6.3 and 6.4).

If the F statistic is significant the least significant difference (LSD) is calculated to determine where the differences among the variants lie. The LSD is the minimum by which the means of any two rows of columns significantly differ from each other. If the F statistic is not significant at the chosen level the LSD is not calculated.

The least significant difference is calculated as follows.

$$LSD = \sqrt{\left(\frac{2}{\text{no. cols. or rows}^a}\right) \times \text{Mean square of error} \times F^b \text{ (0.01 or 0.05)}}$$

(a) Divide by number of rows if LSD for columns is being determined or vice versa.

(b) F (0.01 or 0.05) is obtained from the F table at the level of significance selected in the analysis of variance.

F Value is the same as for obtaining the F-value in the analysis of variance.

The quality control technologist should find these statistical interpretations to be of value in at least two ways during the processing of any given product.

(1) *Machine or process control:* This usually means taking a series of samples from the line and measuring given attributes of quality to determine if such machines as size graders, fillers, closing machines, etc., are within the specified working tolerances.

(2) *Quality determination:* This refers to inspection where man is the factor determining the uniformity of the samples. Inspection may apply all along the line, starting with the receipt of the raw product, during the process, and with the finished product. Such attributes of quality as color, defects, net and drained weight, or maturity are evaluated.

TABLE 6.3
5% Points For The Distribution of F

f1 degrees of freedom (for greater mean square)

f2	1	2	3	4	5	6	7	8	9	10	100	500
1	161	200	216	225	230	234	237	239	241	242	253	254
2	18.51	19.00	19.16	19.25	19.30	19.33	19.36	19.37	19.38	19.39	19.49	19.50
3	10.13	9.55	9.28	9.12	9.01	8.94	8.88	8.84	8.81	8.78	8.56	8.54
4	7.71	6.94	6.59	6.39	6.26	6.16	6.09	6.04	6.00	5.96	5.66	5.64
5	6.61	5.79	5.41	5.19	5.05	4.95	4.88	4.82	4.78	4.74	4.40	4.37
6	5.99	5.14	4.76	4.53	4.39	4.28	4.21	4.15	4.10	4.06	3.71	3.68
7	5.59	4.74	4.35	4.12	3.97	3.87	3.79	3.73	3.68	3.63	3.28	3.24
8	5.32	4.46	4.07	3.84	3.69	3.58	3.50	3.44	3.39	3.34	2.98	2.94
9	5.12	4.26	3.86	3.63	3.48	3.37	3.29	3.23	3.18	3.13	2.76	2.72
10	4.96	4.10	3.71	3.48	3.33	3.22	3.14	3.07	3.02	2.97	2.59	2.55
11	4.84	3.98	3.59	3.36	3.20	3.09	3.01	2.95	2.90	2.86	2.45	2.41
12	4.75	3.88	3.49	3.26	3.11	3.00	2.92	2.85	2.80	2.76	2.35	2.31
13	4.67	3.80	3.41	3.18	3.02	2.92	2.84	2.77	2.72	2.67	2.26	2.22
14	4.60	3.74	3.34	3.11	2.96	2.85	2.77	2.70	2.65	2.60	2.19	2.14
15	4.54	3.68	3.29	3.06	2.90	2.79	2.70	2.64	2.59	2.55	2.12	2.08
16	4.49	3.63	3.24	3.01	2.85	2.74	2.66	2.59	2.54	2.49	2.07	2.02
17	4.45	3.59	3.20	2.96	2.81	2.70	2.62	2.55	2.50	2.45	2.02	1.97
18	4.41	3.55	3.16	2.93	2.77	2.66	2.58	2.51	2.46	2.41	1.98	1.93
19	4.38	3.52	3.13	2.90	2.74	2.63	2.55	2.48	2.43	2.38	1.94	1.90
20	4.35	3.49	3.10	2.87	2.71	2.60	2.52	2.45	2.40	2.35	1.90	1.85
21	4.32	3.47	3.07	2.84	2.68	2.57	2.49	2.42	2.37	2.32	1.87	1.82
22	4.30	3.44	3.05	2.82	2.66	2.55	2.47	2.40	2.35	2.30	1.84	1.80
23	4.28	3.42	3.03	2.80	2.64	2.53	2.45	2.38	2.32	2.28	1.82	1.77
24	4.26	3.40	3.01	2.78	2.62	2.51	2.43	2.36	2.30	2.26	1.80	1.74
25	4.24	3.38	2.99	2.76	2.60	2.49	2.41	2.34	2.28	2.24	1.77	1.72
30	4.17	3.32	2.92	2.69	2.53	2.42	2.34	2.27	2.21	2.16	1.69	1.64
50	4.03	3.18	2.79	2.56	2.40	2.29	2.20	2.13	2.07	2.02	1.52	1.46
100	3.94	3.09	2.70	2.46	2.30	2.19	2.10	2.03	1.97	1.92	1.39	1.30
200	3.89	3.04	2.65	2.41	2.26	2.14	2.05	1.98	1.92	1.87	1.32	1.22

Many Advantages Accrue from a Statistically Controlled Process
(1) Personnel efficiency is increased.
(2) Supervisors are kept alert.
(3) Raw materials are economically used.
(4) Tooling and machine problems are reduced.
(5) Estimates and bids can be made with greater assurance.
(6) Sound predictions can be made.
(7) Work flow becomes smoother.

TABLE 6.4
1% Points For The Distribution Of F

f1 degrees of freedom (for greater mean square)

f2	1	2	3	4	5	6	7	8	9	10	100	500
1	4,052	4,999	5,403	5,625	5,764	5,859	5,928	5,981	6,022	6,056	6,334	6,361
2	98.49	99.00	99.17	99.25	99.30	99.33	99.34	99.36	99.38	99.40	99.49	99.50
3	34.12	30.82	29.46	28.71	28.24	27.91	27.67	27.49	27.34	27.23	26.23	26.14
4	21.20	18.00	16.69	15.98	15.52	15.21	14.98	14.80	14.66	14.54	13.57	13.48
5	16.26	13.27	12.06	11.39	10.97	10.67	10.45	10.27	10.15	10.05	9.13	9.04
6	13.74	10.92	9.78	9.15	8.75	8.47	8.26	8.10	7.98	7.87	6.99	6.90
7	12.25	9.55	8.45	7.85	7.46	7.19	7.00	6.84	6.71	6.62	5.75	5.67
8	11.26	8.65	7.59	7.01	6.63	6.37	6.19	6.03	5.91	5.82	4.96	4.88
9	10.56	8.02	6.99	6.42	6.06	5.80	5.62	5.47	5.35	5.26	4.41	4.33
10	10.04	7.56	6.55	5.99	5.64	5.39	5.21	5.06	4.95	4.85	4.01	3.93
11	9.65	7.20	6.22	5.67	5.32	5.07	4.88	4.74	4.63	4.54	3.70	3.62
12	9.33	6.93	5.95	5.41	5.06	4.82	4.65	4.50	4.39	4.30	3.46	3.38
13	9.07	6.70	5.74	5.20	4.86	4.62	4.44	4.30	4.19	4.10	3.27	3.18
14	8.86	6.51	5.56	5.03	4.69	4.46	4.28	4.14	4.03	3.94	3.11	3.02
15	8.68	6.36	5.42	4.89	4.56	4.32	4.14	4.00	3.89	3.80	2.97	2.89
16	8.53	6.23	5.29	4.77	4.44	4.20	4.03	3.89	3.78	3.69	2.86	2.77
17	8.40	6.11	5.18	4.67	4.34	4.10	3.93	3.79	3.68	3.59	2.76	2.67
18	8.28	6.01	5.09	4.58	4.25	4.01	3.85	3.71	3.60	3.51	2.68	2.59
19	8.18	5.93	5.01	4.50	4.17	3.94	3.77	3.63	3.52	3.43	2.60	2.51
20	8.10	5.85	4.94	4.43	4.10	3.87	3.71	3.56	3.45	3.37	2.53	2.44
21	8.02	5.78	4.87	4.37	4.04	3.81	3.65	3.51	3.40	3.31	2.47	2.38
22	7.94	5.72	4.82	4.31	3.99	3.76	3.59	3.45	3.35	3.26	3.42	2.33
23	7.88	5.66	4.76	4.26	3.94	3.71	3.54	3.41	3.30	3.21	2.37	2.28
24	7.82	5.61	4.72	4.22	3.90	3.67	3.50	3.36	3.25	3.17	2.33	2.23
25	7.77	5.57	4.68	4.18	3.86	3.63	3.46	3.32	3.21	3.13	2.29	2.19
30	7.56	5.39	4.51	4.02	3.70	3.47	3.30	3.17	3.06	2.98	2.13	2.03
50	7.17	5.06	4.20	3.72	3.41	3.18	3.02	2.88	2.78	2.70	1.82	1.71
100	6.90	4.82	3.98	3.51	3.20	2.99	2.82	2.69	2.59	2.51	1.59	1.46
200	6.76	4.71	3.88	3.41	3.11	2.90	2.73	2.60	2.50	2.41	1.48	1.33

(8) The study and analysis of the process develops a wealth of information and leads to many improvements.

TABLE 6.5
Allowance Score Points For Drained Weight Indices
And Minimum Weights Of Tomatoes For
Commonly Used Containers

U.S. Grade Classification	Score Points	Drained Weight Index	Container Sizes						
			8Z Tall	Picnic	No. 303	No. 2	No. 2½	No. 3 cyl.	No. 10
			Minimum Drained Weight in Ounces						
Whole	20	62	5.4	6.8	10.5	12.7	18.5	32.0	67.9
	19	60	5.2	6.5	10.1	12.3	17.9	31.0	65.7
	18	58	5.0	6.3	9.8	11.9	17.3	30.0	63.5
	17	54	4.7	5.9	9.1	11.1	16.1	27.9	59.1
	16	50	4.3	5.5	8.5	10.3	14.9	25.8	54.7
A	20	70	6.1	7.6	11.8	14.4	20.8	36.2	76.6
	19	68	5.9	7.4	11.5	13.9	20.2	35.1	74.4
	18	66	5.7	7.2	11.1	13.5	19.6	34.1	72.2
B	17	62	5.4	6.8	10.5	12.7	18.5	32.0	67.9
	16	58	5.0	6.3	9.8	11.9	17.3	30.0	63.5
C[1]	15	54	4.7	5.9	9.1	11.1	16.1	27.9	59.1
	14	50	4.3	5.5	8.5	10.3	14.9	25.8	54.7

[1] A special limiting rule prevents the assignment of a grade higher than U.S. Grade B to a sample unit (individual can) falling within the C classification; and other provisions are allowed for ascertaining the grade of a lot.—See Section 52.5170.

TABLE 6.6
Squares, Square Roots, And Reciprocals

No.	Square	Square Root	Reciprocal	No.	Square	Square Root	Reciprocal
1	1	1.00	1.000	41	1681	6.40	0.024
2	4	1.41	0.500	42	1764	6.48	0.024
3	9	1.73	0.333	43	1849	6.56	0.023
4	16	2.00	0.250	44	1936	6.63	0.023
5	25	2.24	0.200	45	2025	6.71	0.022
6	36	2.45	0.167	46	2116	6.78	0.022
7	49	2.65	0.143	47	2209	6.86	0.021
8	64	2.83	0.125	48	2304	6.93	0.021
9	81	3.00	0.111	49	2401	7.00	0.020
10	100	3.16	0.100	50	2500	7.07	0.020
11	121	3.32	0.091	51	2601	7.14	0.020
12	144	3.46	0.083	52	2704	7.21	0.019
13	169	3.61	0.077	53	2809	7.28	0.019
14	196	3.74	0.071	54	2916	7.35	0.019
15	225	3.87	0.067	55	3025	7.42	0.018
16	256	4.00	0.063	56	3136	7.48	0.018
17	289	4.12	0.059	57	3249	7.55	0.018
18	324	4.24	0.056	58	3364	7.62	0.017
19	361	4.36	0.053	59	3481	7.68	0.017
20	400	4.47	0.050	60	3600	7.75	0.017
21	441	4.58	0.048	61	3721	7.81	0.016
22	484	4.69	0.045	62	3844	7.87	0.016
23	529	4.80	0.043	63	3969	7.94	0.016
24	576	4.90	0.042	64	4096	8.00	0.016
25	625	5.00	0.040	65	4225	8.06	0.015
26	676	5.10	0.038	66	4356	8.12	0.015
27	729	5.20	0.037	67	4489	8.19	0.015
28	784	5.29	0.036	68	4624	8.25	0.015
29	841	5.39	0.034	69	4761	8.31	0.014
30	900	5.48	0.033	70	4900	8.37	0.014
31	961	5.57	0.032	71	5041	8.43	0.014
32	1024	5.66	0.031	72	5184	8.49	0.014
33	1089	5.74	0.030	73	5329	8.54	0.014
34	1156	5.83	0.029	74	5476	8.60	0.014
35	1225	5.92	0.029	75	5625	8.66	0.013
36	1296	6.00	0.028	80	6400	8.94	0.013
37	1369	6.08	0.027	85	7225	9.22	0.012
38	1444	6.16	0.026	90	8100	9.49	0.011
39	1521	6.24	0.026	95	9025	9.75	0.011
40	1600	7.32	0.025	100	10000	10.00	0.010

STATISTICAL QUALITY AND PROCESS CONTROL

It is the technologist's responsibility to see that (1) *the measurements are taken in a systematic fashion,* (2) *all the data are recorded,* and (3) *the data are properly interpreted.* These functions, if properly carried out, are informative as well as preventive. They should inform management of what is going on at all times. If a machine or an inspection shows variation in the samples they can prevent the product from getting out of control, if the adjustment is properly and promptly made.

The obvious advantage of such a systematic system of interpreting quality control data is that management can easily see what is going on from hour to hour, line to line, year to year, or even season to season. It is a graphic presentation of the statistical interpretation of the observed samples which indicates what is taking place on the production line. If properly used, the quality control chart technique can save management money and assure product uniformity. Moreover, tolerance can be made closer for any particular attribute of quality of machine specification with the ultimate result of improved quality.

The establishment and use of a SQC-SPC program is not just another tool to keep someone busy. It is a tool to force the operator of every unit operation in a food plant to pay strict attention to the process for which he is responsible. It will result in more uniform products produced at reduced costs. Further, the SQC-SPC program has proven to be an effective method of developing the responsibility of plant personnel for the good of a growing organization.

TEST 2.1—TESTING THE SIGNIFICANCE BETWEEN TWO SAMPLE MEANS

Procedure:
(1) Obtain 10 cans each from 2 lots of canned whole tomatoes.
(2) Record the following container information:
   (a) Code mark on container exactly as presented
   (b) External condition of container
   (c) Size of container
(3) Determine and record the gross weight of the container and its contents.
(4) Remove the lid by cutting it out from the top.
(5) Empty the contents on a screen (use 8-in. screen for 2½ size can or smaller and a 12-in. screen for cans larger than 2½ size.)

(6) Let drain for 2 min. and weigh the fruit on the screen without disturbing the contents thereof. Record the drained weight. See Table 6.5 for U.S. grade classification of the whole tomatoes.
(7) Wash the can or package, dry and determine the net weight by subtracting the weight of the empty container from the gross weight.
(8) Repeat the above procedure for all of the cans and record either the gross, drained, or net weight information.
(9) Determine if the two sample means are significantly different, using the procedure devised by Howes. Refer to SQC Form 6.3 and 6.4 for sample calculation.

BIBLIOGRAPHY

HOWES, D.R. 1954. F-Table significance tests for two small-sample means. Ind. Quality Control *11*, No. 3, 25-26.

CHAPTER 7

Variable — Raw Materials and Ingredients

Without question the greatest variable the food processor works with is the materials used for the manufacture of a given product, including all the raw materials, added ingredients, packages, labels, etc. This variable needs to be understood and kept in control for product quality assurance. The use of written purchase contracts with detailed specifications is the only direction to take. Some firms are now only using certified suppliers, that is, suppliers that have agreed to details put forth by the buyer including guarantees of the quality and quantities delivered. Further, the use of Just In Time (JIT) delivery, that is as needed, is of major concern for control of product quality during manufacture.

Man has learned from experimentation that the best cultivar of variety or breed is of utmost importance when processing given quality products. Experimentation has clearly shown the ideal cultivar in terms of best yielder, the most disease resistant, the best for mechanical harvesting and bulk handling, and the one that has greatest acceptance in terms of quality. Further, through years of production experimentation, man has learned the ideal area of production, the better fields, and the crops that can be grown in given areas.

Further, through careful record keeping, man has learned what producers are reliable in terms of producing uniform quality products, promptness in delivery, and up-to-date on production practices. Thus, it should be obvious that we should know how to select and control this all important variable to obtain consistent quality raw materials.

However, Mother Nature has and will continue to make drastic changes that are not always planned for by man. In addition it should be quickly pointed out that man is not willing to operate on the status quo. Therefore, new cultivars, new production techniques, new harvesting and handling practices, new storage know-how, etc. are and will continue to be introduced. As a result it behooves every food processor to be most concerned and aware of the new changes that can immensely benefit the firm and to always plan ahead on how to cope with all the potential of raw material risks.

In the past, the Land Grant Universities have conducted extensive breeding programs and trials to continually upgrade production practices. However, at the present time there is a significant reduction in funds for these programs and there is a very definite shift by many researchers to move to more glamorous areas for their research activities. Thus, the food industry will have to support this much needed research. The food industry must have vision and look to the long range and not the route of the quick fix for their future. Luckily, many organizations and associations see this need and are funding research with a long range view. Sponsored research has merits as proprietary rights can be better protected and the research can become carefully channeled. However, without proper leadership the researcher may be short sighted and the researcher may fail to grasp the long term needs of the industry.

Quality Raw Materials And Ingredients

Raw materials are the life blood of the food industry. The need for improved raw materials is constant and most challenging. Research is greatly needed to improve the quality of the raw material so that uniform high quality finished products can be produced. The perfect raw material is not presently on the horizon for most food products.

One way to establish the raw material need is for all segments of the food industry to write down the specific details of their requirements. In many cases this becomes the specification. This specification then becomes the target for the raw material procurement personnel when working with producers. The producer can then intelligently plan ahead and shoot to meet this requirement. There is no other way for the industry to intelligently move forward. The idea of building beautiful and very functional processing plants that are operating under carefully monitored conditions utilizing modern methods and manned by well trained personnel is all wrong without control of raw materials. Emphasis must be directed at the producer of the raw materials and producers must be required to deliver according to written specifications. This same statement applies to all suppliers of ingredients and packaging materials.

All ingredients used in the manufacture must be of food grade quality and in full compliance with the Federal Food, Drug, and Cosmetic Act of 1938 as amended and all applicable regulations there under. In addition manufacturers must supply the food firm a written detailed list of all individual components used in the formulation of their ingredients and, of course, these must be in compliance with the above law and regulations

Certification Of Suppliers

In many plants today, all suppliers are being certified. This simply means that they have agreed to the detailed set of specifications and that they will deliver what is contracted for at the required time. The vendor must have a well developed quality plan with process control schemes to assure finished materials (ingredients, packaging materials, and/or services to be rendered) with proper documentation to support their plan. They must have and understand the philosophy of continuous improvement. This simply means that they have taken the necessary steps to tighten up their variation in their processes so that the delivered materials are closer to the 6 sigma levels.

Processor/Supplier Relationship

Food processors should develop strong producer-processor relationships. They should share their problems, their concerns and their vision. Ideally, all suppliers of raw materials to a food firm should be considered a partner. Further, every supplier should look at the end products from the food firm and rejoice in the fact that those products are made from my raw materials. He or she should say to themselves, I have great pride in continued improvement of my materials as we both are working together to produce and manufacture finer products today and in the future from my firm for the continued satisfaction of the customer. All food companies must develop a partnership with their vendors by sharing business information to:

(1) Clarify customer quality and service requirements,
(2) Material specification including how the materials are to be used in the process,
(3) Employee site exchange to encourage direct transfer of information between the producer and the user,
(4) Prompt resolution of issues should they arise, and
(5) Promotion of Just In Time (JIT) delivery, thus as the partnership develops inventories of raw materials, ingredients, and packaging materials can be maintained with the ultimate reduction of warehousing costs for all concerned.

When the above is accomplished, the industry can and will have the right materials to manufacture quality products that meet the customers requirements all the time. Sure this is a challenge, but isn't that what all good things are made of. The agricultural production industry can do it, but it will not happen until the food processors establish careful and intelligent specifications carefully describing their needs. then and only then can one expect to operate modern factories to produce uniform quality products.

All other materials coming into a food plant must be accepted only on the basis of their meeting given specification, that is, Acceptable Quality Levels (AQL). All materials must be carefully sampled and analyzed to ascertain their compliance with AQL's. Accepting inferior materials creates many chain reactions within the food plant that are not called for in this day and age. It is mandatory for management to control incoming materials to specific requirements if the firm is to operate successfully today and tomorrow. Written specifications and contracts are the norm today and the supplier should be properly rewarded for his delivery of acceptable merchandise.

Reference
Gould, Wilbur A. 1992. Total Quality Management for the Food Industries. CTI Publications Inc., Baltimore, MD.

CHAPTER 8

Variable — Machines

"You can learn a lot by just watching." Yogi Berra

All machinery in a food plant is designed to perform given functions or unit operatons. These functions or unit operations vary widely with the type of food being processed and, of course, the size of the operation. Together, the unit operations or machines may make up a given process line or product line. All unit operations have given parameters in terms of thru-put capacity, that is, pounds, tons, etc. per hour or shift.

Since machines are obtained from many sources, it behooves management to make certain that the given machine "fits" the line in terms of thru-put. Further, management should make certain that the machine has variable speed or drives to control thru-put. Machines must, also, be operated under specific parameters, that is, dwell time, given temperatures and pressures, amount of vacuum, given nanometers of light, etc. are specific for the intended function to be performed to provide proper effect on the quality of the product.

Machinery should be of sanitary design and all surface in contact with the food or food materials should be inert to the food under the condition of use. All surfaces of machines should be smooth and non-porous, free from pits and crevices, designed to protect the food from external contamination, and free of all dead space. All moving parts should be self-lubricating. Most importantly, all machinery should be easily dismantled for cleaning and where ever possible the machine should be designed for self cleaning using spray balls or other means.

Machines are designed to move products, hold the products, change the style of the products, and/or change the state of the product. For example, cutters may be designed to just cut-in-half, or slice, or quarter, or dice, or strip, or cube, or crush a product or change into other styles.

All machines should be equipped with monitoring devices for speed of operation, volume of thru-put, temperature or pressure and/or vacuum regulators and controllers, etc. The machine should be designed to make it easy for the operator to run the machine and maintain steady-state operation to produce uniform and consistent quality products.

The operators should know the process capability of every machine or unit operation in the process. This information can then be used to know where to put the emphasis on the product during the run for uniform quality control.

In all cases a detailed manual of operation should be available on each unit operation showing the function of the individual unit operation and its relationship to the specific process and, of course, its effect upon the quality of the product. A flow diagram should be developed to show the particular unit operation. In addition a narrative description of each unit operation in the process should be developed with capability data to show set points and effects of queing sequences on material/energy blanace and direct effects on finished product attributes.

Most importantly, start-up and shut-down procedures must be carefully established and clearly elucidated to control quality, improve efficiency and reliability.

Lastly, and most importantly, the operator must be trained to operate the equipment as part of a given process. This includes proper sampling, data recording, charting, and a thorough understanding of appropriate adjustments to maintain an operation in complete control.

Quality assurance must be kept abreast of new changes in machinery and they should contribute to the knowledge of information in making the decision to purchase. Ideally, all new machinery should be evaluated by Quality Assurance as to its capabilities and effects on product qualities (see Table 2.3). Further, QA should make test runs and test the significanc between two sets of sample means, that is, for the new machine versus the old machine (see SQC Data Form 6.3 and 6.4).

Referance
Gould, Wilbur A. 1992. Total Quality Management for the Food Industries. CTI Publications Inc., Baltimore, MD.

CHAPTER 9

Variable — Methods

The methods used in food manufacture are designed to perform given functions to the food. All the methods are unique and have a definite purpose. From a quality control standpoint, the operator must learn all he or she can about the method to better control the unit operation. The operators responsibility is to manage his or her unit operation and control the unit operation within given tolerances or parameters to meet the required specifications.

The actual parameters and control values the operator has to work with to control a given unit operation should be used as his or her guides to allow them to stay within the given specification. These values may be temperature, pressures, vacuums, speed of flow, volumes, concentration, pH, color, texture, consistency, size, shape, numbers of defects, etc. Whatever the value, the measurement must be accurate and reliable of the situation at any given time.

This measured value must be obtained from a random sample taken periodically off the line, batch, or process that is representative of the whole and adequate for the given measurement. Further, it must be an unbiased sample.

The measurement must be appropriate to the unit operation and the given product. It should be a technique that is truly an indicator of the quality of the operation. If the measurement can be an on-line measurement and the results continuously recorded, so much the better. Examples of on-line controllers and recorders include temperature, pressure, vacuum, speed, pH, soluble solids/refractive indice, moisture, oil, viscosity, color, defects, etc.

Regardless of the method of measurement, the operator must see that the data is recorded, analyzed, and proper interpretation of the results. Lastly, and most importantly, the operator should chart or plot the results and let the charted or plotted results do the talking.

Depending on the data, the operator should learn to leave the operation alone when there is no concern, that is, when it is staying within the specifications limits. On the other hand, when the data indicates that appropriate adjustments are necessary, the operator should make appropriate adjustments accordingly. It is most important that the operator does not constantly "tweak" the operation on a hunch. Tweaking may be costly and very untimely. If an operation is right, it is most important to learn to leave it alone and let it run according to the present setting.

The food industry is most complicated and even though the operator may understand his unit operation and how to make measurements, raw materials and method of manufacture are quite varied. Using potato chips as an example, potatoes will vary quite widely in solids content (16 to 24%), the individual chipper may vary his slice thickness from 0.050 to 0.075 (a difference of some 15 slices for every 3 inch in diameter tuber), and the individual chipper may use different fry temperature, that is, a low of 300 degree F to a high of 375 degree F. This wide difference in fry temperature may require a dwell time of the chip in the fryer of over 300 seconds. The resultant oil content of the finished chip can vary as much as 100%, that is, from 25% oil content to above 50% oil content of the finished chip. All this simply says, that the operator of each unit operation, receiving, slicing, and frying in this example must be coordinated and they must communicate carefully if one is attempting to fry chips with a controlled oil content of 30%. This can be done and is being done, but all unit operations must be monitored and the unit operation operator must carefully control his or her unit operation to produce an acceptable product for the customer.

The industry is continually adapting and modernizing to more and more automation, electronic controllers, robotics, and other systematic systems or methods that may be fully computerized, thus requiring much less hand adjustments for control of product quality. These new and modern changes are to be encouraged. They will require better training and operators who are much more knowledgeable. The quality of the finished products will be in greater control with the resultant end products being more uniform and the customer will be the ultimate winner and much more satisfied.

Quality Assurance personnel should contribute to the methods of manufacturing and they must develop the process capabilities of each change in the method as it affects the unit operation. Of course, they are expected to provide feed-back data as they evaluate the product and audit the line. Production personnel must learn to control the process for product uniformity.

Reference
Gould, Wilbur A. 1992. Total Quality Management for the Food Industries. CTI Publication Inc., Baltimore, MD.

CHAPTER 10

Variable — People Power

People are the greatest underutilized resource in the food industry.

Every worker in a food plant should be given a thorough education in what is acceptable product quality and they should be shown how to seek ways to constantly improve existing products. All plant employees must first understand that quality is conformance to requirements. Thus, the very first requirements in a Total Quality Assurance program is to make certain that the supervisors, and all the employees are trained as to what quality is, why control of quality, where to be concerned about quality, when to make changes, who is responsible, and how to identify and acknowledge the consumers wants.

An anonymous executive penned the following:
"To look is one thing,
To see what you look at is a second thing,
To understand what you see is a third thing,
To learn from what you understand is still another thing,
But, to act on what you learn is all that really matters".

People responsible for given unit operations in a food plant must be trained and empowered to control their unit operation to conform to prescribed parameters or specifications. Further, they must always be alert and seek out quality improvement strategies and they must be free to run their unit operation to produce quality products.

People Partners
People need to be a part of the team that is building quality for the future of the food firm. They need to be treated as "associates" or "partners" and not just as a warm body with a strong pair of hands. People in a food plant are not servants or slaves and no person wants to be told to check their brains at the gate.

People want to make a contribution and they want to cultivate their knowledge of the business as they really want to see the food firm survive.

Yes, its true that some employees may not have a very high education, but they do want to grow with the firm and they want to be given a chance to develop through on-the-job training.

Food companies should offer opportunities for the employee to better themselves through in-plant educational programs starting with a better understanding of the firm, its history, its vision, its mission, its goals, its products, its growth plan, and its future. Information on health plans, insurance, retirement, etc. are all part of the orientation process. All this information should be well-packaged and presented verbally, visually, and in a hand book for home study and sharing with family. This is usually part of the Human Resources Office of the Personnel Office.

Training

The rest of the training is really the responsibilities of the Quality Assurance personnel working with the specific supervisors or leaders in each of the given work areas of the firm. Quality Assurance personnel must be good communicators, they must have great patience, they must be most knowledgeable, and they must be great listeners. Their task is not easy, but it is very rewarding to themselves, the employee and, of course the firm.

The individual employee should then be given specific training for the assigned task. This training should not be the big brother approach, but rather details on video or combination of pictures and audio, and/or print media along for their handbook of the assigned area or operation. This information should explain the standard operating procedures (SOP) including all parameters of control, preventive maintenance, cleaning information including schedules, and all safety precautions. Further, the video should detail the records to be kept and the effect any changes may have upstream on product quality. Lastly, the video should spell out his or her responsibilities to the whole line. This should include a flow chart of the food plant line clearly delineating each unit operation with specifics on the assigned unit operation. Thus, the employee can see the whole forest not just their tree. They see the big picture with specifics about their area. This whole aspect of training should detail the 5W's and 1H, that is,

What does this unit operation do to make an acceptable product?

What effect do changes in the unit operation have on product quality?

Where are the controls and the safety switches?

When should changes be made in any of the parameters?

Why is it essential to control within the established parameters?

VARIABLE — PEOPLE POWER

Who authorizes changes in establishd parameters?
How does this unit operation tie into the product flow?
How can I become part of the team that is building better products for the future?

All of these 5W's1H can be greatly expanded for every operation.

The employee should then be given explicit training on how, and when, and where, and why, and the what of sampling and the importance of an adequate, respresentative, and unbiased sample. The sample is used for control of his or her unit operation and for audit of the line by quality assurance personnel.

Next the employee should be carefully instructed on what tests are available to evaluate the sample and the appropriate adjustments for controlling the operation to conform to requirements. These test procedures should be carefully detailed in writing including as many visuals as possible for thorough understanding. Most importantly, the employee should thoroughly understand how to standardize the test instrument or procedure.

Following this training the employee should be taught the principles of data collection, recording, and charting. This information becomes their guide to control of their unit operation. It becomes a "talking picture" of the operation for their specific unit operation. They should thoroughly understand that their data is talking to them about the operation. Most importantly, they should learn to speak with these facts and not "I think" or hearsay.

Training should always continue for every employee and this training should be the start toward team build. Some firms use Quality Circles, Quality Improvement Groups, Study Groups, etc. The title is not as important as the fact that individual employees learn to work together as teams, to help each other, and to continually improve the process. Teams see the big picture, they quickly know the weak links and they work to correct it. Teams can handle absenteeism and performance issues and, yes, they even schedule vacations so as not to disrupt the work. Teams strive for excellence. Teams bring strength to the operation. Teams see their role and they quickly see the long range picture. Effective teams are those teams that have a say in who works with the team and on what problems. Teams may perform their own maintenance. Most importantly, teams find opportunities to improve quality and productivity and they will work to realize these opportunities. The team should establish its own mission and technical training at the teachable moment. Teams can provide people skills for interacting, solving problems, making decisions, and taking appropriate action.

Teamwork

Trained employees must understand their responsibilities and they must be held accountable for their role in the operation of the food plant. If a firm expects quality work, it must train and develop its people. People need to know the "ways" of their food firm. People need to understand the values and the philosophies of the firm they work for. They need to know the policies and procedures. Most importantly, CEO's and Managers should realize that training is the pathway to growth of a food firm, thus the growth of a food firm is directly based on the growth of its people.

Training may come in many forms: (1) Formalized class room with lectures, text assignments, exams, etc.; (2) the use of the laboratory experiments to demonstrate effects of conditions or changes; (3) Staff or Committee Meetings where people feel that they are a part of the whole. Meetings prevent the "grapevine effect" and assures that everyone speaks with the same tongue; (4) One on one counseling or instruction-the personal effect; (5) Personal example, that is dress, manners, practices, etc.; and (6) Private or individual study through readings and talking with those in the know. I believe that everyone can learn something new everyday that will help them if they seek all the opportunities that are available to them.

The most important single factor when working with people is to make certain that the right person is in the right job, at the right time, that is, they are qualified for the job at this time. Next, the employee must have the right tools to do his or her task. Every individual must feel that they belong and that they are part of the team. All employees must work efficiently and they must be concerned about any problems in their area. Most importantly, they must make an effort to grow on the job and show that they are responsible and willing to be held accountable for their actions. Moral may be difficult in some situations and its most important for the team to work to strengthen the moral of every employee. Strong moral ensures enthusiasm and it improves quality and in turn increases productivity.

Every employee must be rewarded for their solid contribution to the long range efforts of the firm. Employees must be paid on the basis of productivity and they must be assured of their future. They must be recognized for their significant contributions, their piece of the action, their personal growth, and they must be given appropriate advancement. The individual employee must feel that their job is a great one and that their firm is a fun place to work because of their freedom, appropriate annual time-off, and adequate recognition for their contribution to the growth and success of the firm. Every employee should feel good about what they do, the products they make, and the satisfaction they bring to the growing list of the firm's customers. Employees should be thankful for their training and they should feel that their actions are a direct reflection of this training and understanding of their role with the firm.

Reference
Gould, Wilbur A. 1992. Total Quality Management for the Food Industries. CTI Publications Inc., Baltimore, MD.

CHAPTER 11

Food Plant Sanitation

Industry's Responsibilities

Persons working in the food industry and firms engaged primarily in the production of food for human consumption have a moral and legal obligation to perform all operations in clean surroundings and with due regard to the basic principles of sanitation. Further, each has obligations to uphold sanitary standards in common practices for food handling establishments.

Sanitation is every person's job in the plant. If a plant is in sanitary shape there is no worry of what to do when an inspection is to be made. However, sanitation should never end with the satisfaction that you have met a municipal, a state and/or a Federal inspection or regulation. It should be a part of the everyday policy of that firm. Only through the individual efforts of each person working in the food plant can one expect to maintain the respect that your product demands at the consumer level. Sanitation is a responsibility that every person handling or working with food must constantly fulfill.

If properly conducted, sanitation removes the worry about spreading of communicable diseases or about the potential of food poisoning. Further, if sanitation is properly maintained, a product free of defects will be produced and waste and spoilage can be eliminated. These are moral obligations.

From a legal obligation standpoint, Section 402a4 of the (Food, Drug and Cosmetic) act states a food shall be deemed to be adulterated "if it has become contaminated with filth, or whereby it may have been rendered injurious to health."

Value of Planned Sanitation:

If a plant has paid proper attention to housekeeping, sanitation problems are over, as sanitation is a way of life. The value of a planned sanitation program can be enumerated into these five basic areas:
    1. A better Product. Competition demands that better products be produced everyday.

2. A more efficient operation. Sufficient studies have been made in recent years to indicate that efficiency in a food plant is directly related to the sanitary conditions.
3. Greater employee productivity. This in and of itself should be reason enough to have a planned sanitation program.
4. Studies have indicated that fewer accidents have been recorded for those plants in top sanitary conditions.
5. Perhaps most important— a sanitary plant serves as a barometer of the factory conditions. "Let me see your home and I'll tell you what kind of a person you are or more importantly, let me see your factory and I'll tell you the kind of product you produce."

PLANTKEEPING

Plantkeeping in a food plant should, if properly done, indicates that a factory is clean. Plantkeeping may be broken down into the following four basic categories: Exterior, Interior, Materials, and the Plantkeeper/Sanitarian.

1. Exterior-Outside

From the exterior standpoint, the plant grounds serve as a reflection of what is inside. The grounds should be properly graded to provide natural drainage. There should be no litter or waste accumulating in or around the factory, the receiving yard, the platforms, etc. grass, weeds, and hedges should be controlled to prevent the harborage of insects or rodents. Roads should be kept free from dust. Gravel, cinders, oil covered or paved roads are to be recommended. The storage of the equipment outside a plant should be neat and orderly. The parking lot should be kept orderly and all debris should be removed daily. The parking spaces should be well-arranged and marked. The buildings themselves, from an outside appearance standpoint, should be clean and well-maintained. All exterior openings should be screened and rodent proof. The roof should be leak proof and there should be no uncovered openings, and the outside of the building should be free from insects, rodents, etc. And lastly, but perhaps most importantly from an exterior standpoint, all spilled or spoiled products should be cleaned up immediately and removed from the premises. Broken containers, etc., should not be permitted to accumulate in the receiving and shipping areas. The outside appearance of a factory is a reflection of the inside.

2. Inside

The interior in itself can be broken down into the different parts, and in making a plantkeeping report, the individual sanitarian should consider each of these accordingly:

a. **Building**—The building proper is perhaps the place to start. All materials of construction in a food processing plant should lend themselves to easy cleaning. The floors should be water tight, smooth surfaced and sloped ⅛ to ¼" per foot to floor drains approximately every 10' apart. The drains should be capable of handling the necessary waste material. It should be covered with removal grates so that they can be kept clean. The walls, doors, partitions, pipes, ceilings, etc. should be kept clean and painted when and if needed. The building should be properly ventilated to prevent any drippage into food or onto food handling equipment, the growth of molds and/or the deterioration of paint or structures. Further, the ventilation should be adequate to provide suitable working conditions for employees. All windows, doors, and openings in the food plant areas should be screened.

The lighting should be adequate for the job being undertaken. This may range from 25 to 150 foot candles and it should be designed according to the quality of light necessary for the job underway. Good lighting promotes cleanliness and makes the sanitation job much easier.

b. **Equipment**— The equipment in use in the food plant should be constructed not only for its functional use, but, most importantly, with due regards to its cleanability and protection from contamination. Materials of construction should be smooth, hard, nonporous and preferably of stainless steel. All pipe lines, fittings, etc., handling food should be of the sanitary type. The elimination of sharp corners in tanks, flumes and other equipment greatly facilitates cleaning and prevents spoilage organisms from building up. All equipment should be directly accessible for cleaning or proper provisions should be provided for cleaning in place (CIP). All open equipment, such as, tanks, hoppers, buckets, elevators, etc., should be covered, and if it is a multi-floor area, there should be a floor curbing to a height of 6" around the shutes, etc. All containers used to transport food materials should be kept clean and not used for other purposes. Nesting of pails, trays, etc. should not be allowed until they have been cleaned. All equipment should be properly guarded, and excess lubricant should be cleaned off after lubricating. The hoses and clean-up equipment should be properly put away after each use, and all unused equipment and equipment repair should be removed from the processing area. The waste in the food plant should be collected in containers properly designed for handling waste. This includes tight fitting lids. All waste materials should be removed daily, and the containers should be kept clean.

c. **Storage**— The storage in the food plant is one of the major areas. The requirements include aisles which should be kept clean and well-marked. The lanes and aisles in the storage area should be marked accordingly. Food materials, packages, etc. should be protected from insects, rodents, dust, dirt, etc. All stored products should be placed away from the walls and in proper storage temperatures.

The management of storage areas is probably one of the secrets to good housekeeping. All storages should be cleaned at least once per week. All incoming goods should be inspected as to damage, rodent or insect infestations. Frequent inventories and evaluation of warehouse products should be constantly taken. The management should also provide adequate space, and this space should be used properly. This includes control of temperature within limits depending on the perishable nature of the commodity. The policy of First In, First Out (FIFO) should be strictly adhered to.

d. **Employees**— Probably no other single factor is as important as employees in terms of housekeeping. The employees, in a food plant should look sharp, they should feel sharp, and, if so, they should be sharp. To help an employee to be sharp, there are certain basic requirements:

    All employees must wear caps and women and men with long hair must wear hair nets and/or hair restraints.

    Pins, curlers, jewelry, sentimental pieces, fingernail polish and other loose attachments shall not be worn.

    Pens, penicls, watches, etc., must not be carried in pockets above the waist line.

    Protective clothing should be worn at all times.

    Gum chewing, smoking, or other use of tobacco and eating shall be confined to designated areas.

Glass bottles etc., shall not be permitted in work areas.

Hands shall be washed and sanitized at the following times:

When reporting for work
After a break point
After smoking or eating
After picking up objects from the floor
After coughing or sneezing and covering mouth with hand
After blowing nose
After using the toilet facilities.

All employees must report any blemishes or break in the skin to the supervisor prior to reporting for work. Band aids or adhesives, which may become loose and fall off during the work time unless covered with gloves, are not to be used.

Safe personal conduct within the food plant should be strictly observed. Running, horseplay, riding on trucks or lifts, taking shortcuts (ducking under conveyors, etc. whether operating or not) are prohibited.

All employees should share responsibilities to maintain lockers and washrooms in neat, clean, and in an orderly manner.

Most importantly, all employees should be required to observe proper habits of cleanliness. A food factory should not employ any person afflicted with infection or contagious disease. Health certificates should be required, and, lastly but most importantly, signs should be used throughout the company's food processing facilities as to smoking, eating habits, washing habits, and general sanitary requirements.

e. **Special Areas**—Special storage areas should be provided for handling of clean uniforms. towels, toilet articles, soiled uniforms and linens, custodian's supplies and equipment, pesticides, employee's personal belongings, and garbage and wastes. Each of these areas should be operated for the storage and handling of the equipment, supplies, etc., stored therein. In some cases it may be necessary to keep them under lock and key, but, most importantly, they reflect the attitude of housekeeping by the neatness, orderliness and maintenance of these facilities in accordance with good housekeeping practices.

Other areas that many consider to be highly significant in housekeeping practices are the toilet, lockers and restrooms. These facilities will determine in great part the attitude of personnel about the factory conditions. They are intangible factors, but they must be part of any sanitation program. They may pay the greatest dividends from the standpoint of plant sanitation, quality improvement, and employee morale. Adequate facilities include liquid soap and throw-away towels in the restrooms and within reasonable distances from each work station are most vital. Drinking fountains or taps with throw-away cups should be used. The fountains should be provided with side-outlets of water. In addition to these facilities, first aid rooms and locker facilities for permanent employees are a must. Toilets should be provided with double doors and never open directly into any room where food is being processed. They should be constructed of sanitary materials, adequately ventilated with all openings screened. The toilet facilities, toilet areas, restrooms, etc., should be kept scrupulously clean. There are minimum requirements depending on the numbers of employees for any given operation and obiviously there should be facilities for both men and women, and they must be plainly marked. The area that is important as part of the interior of a modern food plant is the lunchroom. This, of course, will vary depending on the size of operation and the general facilities in the area. Regardless, minimum lunch room facilities if no more than vending machines should be provided, even for those workers that carry their lunch. Of course, the full service lunch room is the most ideal. It should be a place where people want to eat and this can be maintained as such if it is kept in good housekeeping order.

The quality of the food may be secondary to the type of facility in terms of furniture, utensils, floors, etc.

f. **Materials**-The handling of materials in a processing plant for both incoming and outgoing warehousing is most important. Incoming materials should be carefully inventoried and evaluated as to assurance quality standards. Likewise, outgoing materials must be properly identified in terms of shipments and qualities of products. Internally, the warehousing of same is quite obvious in terms of housekeeping. Great losses occur in warehouses by improper housekeeping practices in terms of breakage, pilferage, looting, etc. Materials must be handled orderly for proper housekeeping practices to be maintained.

g. **Sanitary organization**—First, management must be responsible for sanitation. The plant manager directs all activities and determines policy. Since management cannot be aware of all the details, authority must be delegated to some individual.

The authority to uphold standards of sanitation usually falls to the plant sanitarian. In the small plant, the sanitarian may be one of the Quality Assurance Technologists or even a production supervisor; however, this individual should not be so overloaded with other activities that he cannot effectively control sanitation. The sanitarian should have understanding of microbiology, chemistry, entomology, parasitology, and sanitary engineering.

The sanitarian should be directly responsible to management. If he is part of the quality assurance department, his reports should be forwarded to management. Since the sanitarian frequently deals with matters under the supervision of other technical personnel, he must be tactful to avoid conflicts of authority.

1. The responsibilities of the sanitarian are:
    a. To develop a sanitation program.
    b. To secure the support of management and employees.
    c. To strive to constantly improve the program.
    d. To study sanitary problems and evaluate results.
    e. To keep informed of new developments.
    f. To report to management.

2. The duties of the plant sanitarian and his staff:
    a. To supervise matters of personal hygiene.
    b. Maintenance of adequate cleanup.
    c. The elimination of rodents and insects.
    d. The supervision of water supply, sewage, and waste disposal.
    e. To supervise sanitation and health in the company owned lunch room.

f. Maintain the sanitation of rest rooms and toilets.
g. Check general plantkeeping.
h. Supervise sanitary storage of raw and finished products.
i. Take correct action to prevent any contamination.
j. Conduct organized training programs for plant personnel.
k. Make individual inspections of the plant and report to management.
l. Participate in general inspections with management and supervisory personnel.
m. Cooperate with local, state, and Federal inspectors.

3. The sanitarian acquires the support of management by showing cost saving possibilities of good sanitation, by emphasizing the public relations aspect of a clean plant, equipment, grounds, and by pointing out good employee reaction to sanitary conditions.

The plantkeeper or sanitarian is the key to the success of plantkeeping in the food plant. His position in the company and the support which he receives from top management is most vital. This head plantkeeper or sanitarian must have native intelligence. He must have the executive ability to sell the company and its employees on why and how to keep the plant in sanitary condition. He should be trained in the how, what, where, and when of cleaning. This education will require considerable science and great knowledge of the food industry. The success of his job will depend in great part on the assistance which he has. In a large operation, this will necessitate a foreman with personality, tact, and enthusiasm for his job. In a smaller company, this would be a requirement in addition to the above Plantkeeping. The tools with which he works are many and varied depending upon the type of plant in which he is operating. In general, it would include the following:

a. Ample supply of water and of desired quality.
b. Different types of brushes (nylon preferably, but these will depend to some degree on the type of equipment to be cleaned).
c. Detergents and knowledge of how to use them.

d. Chlorination and chlorinating equipment to control bacterial counts.
e. Steam and/or high pressure air/steam equipment fitted with proper nozzles for hard to get at areas.
f. Flashlights for inspection of out of way places.
g. Black light for detection of rodents, etc.
h. Camera (Polaroid)
i. His attire: white cap, clean white overalls or white shirt and pants.

The success of the job will depend in great part upon the training courses, seminars, etc., which he conducts for the employees. These educational programs should point up the methods to be used and the responsibility of each individual in practicing proper housekeeping in the modern food plant.

The success of the plantkeeper, as previously mentioned, will vary with the lines of communication between management, between production employees, and between the staff personnel. Sanitation must be wanted by all, including management. Sanitation programs never cost, they pay for themselves.

3. **Plant Inspection**

The final aspect in a sanitary evaluation of a factory is the inspection itself. The inspection may be conducted by company personnel from either the home office or the local plant, or by outside agencies or groups. The inspection at the plant level should be made daily. The most important part of the inspection is a written report filed with plant management at the local level and the home office to give them the concepts and knowledge of what is going on. Perhaps more important than either of these, the home office can exert help where needed to correct conditions. By making inspections and particularly by outside groups, areas that may have been overlooked by local personnel can be noted. The inspection should include the outside as well as the inside of the factory.

A written report should be made in all observed conditions listed as satisfactory, needs improvement or unsatisfactory. There should be a manual containing the minimum standards for each of the plant areas. After the report is read, it should be acted on accordingly. The report in itself is worthless unless the information is used advantageously.

4. **Sanitation Evaluation:**

The evaluation survey of a factory will depend in great part upon the standards of cleanliness which are established. These standards can be divided into 3 basic areas:

1. **Physical cleanliness**—This is the absence of visual product waste, foreign matter, slime, etc.
2. **Chemical cleanliness**—This is the freedom from undesirable chemicals. Contamination could occur from cleaning compounds, germicides, pesticides, etc., which might be left on the product and on the equipment. In general, these can be easily corrected by proper washing and rinsing.
3. **Microbiological cleanliness**—This is probably the most dominant factor in sanitation today, and, of course, it is controlled by the amount of microorganisms that may be present on the product, in the equipment, building and people. These can and must be controlled through proper sanitary cleanups.

A laboratory equipped to assist the housekeeper or sanitarian is a must, and the standards of cleanliness must be a part of the minimum specifications for the operation of a factory. These standards of cleanliness can be established through the quality assurance personnel.

5. **Suppliers**—You have a right to expect certain things from suppliers. These are necessary services and should not be taken lightly. Your chemical supplier can survey your plant to begin the organization of your cleaning program. He can outline specific cleaning methods, products and exact amounts, times, and temperatures. He can set up portion control and help outline who does what job. In most cleaning programs, labor amounts to more than 90% of the total cost. A cleaning material supplier can give good advice on efficient use of the cleaning crew.

The supplier representative can be your quality control "watchdog" and cost "controller" with his regular and on-call service. Remember, the two important items are highest quality results at proper and reasonable total cost. The supplier can provide control and feeding equipment, engineering services, technical assistance, laboratory services, and sound planning for future needs.

Program

A food firm should organize its sanitation program with firm guidelines. They should use cleaning surveys and assign well defined responsibilities to the sanitarian. The accomplishments should be measured regularly and a plant sanitation committee should meet monthly. A regular schedule of meetings will help to enforce the importance of the sanitation program and to keep it in the forefront and on the minds of key people in the organization.

A sanitation committee might well consist of the following: Plant Manager, Production Foreman, Quality Assurance Supervisor, Food Technologist, Maintenance Engineer, Personnel Supervisor, Plantkeeper/Sanitarian.

The fact that management expects certain things to be done, and it is known that they will follow up, is a great motivating force for a sanitation program. The sanitarian should be inquisitive. He should get to know why certain things are done certain ways and when certain things are not being accomplished. Regular performance reviews for the sanitation crew will help set high goals of quality.

As always in the management of people, training in the sanitation program is of utmost importance. Training and retraining should be undertaken continuously. The trainer should also be trained on, at least, an annual basis.

A few items to follow in any training program:

Break down each job into components to instruct,

Take the important things first,

Stress the repetitive situations so that people will be able to do best the things that they are called upon to do the most, and

Try to get people to think about what they are doing. Get people into the habit of checking to see that things are operating as they should be and looking to see if the equipment is really clean.

The difference between teaching a person what to do and helping him/her acquire the skill takes practice, where learning the "what to do" may require only a good memory. The supplier can be a great help in developing and training sanitation people. Everyone, including the owner through the lowest paid person in the organization, has a moral and legal obligation to perform all operations in clean surroundings and with due regard to the basic principles of sanitation. Further, the plant manager has obligations to uphold sanitary standards in common practices for food handling establishments.

SUMMARY

In summary, sanitation is of utmost importance to the food business not only because it is good economic sense but because the law requires sanitary practice.

Each food plant should have some individual, sanitarian or otherwise, to oversee matters of sanitation. That person should be directly responsible to management and must be able to get along with and influence people in order to make a sanitary program effective. It is a most important job in the food plant.

Reference

Gould, Wilbur A. 1990. CGMP's/Food Plant Sanitation. CTI Publications Inc., Baltimore, MD.

CHAPTER 12

Fundamentals of Research and Development

Research and Development in many food firms is assigned to the Quality Assurance Department. A.S. Gyorgyi stated that "Research is to see what everybody has seen and to think what nobody has thought" and Boss Kettering stated "if you don't do research until you have to, its too late". These two statements pinpoint the essence and the need for a food firm to establish an R&D group.

The role of research in a food firm can be very broad because progress as we know it today would be impossible without research. We learn new facts and ideas by chance, or by trial and error, or by generalization from experience, and/or by logic. Logic means reasoning things out or by using the scientific method of inquiry. The scientific method is an orderly system of searching for the truth. Most people agree that there are six steps to be followed when conducting research:

1. Identify the problem and establish the objectives, that is, the whys.

2. Research starts in the library, that is, find out what has been done. Basic sources of information may be found in text books, scientific journals, abstracts, reviews, nomographs, annual reports, symposia, thesis and dissertations from universities, translations, trade publications, and association bulletins and newsletters.

3. Establish the approach to be undertaken in solving the problem, that is, the development of how you will proceed or more scientifically, the experimental design including the replications of the study.

4. Gather the data, that is, work the proposed plan by recording the facts.

5. Test the hypothesis, that is, answer the objectives by interpreting the data.

6. Summarize the study, draw conclusions, and make recommendations based on the facts.

The R&D group has many different functions or objectives depending on the size of the firm and the diversity of the product lines. The following are some broad areas that may be a part of the objectives or reasons for establishing an R&D program:

1. To cure existing troubles and nuisances in connection with materials, processes, products and services and to anticipate and prevent such troubles from occurring in the future.
2. To reduce costs involved in the use of materials, processes, products, and services.
3. To improve the quality of existing materials, products, processes, and services.
4. To develop new products or new uses for existing materials, products, processes and services.
5. To develop suitable substitutes for existing materials, products, processes, and services.
6. To develop uses for by-products that may be otherwise wasted.
7. To keep abreast of scientific and technical knowledge in areas of vested corporate interests.
8. To establish, issue, and maintain standards and specifications for raw materials and finished packaged products bought and/or sold by the firm.
9. To represent the firm in matters involving technologies.
10. To deal with governmental agencies in relation to technical matters including labeling, nutrition, standards, waste disposal, etc.
11. To increase sales appeal of the product and improve customer and public relationships.
12. To increase the profit on invested capital and on unit sales volumes by increasing product quality at the lowest possible costs.
13. To keep the firm up-to-date with competitive products in the market place.
14. To assist management of the firm in the development of long range plans for the firm.

To accomplish the above and to keep a firm growing, an R&D Manager must be employed. The qualifications of an R&D Manager should include a person with bold imagination, one that is enthusiastic but balanced with good judgement, one that is creative and able to solve original solutions, one that has experience in interpreting food processing operations and

FUNDAMENTALS OF RESEARCH AND DEVELOPMENT 115

other research, and one that has the ability to work cooperatively with all personnel in the firm. He should be a good manager first and have the know how to make timely decisions and how to influence the right people about budgets and technical matters. He should be an expert in listening and understanding with great patience and insight. Most importantly he should identify with the firm he works for and have a broad overall knowledge of their business. Further, a good R&D Manager must have a sound education in the foods areas, that is, processing, technology, and science. As the firm grows and R&D proves its worth, specialists can be brought on board, that is, flavorists, sensory analysts, microbiologists, chemists, engineers, and technologists in the areas of processing/preservation/packaging.

There are many ways to organize an R&D program. The following organization charts illustrate four such practices. Figure 8.1 is an organizational plan for a small R&D set-up. Here, the Director, Manager, or Project leader may be one and the same person with the Technologist,

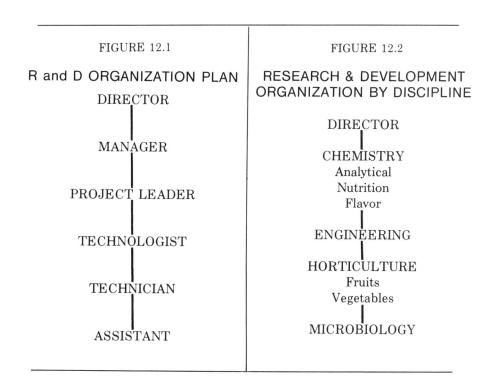

FIGURE 12.1

R and D ORGANIZATION PLAN

DIRECTOR
|
MANAGER
|
PROJECT LEADER
|
TECHNOLOGIST
|
TECHNICIAN
|
ASSISTANT

FIGURE 12.2

RESEARCH & DEVELOPMENT ORGANIZATION BY DISCIPLINE

DIRECTOR
|
CHEMISTRY
Analytical
Nutrition
Flavor
|
ENGINEERING
|
HORTICULTURE
Fruits
Vegetables
|
MICROBIOLOGY

Technican and Assistant as support personnel. Obviously, in a one person R&D operation there would not be the luxury of three support personnel. However, its costly to waste the talent of the Director, Manager, or even Project leader in the set-up and clean-up after a given study is made. That is, hourly help may be used at the level of the Technician and/or the Assistant, thus, utilizing the talents of the trained personnel more effectively.

FIGURE 12.3

RESEARCH & DEVELOPMENT ORGANIZATION BY PRODUCT

DIRECTOR

ANIMAL PRODUCTS
Milk
Meat
Poultry
Eggs
Fish
Etc.

PLANT PRODUCTS
Grains
Pulses
Fruits
Vegetables
Tubers
Nuts
Sugar Crops
Oil Seeds
Sea Weeds
Beverage Crops
Etc.

FIGURE 12.4

RESEARCH & DEVELOPMENT ORGANIZATION BY PROCESS

DIRECTOR

BAKING

CANNING

CHEMICAL PRESERVATION

CHILLING

CONCENTRATING

DRYING

FERMENTING

FREEZING

PASTEURIZATION

PACKAGING

RADIATION

In Figures 12.2, 12.3, and 12.4 the organizational plan stresses disciplines, products, and processes, respectively. The organizational plan for anyone of these may be appropriate depending on the size of the firm and the lines of products the firm produces. Each of the plans have their advantages.

The ideal organizational plan is shown in Figure 12.5. This plan utilizes much that is in each of the other plans and stresses the team effort approach. Personnel working on any project may be drawn from several areas of the firm and may not always be in the R&D group. However, its a team effort and the team has the objective of improving a process, or products, or services that can be put together to accommmplish the overall objectives. This format calls for strong project leadership under the direction of a capable director. This plan works and is used in small firms as well as large firms. The total objective of the plan is to build on where the firm is and in most cases it can only be accomplished by a total team effort.

FIGURE 12.5

RESEARCH AND DEVELOPMENT

IDEAL ORGANIZATION PLAN (TEAM WORK)

The establishment of what research to work on is usually a management decision. Management must establish the urgency, the potential tangible benefits, the prospects for success, the costs to implement the study, and the time schedule. In making their decisions, management must consider the market needs, that is, size, stability, consumer prejudice, legal aspects of the product, the process, the patent, etc. The Return on the Investment (ROI), the potential sales life of the product, and what kind of competition is out there are other factors to consider. Management decisions must be up front and be clearly spelled out to the R&D personnel. Once management has established the criteria and the factors to be considered and communicated these to the R&D personnel, the R&D personnel must be held accountable for their efforts in behalf of the firm. An R&D function is only as strong as management makes it.

One of the chief functions of an R&D group is the development of new products. The first prerequisite in the development of a new product is clear cut objectives by management. These objectives should be in writing as part of the policy statement of the firm and approved by the Board of Directors of the firm. Thus, the decision to develop a new product is the decision of the Board. The following are a few of the many questions that must be answered:

Is there a stable source of supplies or raw materials?

Does the proposed product fit into the line we now manufacture?

Do we have the necessary equipment to manufacture the new product?

Do we have the capability, that is, in R&D, Engineering, Production, Marketing, etc.

Do we have the capital available for the needed investment?

Is this the best alternative use of present capital?

What is the estimate on ROI?

Are there competitive products on the market and, if so, what is their price and can we be cost competitive?

Have we determined the projected market?

What are the chances that our competitors will scoop us?

How can we protect our vested interest? Is the proposed product or process patentable? Are there trade secrets involved? How much lead time do we have?

Have we determined the projected market?

If it is agreed to develop a new product, the following steps should be followed:
1. After the goal has been defined by management, analyze the competition and the literature for any similarities.
2. Bench top, that is, kitchen design the new product with appropriate quality tests to insure uniformity in repeat products. The quality tests include physical, chemical and sensory analysis with procedures clearly spelled out for the development of the defined specifications.
3. Move the proposed product from the bench top to the pilot plant for a study of manufacturing inputs and their evaluation. If satisfactory, a test run on the production line follows. This product should be evaluated to insure uniformity and compliance with specifications and for repeatability in manufacture.
4. Test market the new product in the market place using focus groups, in-store evaluation of the product, and actual sales.
5. Test the new product in service; find out what the user thinks of it, and why the non user has not bought it.
6. If necessary in light of consumer reactions, redesign the product as in 1 above.
7. Repeat steps 2 to 4 of the new product followed by consumer evaluation and home use. If the newly designed product is satisfactory, then proceed to start up a production run followed by market introduction.

Research and Development in a food firm is an essential arm whether operated as an independent unit or as a part of the Quality Assurance Department. The measure of the R&D value to a firm is based on the net profit generated from the newly developed processes or products or improvements in present products. In some cases, their measure of success is based on the number of patents or the number of scientific publications or presentations made. R&D personnel can be most visible to a firm and in many cases the R&D personnel will be transferred to key position in other phases of the firm to continue their productivity for the overall growth of the firm.

Reference

Gould, Wilbur A. 1991 Research and Development Guidelines for the Food Industries. CTI Publications, Inc., Baltimore, MD.

CHAPTER 13

Communicating Quality and the Cost of Quality

The art of communicating is perhaps the most important area for the Quality Assurance personnel to be involved. The first way to communicate quality mindness is by the behavior and appearance of the quality assurance personnel. They should always set the perfect example. Their hair (male or female) should be under control and the men should be clean shaven. Their clothing should be clean and preferably white from the hard hats to their shoes. They should be pleasant and have the ability to speak so that they are completely understood.

The first essential about communicating quality is to be able to be a good listener. Listening helps us to know and understand the people we are working for and with. Listening helps to understand the motivational wants and needs of the people you work with. Listening will tell whether or not the person you are working with and for understands your message— did they really hear what you thought you said.

Quality assurance personnel must be able to lead discussions and make demonstrations of the "Whys" and "Hows" of quality control practices and costs. They should know how to use visual aides, models, slides, videos, movies, and charts in their demonstrations. They need to be positive in their mannerisms and their activities to lead the way to maintain the desired standards and expectations of management.

Individual Quality Assurance personnel must keep laboratory notebooks and records to accurately record all their efforts in their quality assurance work. A bound notebook is the preferred type. The notebook provides the daily information and validates the quality of the product or the process. The notebook should have numbered pages and all entries should be in ink with each page dated and signed at the end of the day.

The notebook is the property of the firm and should be reviewed by peers on a regular basis. The notebook serves as the basis for writing the monthly and annual report. It may be most vital to prevent loss of data which could be needed if a recall is made. It is, also, important to protect discoveries, ideas, and the prevention of duplicating the work at some later time.

If information is taken from other sources or drawings, computer printouts, data sheets, photos, etc., and they are part of the day's report, they should be carefully noted and they should be filed as exhibits of the day's activities. After the notebook is completely full or it has covered a given topic, it should be filed in the firms library for future reference. The notebook is the first fundamental in communicating quality.

Reports from a quality assurance department are expected by other departments within the production and management areas on a regular basis. The reports are the message to management that the processes or the products are in control or that there is a need for action. The reports should be as short as possible and directly to the point. All reports should carry a summary and conclusion as to recommendation relative the matter or concern. A report is really a feed-back method of covering pertinent quality information to cooperating departments and the management.

The first essential point about a report is that the report must be honest and it must be accurate. Secondly, the report must be always on time. After the fact is of no value to control processes or product quality. Thirdly, the Quality Assurance manager should prepare an annual report to cover changes made in the operation and how these changes have saved dollars and/or improved the process or products quality. The report should emphasize the cooperation of all personnel involved and may be jointly prepared for management.

Reports vary as to types, that is, the daily report with numbers and charts of the process showing measurements that indicate any variation from the expected. Remember a picture (chart) is worth 10,000 words and it is much easier to understand.

The chart could be a Pareto Chart, that is, the ranking of all potential problems or data or sources of variation wherein the points are prioritized and the trivial many causes are separated from the vital few.

Another chart is the Histogram. This is used to show the frequency distribution of a particular attribute that is measured in the process or of the product.

A Run Chart is the graphic presentation where one measured characteristic of a process is plotted over time. Run Charts give a visual picture of what is happening during a given period of time or a run.

The Control Chart is a graphic presentation of a measured characteristic of a process showing actual data. The Control Chart has a Central Line, the Average designated X Bar, and Control Limits representing what the process can do when operating consistently. The Control Limits (Upper and Lower Control Limits) are calculated statistically using actual data from the process. The objective of Control Charts is to show how the process is doing, that is, is it in control and are any trends taking place that may mean adjustments to the process. Any adjustments should be noted on the Chart to help in interpreting future actions. Generally, Control Charts are for specific unit operations in the process and they are maintained by the operator of that unit operation. Quality assurance personnel must monitor these charts and help the operator if the process should get out of control.

Another class of reports would include those used for problem solving using the scientific method approach. The following is a brief description of the how to proceed:
1. Identify the problem, that is, what is the problem?
2. Gather all the facts and data in a systematic manner.
3. Statistically analyze the data using Pareto, Control Charts, Correlation Charts, and/or ANOVA.
4. Determine possible solution to the problem.
5. Recommend the best solution and put it into effect.
6. Follow-up at a later date to see if the solution is working, if not, use the second best solution as determined in 4 above and put this one into effect, etc.
7. Document the solution to the problem by answering the following questions:
   A. What was done?
   B. Why was it done?
   C. How was it done?
   D. When was it done?
   E. Where was it done?
   F. Who was involved?

A third type of report deals with the quality costs. The cost of quality is the cost of doing things wrong. Crosby states that "Quality is Free", its the non-quality things that cost. The cost of quality is the price of non-conformance. Price of Non-Conformance (PONC) includes all the expenses involved in doing things wrong. The successful busines firm must show a favorable return on the investment (ROI). Studies have clearly shown that there is a direct relationship between quality performance and the return on the investment.

Quality assurance personnel must not only be able to effectively communicate in terms of process control systems and product attributes to management, but they must also be able to articulate in management's language their contribution to the bottom line, that is, the Return on the Investment (ROI). No effective quality assurance program is complete without an understanding and an identification of the costs of quality.

Quality costs are defined as costs which exist in the design, development, manufacture and distribution of a product, a process, or a service, because a quality deficiency either does or might exist. These costs would disappear if all possible quality deficiencies disappeared and if perfect control of materials (ingredients, products. packaging materials and utilities, etc.), people and processes where possible. The objective of such a system is to quantify the opportunity dollars available; thereby, pinpoint problem areas for corrective action, reducing manufacturing costs and improving product quality. In short, it is designed to recognize the costs of "doing things over" as a significant addition to the bottom line.

Quality costs are the separation of accounting figures into four categories:

Prevention—"Costs incurred for planning, implementing and maintaining a quality system that will ensure conformance to quality requirements at economic levels" or costs to ensure "We do it right the first time", that is, costs of quality planning, training programs, and vendor surveys.

Appraisal—"Costs incurred to determine the degree of conformance to quality requirements" or the costs of auditing to see if we actually "Did do it right the first time", that is, the operator use of statistical control charts, package integrity checks, and quality audits of incoming materials.

Internal Failure—"Costs arising when products, components and materials fail to meet quality requirements prior to transfer to ownership to the customer", or the price "We pay when we find we didn't do it right the first time," that is, costs associated with a product which must be reprocessed, recouped, rejected, or lot by lot acceptance sampling.

External Failure—"Costs incurred when products fail to meet quality requirements after transfer of ownership to the customer", or the price we pay and will continue to pay for failure to discover "We didn't do it right the first time", that is, any customer complaint or any product recall.

To achieve a favorable return on the investment (ROI) of materials, processes and people the four categories of quality costs must be determined to establish a benchmark for comparison. As in most companies there will be a considerable inbalance between the dollars spent on failure costs versus those dollars spent in the value added areas of prevention and appraisal. This type of dollar breakdown will focus where improvement projects should be undertaken.

Total quality costs alone do not provide a clear picture to management. They should be compared to various parts of the business that identify change, that is, a labor base, production costs, sales or per unit produced, in order to aid in providing a realistic review of present performance and/or assist in appraising future potentials.

By clearly identifying and communicating the financial impact of the cost of quality, the quality assurance manager will help top management set objectives and develop a strategy to achieve consumer enthusiasm.

Problems in a food plant may be many and varied. They start with the raw materials (cultivars, growing areas, maturities, harvesting, handling, storage, pesticides etc.), preparation operations (receiving, sorting, separating, slicing, washing, disintegrating, blanching, mixing, filling, etc.), packaging (container types and their effect on flavor migration and pickup), processing (time, pressures, temperatures, agitation, etc.), and storage-shelf life (time, temperature, humidity etc.). To solve the many problems relative quality retention or product improvement, the quality assurance personnel must learn to communicate and they must use the scientific method in reaching their conclusions. Most importantly, they must report the results of their findings promptly to prevent recalls, consumer complaints, non-conformance costs of manufacture, reliabilities of vendors, changes in government regulations, and product behavior in the market place. Communicating to all levels of management by the quality assurance personnel is a must for continued growth of a firm and a favorable return on the investment for the stockholders.

References

AMA Course notes "Managing Costs of Quality".
Anon. Quality Costs-What and How. Am. Soc. for Quality Control. 1971.
Crosby, Philip B. Quality is Free. McGraw Hill Book Co., NY, 1979.
Crosby, Philip B. Quality without Tears. McGraw Hill Book Co., NY 1984.

CHAPTER 14

Specifications and Quality Standards

Quality assurance and product control follows the establishment of product and process specification. No other facet of quality assurance is more important than the establishment of specifications and the development of standards of quality for product evaluation. Remember, "Before you can control–you must first be able to measure to some standard."

With the new information that is available on quality evaluation methods and specific processing procedures for the different products, management in cooperation with production, sales, and quality assurance must draft specific process procedures, specifications for all incoming materials, and standards of quality for all finished products. These specifications and standards are developed for the primary purpose of providing information and parameters of unit operation to production personnel for the packing of desirable qualities of finished products. They become the Standard Operating Procedures (SOP) for each unit operation in the process.

The quality assurance technologist's major function is to train the unit operators and line process operators and then audit and evaluate the product's deviation from these specifications and when necessary, make the needed changes to control the quality of the product at the desired level. Thus, management always knows what quality is being packed. Obviously, the personnel in the production department and the quality assurance department must cooperate to the utmost to produce a given quality product.

Please The Customer

One must always remember that a food company is in the business to make a product– a product intended for sale to a customer– from which the firm hopes to make a profit. The key word is the customer, he is the one who must be satisfied. Therefore, the customer establishes the level of quality. Food companies must learn to listen to their customers through focus groups, in-store observations, and actual home surveys. What is being reported is most important to what should be manufactured.

Specifications

A specification is basically a communication tool to define reasonable expectations. A specification serves as the body of rules for the manufacture and sale of food products. The specification must be established to satisfy the customer. A food firm should strive to do more than just meet a specification. The ultimate goal should be to reduce the variability of the process output to zero. This is accomplished by determining the process capabilities of each unit operation and the total process. The goal is to improve the process capability of each unit operation to acceptable limits.

Juran states "the purpose of a specification is to define the subject matter of quality. Specifications may include: (1) materials, (2) processes, (3) products, (4) methods of test, (5) criteria for acceptance or rejection, and (6) method(s) of use."

Material specifications originate with firms or governmental agencies or even professional societies. The material specification describes and defines the product, that is, cultivars or varieties, quality characteristics and tolerances (color, maturity, freedom from defects, consistency, pH, etc.), size and their limitations, handling practices, pesticides and their application, etc. A material specification, among other things, establishes the method of quality evaluation, sampling procedures, and the lot identification.

A process specification serves "to instruct personnel on how to make the product successfully and economically" according to Juran. It serves as an "indirect means for specifying product quality, that is, evidence of conformance to the process specification". A process specification provides the "know-how" to operate each unit operation in conformance to product quality requirements. Process specifications include times, temperatures, pressures, flow rates-velocities, specific gravities, heat and mass transfer, etc.

Test specifications are detailed descriptions of how to measure and evaluate quality characteristics of incoming materials, foods in process, or the quality characteristics of the finished products. (see Chapters 16 through 31).

Specifications provide for acceptance or rejection of materials, processes or products. In other words, the standard of acceptance or rejection, that is, acceptable quality level (AQL). A standard is a basis for comparison, a guide or a level of excellence, a pattern or an authoritative measure.

Specifications even provide for methods of use. Many food firms publish their recipes of use giving details of preparation and recommended cook times. An excellent example is those details stated on the package of many microwavable products with the times for heating the specific product specified on the container.

All specifications should be in writing so that the information can be conveyed to all parties concerned. Harriman states that the specification writer should either specify or omit, that is, do not put anything in the specification that cannot be enforced. He suggests the use of simple words followed up by technical words for exact meanings. Further, he suggests the use of nouns, short sentences and the use of commas sparingly. He states that specifications should give directions– not suggestions, that is, specifications should be specific.

An example of a specification for potatoes for use in chip manufacture and potato salad use follows:

Characteristic	Chip Manufacture	Potato Salad
Cultivar	Norchip, Atlantic Denali	Red Pontiac Sebago
Maturity	No Feathering	Skin Set
Sugars		
Reducing	Less than 0.20%	No Limit
Total	Less than 2.5%	No Limit
Size	Min. 2" to Max. 3"	Min. 2"
Shape	Round	Round
Peel Color	White or Brown	White, Brown or Red
Eye Depth	Shallow	Shallow
Peel Thickness	Less than 1/8"	Less than 1/8"
External Defects	Max. of 4% No soft rot	Max. of 2%
Flesh Color	Creamy white	White
Specific Gravity	Greater than 1.080	Less than 1.070
Internal Defects	None including Hollow Heart	None including Hollow Heart

Standards

Standards by definition are anything taken by general consent as a basis of comparison, a grade or level of excellence, a serving as a basis of weight, measure, value, comparison or judgement. A standard is an authoritative model or measure or a pattern for guidance.

The United States Standards for Grades by given food products have been in wide use for product evaluation since the early 1930's. The fruit and vegetable Grade Standards are voluntary and are used by both buyers and sellers. Standards for grades of meat poultry, dairy foods, fish, etc. have been established by governmental agencies and in many cases are mandatory. The Food and Drug Administration has established Minimum Standards of Quality, Fill of Container, and Standard of Identity for several food products. All of these standards are available from the Superintendent of Documents. Further, many are available in an abstracted form from the ALMANAC.

Food firms have their own label standards of quality which may be similar to the U.S. Standards for Grades. Generally, a food firm's label requirements may be more restrictive for certain quality characteristics. That is, the U.S. standard may allow a given consistency for Grade A Catsup (Bostwick of 9 cm.) while a given food firm may allow only a Bostwick of 6 cm. The consumer recognizes the firm's standard of quality by the label. This, perhaps, is the most common system and it does work if a firm has a reliable quality assurance program.

Codex Alimentarius

In 1962 the first sessions were held with some 30 countries participating in the development of CODEX ALIMENTARIUS as outgrowth of World Health Organization (WHO) and the Food and Agricultural Organization of the United Nations (FAO). These world wide standards could be fully accepted by the participating countries, that is, products complying with the standard can be freely distributed under the name and description laid down in the standard. Or, the participating country could accept as a target, that is, intent to accept after a stated number of years and the country will not hinder the distribution of products complying with the standard. Or, the participating country could accept with minor deviations, that is, the participating country states what the deviations are, the reasons for them, and indicates whether products fully conform may be distributed freely and whether it expects to be able to give full acceptance and, if so, when. To date many standards have been approved and published. The following countries are assigned these areas for the development of standards:

SPECIFICATIONS AND QUALITY STANDARDS

Country	Standards responsible for
Canada	Food labeling
Germany	Analysis & sampling, meat & meat products, and dietetic foods
Netherland	Food additives, pesticide residues
Norway	Fish
Switzerland	Cocoa products, natural mineral water
United Kingdom	Sugar, fats & oils
UN EC/Europe	Fruit juices & frozen foods
United States	Food hygiene, processed fruits & vegetables, poultry and poultry products
WHO/FAO Com.	Milk and milk products

ISO Standards

In 1946 an International Organization for Standardization (ISO) was founded to promote the development of international standards and to facilitate the exchange of goods and services throughout the world. ISO is composed of member bodies from over 90 countries, the U.S. member body being the American National Standards Insitute (ANSI). In 1987, ISO published five International Standards (ISO 9000, 9001, 9002, 9003 and 9004) developed by ISO Technical Committee 176 on quality systems. These standards along with the definitions contained in ISO standard 8402 provide guidance in the selection of an appropriate quality management program for a supplier's operation.

ISO 9000 (ANSI/ASQC Q90), Quality Management and Quality Assurance Standards-Guidelines for Selection and Use, explain fundamental quality concepts and define key terms and guidance on selecting, using, and tailoring ISO 9001, 9002, and 9003. ISO 9000 has been adopted in the United States as the ANSI/American Society for Quality Control (ASQC) Q90 Series.

ISO 9001 (ANSI/ASQC Q91), Quality Systems-Model for Quality Assurance in Design/Development, Production, Installation and Servicing, is the most comprehensive standard in the series. ISO 9001 covers all elements listed in ISO 9002 and 9003. In addition, it addresses design, development, and servicing capabilities.

ISO 9002 (ANSI/ASQC Q92), Quality Systems-Model for Quality Assurance in Production and Installation is a standard related to the preventive, detection, and correction of problems during production and installation.

ISO 9003 (ANSI/ASQC Q93), Quality Systems-Model for Quality Assurance in Final Inspection and Test is the least comprehensive standard. It addresses requirements for the detection and control of problems during final inspection and testing.

ISO 9004 (ANSI/ASQC Q94), Quality Management and Quality Systems Elements-Guidelines provides guidance for a supplier to use in developing and implementing a quality system and in determining the extent to which each quality system element is applicable. ISO 9004 examines each of the quality system elements (cross referenced in the other ISO 9000 standards) in greater detail and can be used for internal and external auditing purposes. A copy of the ISO Standards are available for a fee from American Society of Quality Control, Milwaukee, WI.

All specifications and standards must be kept up-to-date. They must be realistic. They must be objective. According to Hi Pitt "they must be treated as laws and one must accept all of the difficult responsibilities attendant upon that decision. The reason for this conclusion is that the alternative to it may well destroy a firm's reputation and put them out of business."

Reference
Anon. ALMANAC. Edward E. Judge & Sons Inc. 1987
Breitenbery, Maureen. 1991. Questions and Answers on Quality, the ISO 9000 Standard Series, Quality System Registration, and Related Issues. U.S. Dept. of Commerce NISTIR 4721.
Cary. Mark et al. The Customer Window. Quality Progress June 1987.
Harriman, H.F. Standards and Specifications. McGraw Hill Book Co. 1928.
Juran, J.M. Quality Control Handbook. McGraw Hill Book Co. 1962.
Pitt, Hi. Specifications: Laws or Guidelines? Quality Progress July 1981.

SPECIFICATIONS AND QUALITY STANDARDS

TABLE 14.1 — FDA Food Standards By CFR Part No.

CFR Part	Title	# of Products
131	Milk & cream	31
133	Cheeses & related cheese products	73
135	Frozen Deserts	5
136	Bakery Products	5
137	Cereal flours & related products	33
139	Macaroni & noodle products	15
145	Canned fruits	27
146	Canned fruit juices	25
150	Fruit butters, jellies, preserves & related products	5
152	Fruit pies	1
155	Canned Vegetables	11
156	Vegetable juices	2
158	Frozen vegetables	1
160	Eggs & egg products	10
161	Fish & shellfish	17
163	Cacao products	14
164	Tree nut & peanut products	3
165	Nonalcoholic beverages	1
166	Margarine	1
168	Sweeteners & table sirups	9
169	Food dressings & flavorings	11

Part II —

Process and Food Product Evaluation

CHAPTER 15

Packaging and Container Integrity Evaluation

Introduction

Foods are packaged for transportation for further manufacture (tote bins, barrels, tankers, railcars, etc.) or transportation for direct ultimate use. They are packaged to prevent damage en route due to mechanical or biological elements (water vapor, oxygen and other gases, light, rodents, insects, and microorganism invasion). Foods are, also, packaged for dispensing, identification, or for carrying directions and instructions for use, and for attracting the potential buyer in the market place.

A good consumer package should be printable; easy to open and reclose; easy to dispose of preferably by recycling; light in weight; inexpensive; tamper proof; strength to protect the product; and a barrier to water vapor, gases and liquids.

Consumer packages may be catergorized into the following categories: metal cans, glass jars and bottles, paper and wood products, films and plastics, edible coatings, and composites. Metal cans, glass jars and bottles, and most composite containers are rigid while paper and plastic trays, boxes, and cartons are semi-rigid, and films in form of bags, envelopes, pouches, tubes, and wraps are flexible.

The metal can has several advantages for use as a food container: 1. Capable of withstanding thermal temperatures during processing, 2. Great strength, 3. Light in weight, particularly aluminum containers, 4. Superior barrier properties for gases, liquids, and water vapor, 5. Relatively inert to most foods, depending on coatings, and 6. Recylcable, particularly aluminum containers.

Glass containers may be extruded, blown or pressed from 72% silica, 12% calcium carbonate or calcium oxide, and 2% aluminum oxide along with substances that changes its color and clarity properties. Glass containers finish or the closure part of the container may be classified as plug, crown, twist off, roll on, continuous thread, interrupted (lug) thread or press on-twist off (PT) caps. Glass containers are an excellent barrier to water vapors, gases and odors, and chemical substances. Glass containers have great strength, they are resealable, recyclable, and very rigid. However, glass containers are usually quite heavy and they are very fragile.

Paper and paper products are made from wood by using chemicals (soda, sulfite or sulfate) to dissolve the binding agents between the cellulose fibers. Paper (less than 0.009" thick) is used for overwraps or single or multiwalled bags after laminating with other materials to provide greater barrier and tensile strength. Some papers are coated or impregnated with waxes, lacquers or plastics to add strength and odor barrier properties. Cellophane is a cellulose product and must be treated with plasticizers to give it flexibility and coated to give it barrier properties. Paperboard products (greater than 0.009" thick) provides structure and stiffness for liquid containers. Paper products are light in weight, can be manufactured to given thickness for rigid control. They are easy to print on and can be constructed of given densities, they are easy to dispose of and can be recycled many times.

Aluminum foil has had much use as a flexible film for snacks and other foods because of its light resistance properties. It is generally laminated to plastic films to improve its barrier properties.

Plastics are easy to fabricate, they are light in weight, they resist corrosion, they can be colored, and generally they have good barrier properties to water, water vapor, gases, odors, oils and grease. Most plastics are relatively easy to dispose of although not always recyclable, and they are easy to heat seal although they are not always resealable. The disadvantages of plastic include varying thermal qualities, and lack of strength (tensile, impact, stiffness, tear, etc.). The following are some of the basic types of plastics.

Polyethylene with several types of molecular structures, that is, HDPE (High Denisty Polyethylene) with a density of 0.94-0.965 g/cc and the molecules are more linear. This product is a rigid plastic and has low water vapor permeability, greater inertness, higher gas transmission, limited odor barrier, and a narrower heat sealing range than other polyethylenes; (LDPE) Low Density Polyethylene has a density of approximately 0.91-0.925 g/cc with a high degree of branching. It is used as film and as a coating. It imparts flexibility and heat sealability to films. It provides water vapor and liquid resistance, but is a poor oxygen barrier; and LLDPE (Linear Low Density Polyethylene) has a more linear structure than LDPE, but exhibits similar properties and is used as a film primarily.

Polyesters are reaction products of polyethylene and terephtalic acid (PET) and are transparent, chemically inert, good gas barrier, high tensile strength products, but have only moderate water vapor properties. They are difficult to heat seal, but can be used for a wide range of temperatures, such as boil in bag containers. They are, also, used in blow molded carbonated beverage containers.

Polypropylene is a highly crystallene polymer exhibiting similar properties to polyethylene, but is stiffer and has better tensile strength. It has good transparency and is grease resistant. It is used extensively in snack food packaging and can be blow molded for simi-rigid containers.

Polyvinyl Chloride (PVC) is used for semi-rigid, blow-molded or thermoformed food packages. It is generally transparent, heat sealable, abrasion resistance with moderate barrier properties to water vapor, gas, and grease. Plasticized PVC is commonly used as a meat wrapper.

Polyvinyliden Chloride (PVDC) is an excellent water vapor and gas barrier, but because it is expensive it is generally used as a copolymer, such as with PVC to produce Saran. PVDC has additional properties such as stiffness, grease and alcohol resistance and in the shrinkable form for vacuum packaging.

Ethylene Vinyl Alcohol (EVOH) is used in conjunction with other materials to produce rigid containers through thermoforming or injection molding.

Polyamides (Nylons) are used because of their heat resistance, toughness, inertness, and good mechanical properties. They are used in retort pouches after laminating to other water vapor and gas barrier films.

Some other heat sealant copolymers include Ethyl Vinyl Acetate (EVA), Ethyl Methyl Acrylate (EMH), and Ethyl Arcylic Acid (EAA). Ethylene Vinyl Alcohol polymer (EVAL) is a polymer that is a high barrier polymer to oxygen. It is available in three forms (EVAL-E, EVAL-F, and EVAL-H with differing ethylene contents.) EVAL E and F are the most common.

Edible Coatings have found very limited use in processed foods, but do have a niche in fresh foods, candies, etc. They are generally made from wax, starch, or other complex carbohydrates.

Composite containers are considered the container of the future for many food packaging operations. Materials are combined to give desired qualities and structures by converters or they may be manufactured in-house by the processor from assembled materials. They may provide difficulties in recycling due to types of laminations used.

Regardless of type of packaging material being contemplated for use, the user must require and assure themselves that the container complies with all regulations as to safety and freedom from toxic substances, free from odors and foreign materials, and free from any infestation or heavy metals.

One of the main areas from a control standpoint is how the packaging materials are shipped to the plant and how the processed product is shipped to the customer. All pallets should be clean and free from any infestation. The pallet should have a 4 way entry and be in good repair (no broken slats, no protruding or bent nails), and none of the pallet load protrudes over the side of the pallet (length or width).

Container Evaluation

The evaluation of a container as to size, net weight, seal, vacuum, headspace, drained weight and external/internal conditions is the first consideration in a quality evaluation/assurance program.

SIZE

Size for metal containers is determined by measuring the diameter and height in sixteenths of an inch. These measurements are reported as follows: 303 × 406 measures $3\,^{3}/_{16}$ in. in diameter by $4\,^{6}/_{16}$ in. in height. Table 15.1 shows the name of container, diameter and height, volume fill and net quantity for statements for various can and glass containers.

CODE

The coding of processed food packages is a requirement as specified in the Food and Drug Administration Good Manufacturing Practices which provides a means of can or lot identification for:
(1) Quality control
    (a) Product evaluation
    (b) Machine efficiency
(2) Legal protection
(3) Segregation of defective lots
(4) Answering of complaints and in case of seizures
(5) Product flow through the plant, warehouse or distribution

The code should be complete, legible and indicate:
(1) Product packed
(2) Quality
    (a) Raw product
    (b) Contemplated grade

TABLE 15.1 — Container Size Conversion — Tin and Glass

Can Name	Diameter × Height	Vol. Fill Cubic Inches	Total Capacity[1]	No. 303 Can Equiv.	No. 2½ Can Equiv.	No. 10 Can Equiv.
6z Jitney	202 × 308	9.42	6.00	.404	.229	.062
211 Baby Food	211 × 200	10.38	4.90	.395	.223	.061
8z Short	211 × 300	12.34	7.90	.469	.266	.072
No. 1 Picnic	211 × 400	17.06	10.90	.648	.367	.100
211 Cylinder (12r)	211 × 414	21.28	13.55	.809	.455	.125
No. 300	300 × 407	23.71	15.20	.901	.511	.139
No. 300 Cylinder	300 × 509	30.17	19.40	1.147	.651	.177
No. 1 Tall	301 × 411	25.99	16.60	.988	.561	.152
No. 303	303 × 406	26.31	16.85	1.000	.566	.154
No. 2 Vac. (12z Vac)	307 × 306	22.90	14.70	.870	.493	.134
No. 2	307 × 409	32.00	20.50	1.216	.689	.187
No. 2½	401 × 411	46.45	29.75	1.765	1.000	.272
No. 3 Vac.	404 × 307	37.19	23.85	1.414	.801	.218
No. 3 Cyl. (46z)	404 × 700	80.54	51.70	3.061	1.735	.472
No. 10	603 × 700	170.71	109.45	6.488	3.673	1.000
		GLASS				
16z (No. 303 or 1 lb jar)		27.97		1.063	.602	.164
30z No. 2½		48.06		1.827	1.035	.282
64z		115.20		4.390	2.487	.677
128z (1 gal. Jug)		231.00		8.780	4.973	1.353

[1] Avoir of Water at 68°F.

(3) Plant in which processed
(4) Line or machine within plant
(5) Date processed and, where applicable, time of day
(6) Period packed

The packing period code shall be changed with sufficient frequency to enable ready identification of lots during their sale and distribution. Codes may be changed on the basis of one of the following:

(a) Intervals of every 4–5 hr
(b) Personnel shift changes, or
(c) Batches, provided the containers comprising such batch do not extend over a period of more than one personnel shift

The code may be in numbers or letters. In either case, it must be logged in a Closing Machine Record Book to show:

(1) Code
(2) Number of containers packed with any code
(3) Time of start and stop of any run under that code
(4) Seam evaluations for each closing machine on a regular, routine basis using representative samples

(5) Closing machine vacuum, if using a steam vacuum closure, and/or center or fill temperature of product

The technologist should record the code exactly as embossed or printed on the package.

VACUUM

"Vacuum" is the term used to denote the pressure conditions inside a hermetic food container and is a measure of the extent to which air has been eliminated from the container prior to processing. The food industry measures vacuum in terms of inches of mercury. A zero vacuum indicates that the pressure in the headspace is equal to atmospheric pressure, whereas a vacuum of 30 in. of mercury would indicate that all gas had been removed from the container. A vacuum gauge measures the difference between atmospheric pressure and the pressure in the container and is usually calibrated to read in inches of mercury.

Many desirable effects have been attributed to vacuum in canned foods:
(1) It maintains the can ends or jar closures in a concave position during normal storage.
(2) It reduces the quantity of oxygen in the container.
(3) It prevents permanent distortion of can ends and helps hold the closure on glass-packed products during thermal processing.

The vacuum should be taken on all thermal processed foods. For metal cans, the reading should be made as close to the side seam as possible so as not to draw brine or sirup up into the gauge, particularly if taking the vacuum in the center of the lid and evaluating a full container. For glass, the gauge should have a longer stem to penetrate through the cap and gasket. Readings should be recorded in inches.

CAN SEAM EVALUATION

The can seam is referred to as a double seam. The double seam is formed by joining the can body and can end components in an interlocking formation. The double seam consists of three thicknesses of the end component and two thicknesses of the body component. There also exists a rubber compound in between the layers to complete the hermetic seal.

Visual examinations should be made at intervals of not more than 30 min. Also, visual seam inspections should be made after a can-jam at the closing machine or after startup of a closing machine following prolonged shutdown.

External or visual inspection of a seam and can should be made for the following defects:
(1) Cut-over or sharpness
(2) Skidding or dead-heading
(3) False seam
(4) Droop at crossover at lap
(5) Condition of inside of countersink wall for evidence of broken chuck
(6) Dents or scratches on body or double seam

Double seams should be evaluated to determine the integrity of the seal. The minimum length of time that should be considered for a teardown inspection is one can once every four operating hours from each seaming station. Teardown examinations should also be made following a severe jam at the closing machine, and after a closing

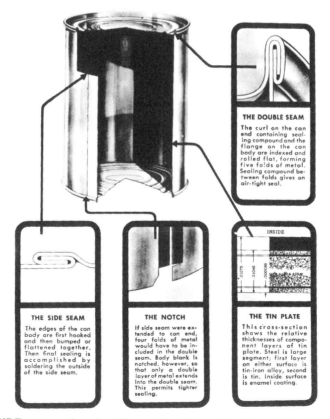

FIGURE 15.1 — Standard Terms Identifying The Basic Parts Of A Can

machine has been changed over from one can diameter or height to another.

External measurements should be made and recorded. These include the thickness, the width (length or height) and the countersink measurement.

The seam should then be taken down and the following measurements taken: coverhook, wave wrinkle condition, fractures, condition of crossover area (side seam), body hook, and tightness rating.

Figure 15.1 illustrates the parts of a can and Figure 15.2 illustrates the double seam and its terminology. The data in Table 15.2 illustrates double seam dimensions of common can lid sizes.

FIGURE 15.2 — Required Measurements For Evaluating The Quality Of Double Seams

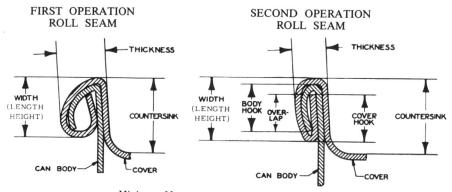

Minimum Measurements

Width* (Not essential if overlap is measured optically)
Thickness*
Countersink (desirable but not essential)
Body hook*
Cover Hook* (required if micrometer is used)
Overlap* (essential if optical system used)
Tightness* or wrinkle
　　　　*Essential Requirements

Calculation of Overlap Length

Overlap length = CH+BH+T-W
Where CH = cover hook
　　　 BH = body hook
　　　 T** = cover thickness, and
　　　 W = seam width

**In general practice 0.010 may be used for the tin plate thickness

TABLE 15.2 — Double Seam Dimensions Of Common Can Sizes
(Dimensions In Inches)

Can Size	Thickness	Length (Countersink)	Body Hook	Cover Hook
211 × 400	0.051-0.056	0.112-0.120	0.075-0.082	0.075-0.082
300 × 407	0.051-0.057	0.114-0.122	0.075-0.082	0.075-0.082
303 × 406	0.053-0.060	0.114-0.122	0.075-0.082	0.075-0.082
307 × 409	0.054-0.061	0.117-0.125	0.075-0.085	0.075-0.085
401 × 411	0.054-0.061	0.117-0.125	0.075-0.085	0.075-0.085
404 × 700	0.056-0.063	0.117-0.125	0.075-0.085	0.075-0.085
603 × 700	0.065-0.071	0.120-0.130	0.075-0.085	0.075-0.085

GLASS CONTAINERS AND CLOSURES

Glass containers amount to approximately 40 billion units or 163 glass containers per capita. Glass containers for food represent 31.1%, beer 30.7%, soft drinks 21.1%, and liquor and wine 9.8%.

As a packaging material, glass possesses qualities unequaled by any other material. It is chemically inert, impermeable, transparent, nonporous, sanitary and odorless. Further, it may be formed in many colors and into an infinite variety of shapes and sizes.

Clear or flint bottle glass is made essentially from sand (almost pure silica), soda ash (sodium carbonate), and limestone (calcium carbonate or calcium magnesium carbonate) which gives the glass hardness and chemical durability. If color is desired, it is obtained by adding small quantities of chromium, cobalt, iron or nickel. In the case of amber glass, carbon and a sulfide, or a sulfate compound are added.

Figure 15.3 illustrates the basic parts of a glass container. From this illustration it can be seen that there are three major portions of a glass container: the finish, the body and the bottom.

The finish is the area through which the product is filled into the glass container and where the hermetic seal is applied by the proper closure. Every type of closure for glass container sealing has a special glass finish or finishes with which the closure has been designed to function. However, glass finishes are standardized and the specifications and tolerances have been established by the Glass Container Manufacturer's Institute.

FIGURE 15.3 — The Basic Parts Of A Glass Container

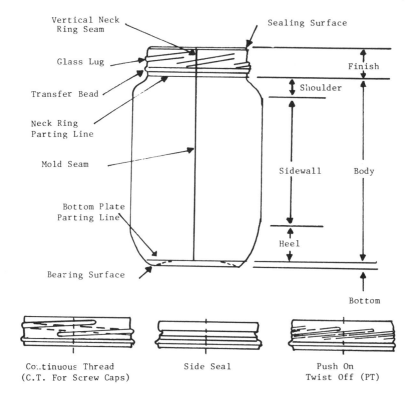

Closures

Closures are an indispensible component of glass packaging. Closures must meet two demanding and contradictory requirements. First, the seal must be so positive that the contents cannot escape and no outside substance may enter. Secondly, consumers must be able to break the seal quickly and easily and, often, to reseal the jars or bottles many times.

One of the principle types of closures is the metal screw cap. This type of cap is threaded to intermesh with complementary threads made on the neck of the glass container.

Almost all low-acid foods packed in glass containers are sealed with vacuum-type closures. There are four principal vacuum closure types:
 (1) Pry-off side seal cap—This cap is pressed straight down onto the glass finish by the sealing shoe of the capper.

PACKAGING AND CONTAINER INTEGRITY EVALUATION

(2) Regular lug-type or twist caps—These caps are applied to the glass finish by turning or twisting the cap onto the finish to seat the lugs of the cap under the threads on the glass finish.

(3) Lug-type cap for baby foods—This cap is applied the same as the regular lug-type cap except that production lines move at much higher speeds.

(4) PT (press-on twist-off) cap—This cap is applied by simply pressing the cap down on the glass finish, with the gasket heated prior to application.

The crown closure is one of the most important. This closure has a resilient liner that forms the actual seal. A metal skirt with flutes that are forced against a projecting ring on the neck of the bottle provides a strong grip to hold back the internal pressure of products like carbonated beverages or beer.

Evaluation of the Integrity of the Glass Closure

Since the closure on a glass jar is based on one of the three types of caps previously described, the evaluation of the integrity of the seal must correspond to the type of closure. However, since the type of closure is not known until the seal is broken, the technologist should evaluate the closure by an external inspection to make certain that the cap is level and seated well down on the finish. It should not be cocked or tilted. In most cases there should be a vacuum and the panel (center of the cap) should be dished-in or concave indicating the presence of a vacuum.

The vacuum should be read using a vacuum gauge as in evaluating the vacuum on metal containers. However, due to the design of the cap, the rubber gasket on the vacuum gauge must be more flexible to allow greater penetration of the gauge. After the vacuum has been taken the cap should be removed and the gasket examined to ensure a secure seal and that the gasket was not pulled or broken at any point.

TEST 15.1 — Container Size, Vacuum, Fill And Drained Weight

General

Before a processed item can be evaluated or graded for quality, a number of preliminary measurements and recordings must be taken; that is, product style, type and code are necessary for proper identification of the sample while grading it. Drained weight, net weight, container condition, vacuum, and headspace directly and indirectly affect the quality and acceptability of the product. Drained weight may or may not be one of the attributes that contributes to the final quality grade of a product, but for many products there is established a minimum drained weight and fill weight for that particular container.

Procedure

Record or measure the following items where necessary and applicable.
(1) Label information
   (a) Name of the product and list of ingredients
   (b) Style or method of preparation as affected by formulation and processing
   (c) Type—differences that are due to variety or cultivar
   (d) Vendor
   (e) Other information—artificial coloring, artificial flavoring, preservatives, packing media including sirup concentration, etc.
(2) Record the following:
   (a) Code mark on container and label information exactly as presented.
   (b) External condition of container; note seams and signs of abuse such as dents, panelling, rust, etc.
   (c) Size of container—the dimensions of a can are expressed as diameter and height. Thus, a number 303 means that the diameter of the lid is 3 and $3/16$ in. while the height of the can is 4 $6/16$ in. The first digit is expressed in inches while the last two are expressed in sixteenths of an inch. If the size of the container is in doubt, it can be measured and expressed in proper terms by measuring with a suitable ruler.
(3) Determine the vacuum with the aid of a vacuum gauge. To determine the vacuum moisten the tip of a vacuum gauge, hold it firmly in one hand while holding the can firmly in the other hand on a flat level surface and insert the needle of the gauge into the lid by penetrating the can with the gauge needle. The reading, expressed in inches, should be taken as near to the outer edge of the can lid as possible to prevent the product from being drawn up into the gauge. The vacuum gauge and can must be held firmly while taking the reading.
(4) Net Weight—the weight of the contents of the container exclusive of the container weight. This may be determined by using a can which is representative of the can being examined as a tare weight or by washing, drying and weighing the emptied can and subtracting the can's weight from the gross weight.

PACKAGING AND CONTAINER INTEGRITY EVALUATION   149

FIGURE 15.4 — Determination Of Net Weight

(5) Remove the lid by cutting it out from the top.
(6) Headspace—determine the headspace using a headspace gauge by lowering the ruler to the surface of the product. In some products the surface is uneven and it may be necessary to make more than one measurement, taking the average. Record this measurement in 32nds of an inch and allow 3/16 in. for the seam of a double seam container.
(7) Pour the sirup or brine into a cylinder. If a sirup pack, determine the sirup strength with the aid of a hydrometer or refractometer and express as degrees brix. If a brine pack item, note the clearness, color and/or general appearance of the brine. If part of the grade, score in accordance with the applicable USDA Grade Standard.
(8) Determine the fill of the container using one of the following five methods:
 (a) The term "general method for water capacity of containers" means the following method:
  (1) In the case of a container with lid attached by double seam, cut out the lid without removing or altering the height of the double seam.

FIGURE 15.5 — Determination Of Headspace

(2) Wash, dry and weigh the empty containers.
(3) Fill the container with distilled water at 68°F to $3/16$ in. vertical distance below the top level of the container, and weigh the container thus filled.
(4) Subtract the weight found in step (8) (a) (2) from the weight found in step (8) (a) (3). The difference shall be considered to be the weight of water required to fill the container.

In the case of a container with lid attached otherwise than by double seam, remove the lid and proceed as directed in steps (8) (a) (2) to (4) inclusive, except that under step (8) (a) (3) fill the container to the level of the top.

(b) The term "general method for fill of containers" means the following method:
(1) In the case of a container with lid attached by double seam, cut out the lid without removing or altering the height of the double seam.

(2) Measure the vertical distance from the top level of the container to the top level of the food.
(3) Remove the food from the container; wash, dry, and weigh the container.
(4) Fill the container with water to 3/16 in. vertical distance below the top level of the container. Record the temperature of the water, weigh the container thus filled, and determine the weight of the water by subtracting the weight of the container found in step (8) (b) (3).
(5) Maintaining the water at the temperature recorded in preceding step (4), draw off water from the container as filled in step (4) to the level of the food found in step (8)(b)(2); weigh the container with remaining water, and determine the weight of the remaining water by subtracting the weight of the container found in step (8) (b) (3).
(6) Divide the weight of water found in preceding step (5) by the weight of the water found in step (8)(b)(4), and multiply by 100. The result shall be considered to be the percentage of the total capacity of the container occupied by the food.

In the case of a container with lid attached otherwise than by double seam, remove the lid and proceed as directed in steps (8)(b)(2) to (6) inclusive, except that under step (8) (b) (4), fill the container to the level of the top.

(c) The drained weight of the food product measured against the water capacity of the can by weight is determined as follows:
  (1) In the case of a container with lid attached by double seam, cut the lid without removing or altering the height of the double seam.
  (2) Tilt the open container so as to distribute the contents evenly over the 8-mesh 8 in. screen for 2½ can or smaller and a 12 in. screen for can larger than 2½ which has been previously weighed. Without shifting the material on the sieve, incline the screen at a 5–10° angle by resting the screen on the edge of the tray so as to facilitate drainage. Two minutes from the time drainage begins, weigh the screen and drained food. The weight so found, less the weight of the screen, shall be considered to be the total weight of drained food.
  (3) Wash, dry, and weigh the empty container.
  (4) Fill the container with distilled water at 68°F to 3/16 in.

FIGURE 15.6 — Determination Of Drained Weight

vertical distance below the top level of the container, and weigh the container thus filled.
(5) Subtract the weight found in step (8) (c) (3) from the weight found in step (8) (c) (4). The difference shall be considered to be the weight of water required to fill the container.
(6) To obtain the percentage of drained food in relation to the water capacity, divide drained weight (step (8) (c) (3)) by water capacity (step (8) (c) (5)) and multiply by 100.

In the case of a container with lid attached otherwise than by double seam, remove the lid and proceed as directed in steps (8) (c) (2) through (6), except that in step (8) (c) (4) fill the container to the level of the top thereof.
(d) The percentage of the total capacity of the can is determined as follows:
(1) In the case of a container with lid attached by double seam, cut out the lid without removing or altering the height of the double seam.

PACKAGING AND CONTAINER INTEGRITY EVALUATION

(2) Measure the vertical distance from the top level of the container to the top level of the food.

(3) Remove the food from the container. Wash, dry, and weigh the container.

(4) Fill the container with water to $3/16$ in. vertical distance below the top level of the container. Record the temperature of the water, weight of the container thus filled, and determine the weight of the container found in step (8)(d)(3).

(5) Maintaining the water at the temperature recorded in preceding step (4), draw off water from the container as filled in step (4) to the level of the food found in step (8)(d)(2). Then weigh the container with remaining water, and determine the weight of the remaining water by subtracting the weight of the container found in step (8)(d)(3).

(6) Divide the weight of water found in step (8)(d)(5) by the weight of water found in step (8)(d)(4), and multiply by 100. The result found shall be considered to be the percentage of the total capacity of the container occupied by the food.

In the case of a container with lid attached otherwise than by double seam, remove the lid and proceed as directed in step (8)(d)(2) through step (8)(d)(6) above, except that in step (8)(d)(4) above, fill the container to the level of the top thereof.

(e) The volumetric determination follows:

(1) In the case of a container with lid attached by double seam, cut the lid without removing or altering the height of the double seam. (In the case of a container with lid attached otherwise than by double seam, remove the lid and proceed as for double seam containers.)

(2) Remove the product and liquid from the container and return them thereto.

(3) Fifteen seconds after the product and liquid are returned, measure the headspace from the top of the double seam or the top of the glass container.

For double seam containers, the product and liquid fill the container to the level of $3/16$ in. vertical distance below the top of the double seam. For glass, fill the container to the level of $\frac{1}{2}$ in. below the top of the container. Use a headspace gauge for ease and uniformity of measurement.

(9) Examine the interior of the container. Look for corrosion, discoloration or breakage of enamel and solder splash.

(10) Record the count or size of the commodity, where applicable.
(11) Check the odor to see if it is typical and normal for the product. Only taste fruit or acid products, as removed from the container if normal odor. For vegetable and low acid products the product should be warmed 10 min. prior to tasting.
(12) Determine the score points for each factor (attribute) of quality in accordance with the applicable USDA Standards for Grades for the particular commodity and record the score points on the appropriate score sheets for the specific commodity.
(13) Ascertain the grade by totaling the scores for each factor of quality. Keep in mind the limiting rule (stopper) for any particular factor of quality, where applicable.
The U.S. Grades for fruits are:
  Fancy or Grade A
  Choice or Grade B
  Standard or Grade C
  Substandard or Grade D
The U.S. Grades for vegetables are:
  Fancy or Grade A
  Extra Standard or Grade B
  Standard or Grade C
  Substandard or Grade D
(14) Under "Remarks" indicate any other information which may be helpful about the product, cultivar or label. Example includes juice, nutritive information or abnormal attributes of quality.
(15) All records should be signed and dated by the technologist evaluating the product and cosigned if more than one technologist evaluates the product.

TEST 15.2 — Can Seam Evaluation

Equipment

(1) Can opener—heavy-duty unit which is adjustable to the can diameter and removes the center panel of the cover without damaging the seams or the body
(2) Standard No. 5 nippers
(3) Seam micrometer
(4) Countersink gauge

Procedure

(1) Obtain two cans and note any visual external defects.

PACKAGING AND CONTAINER INTEGRITY EVALUATION

QC DATA FORM 15.1 — Daily Double Seam Report

Date_____ Line No._____ Can Size_____ Can Code_____

Time	Head No.	Width		Body Hook		Overlap		Coverhook		Tightness		Countersink		REMARKS
		Min.	Max.	Min.	Max.	Min.	Max.	Min.	Max.	Min.	Max.	Min.	Max.	

Technologist_____

(2) Measure the thickness of double seam by using a seam micrometer. A minimum of three measurements should be taken at points approximately 120° apart, excluding the side seam.
(3) Measure and record the width (length or height) of the can seam for at least three measurements.
(4) Countersink measurements should also be taken—making sure that the point of the depth gauge pin is positioned so that the reading is taken at the bottom of the countersink radius with the bar of the gauge positioned across the diameter of the can.
(5) The lid is then removed using the can opener making sure not to damage the seams or the body.
(6) The nippers are then used to tear off the remaining strip of cover left by the can opener.
(7) The cover hook is then measured recording the lowest and highest readings. The cover hook should also be visually examined.
(8) The body hook is measured and the maximum and minimum values are recorded.
(9) Determine the tightness rating (wrinkle) of the cover hook according to the following classifications:
0—No wrinkles to a slight trace of a wrinkle.
1—Slight wrinkle up to about ⅓ the depth of the cover hook.
2—Somewhat heavier wrinkle, more pronounced and extending approximately ½ the depth of the hook.
3—Large wrinkle, extending down about ⅔ the depth of the hook. Cover hook is slightly rounded.
(10) Record all data on the SQC Data Form 15.1, Daily Double Seam Report.

(11) Compare results obtained with the ranges for the various can sizes (See Table 15.2) or with the processor's end versus the can maker's end.

TEST 15.3 — Permeability of a Barrier Film- Absolute Pressure Method

The absolute pressure method for determining the steady-state rate of transmission of gases through barrier films is carried out according to the standard method of test ASTM D-1434 (Gas Transmission Rate of Plastic Film and Sheeting). According to the method used, the sample is mounted between the two chambers in a gas transmission cell. The cell and sample are sealed so that the only path for gas to move from one chamber to the other is through the sample. Two methods are provided for determining the gas transmission rates: Method I Manometric (Pressure Differential Method) and Method II-Volumetric (Volume Differential Method). The results are usually expressed as cc's of oxygen per 100 square inches per 24 hours per atmosphere (temperature and relative humidity). Good oxygen barriers have oxygen transmission values less than 0.05 cc/100 square inches/24 hours at 86°F at 65% R.H.

TEST 15.4 — Water Vapor Transmission Rate (WVTR)

The most commonly used method for measuring water vapor transmission rates of flat sheet samples is the cup test (ASTM E96 and TAPPI T464). A circular sample (creased or uncreased), usually approximately 20 square inches in size is sealed with wax to a metal cup containing a desicant. The test cell is then placed in a chamber in which the temperature and relative humidity can be maintained constantly (usually at 90% R. H. and 100 degrees F.). The test cell is weighed periodically and the weight is observed. Since this is a static test method, the slope of the steady state portion of this curve is related to the transmission rate of water vapor into the cup through the sample. The permeability can be calculated by normalizing for sample thickness and surface area. This test has the advantage that many samples can be tested concurrently, limited only by the size of the environmental chamber and the number of test cells available. It may take several days to several weeks to complete the test if the samples are very good barriers to water vapor. However, it is necessary to wait until enough water vapor has passed through the sample so that there is a measurable change in weight. Lengthy test periods may be important if only a limited number of samples are to be tested. The results of the tests are usually expressed in terms of grams of water/100 square inches of packaging materials/24 hours at a specified temperature and relative humidity (normally 100 °F. and 90% R. H. Thus, a MVTR of 0.1 means 100 square inches of the packaging material will allow 0.1 grams of water vapor to pass through it in 24 hours if the environment is 100°F and 90% R. H.

TEST 15.5 — Burst Test

Burst strengths and/or pressure tests are used to establish, gross leaks in heat sealed packages and the integrity of seams in heat sealed packages. They are used to detect leaks in the weakest part of the package. A leak is defined as any opening in a flexible package that contrary to intention either lets contents escape or permits substances to enter.

TABLE 15.3 — Physical Properties Of Some Films

Properties	Cellophane Lacqured	Cellophane Coated	Cellulose Acetate	LDPE	Polyethylene MDPE	HDPE	Pliofilm
BARRIER							
- Water vapor(1)	3-15	18	Very high	18	8-15	5-10	8
- Oxygen (2)	2-80	1-9	1800-3100	3900-13000	2600-5200	520-3900	130-1300
- Carbon Dioxide	15-95	6-9	4700-5200	7700-77000	7.00-13000	3900-10000	520-5200
- Oils & Greases	Impervious	Impervious	Impervious	Swells	Good	Excellent	Excellent
CLARITY (3)	Transparent	Transparent	Transparent	Transparent Translucent	Transparent Translucent	Transparent	Transparent
STRENGTH							
- Tensile (4)	7000-18000	7000-18000	7000-12000	1000-3500	2000-5000	3000-10000	3000-4100
- Tear (5)	2-10	7-15	2-15	100-400	50-300	15-300	60-1600
- Stretch %	15-25	25-50	15-50	225-600	225-500	10-500	10-20
PERMANENCE							
- Max. Tempt. (F)	375	375	250	150	180-220	250	200
- Min. Tempt. (F)	-	0	0	-60	-60	-60	-
- Flamability	burns	burns	burns	slow	slow	slow	no (5)
HEAT SEALABILITY (degree F.)	200-300	200-250	350-450	250-350	260-310	275-310	240-300
HEAT SHRINKABILITY	no	no	no	some	some	some	some
DIMENSIONAL CHANGE AT HIGH RH IN %	3-5	2-3	0.2-0.6	none	none	none	none

TABLE 15.3 (continued) — Physical Properties Of Some Films

Properties	Polypropylene Oriented	Polypropylene Unoriented	Polyester Uncoated	Polyester Coated	PVDC	PVC	Polyamide
BARRIER							
- Water vapor(1)	4	8-10	15-30	1-2	1.5-5	8	very high
- Oxygen (2)	2400	1300-6400	52-130	9-15	8-26	77-7500	30-110
- Carbon Dioxide	8400	7700-21000	18-390	20-35	52-150	770-5500	Impermeable
- Oils & Greases	Excellent	Excellent	Excellent	Excellent	Excellent	Excellent	Impermeable
CLARITY (3)	Transparent	Transparent	Transparent	Transparent	Transparent	Transparent	Transparent
STRENGTH							
- Tensile (4)	25000-30000	3000-6000	17000	17000	8000-20000	2000-19000	10000-18000
- Tear (5)	4-6	40-330	13-80	10-20	10-20	-	20-150
- Stretch %	70-100	200-500	70-130	80-180	40-80	5-500	250-500
PERMANENCE							
- Max. Tempt. (F)	275	250	250	190	290-310	200	350-475
- Min. Tempt. (F)	-60	-	-80	-60	0	-	-50
- Flamability	slow	slow	slow	slow	slow	slow	slow
HEAT SEALABILITY (degree F.)	needs	325-400	275-400	275-400	280-300	200-350	350-500
HEAT SHRINKABILITY	some	no	some	some	some	some	no
DIMENSIONAL CHANGE AT HIGH RH IN %	none	none	none	none	none	none	1-3

(1) g/24 hr/m   (2) cc/mil/m /24 hr/atm.   (4) lb/in   (5) g/mil.

Equipment

Any suitable transparent container capable of withstanding approximately one atmosphere pressure differential, fitted with a flat vacuum tight cover. A vacuum gage, an inlet tube from a source of vacuum, and an outlet tube to the atmosphere shall be sealed into the cover. The inlet and outlet tubes shall be equipped with hand valves. Attached to the underside of the cover shall be a transparent plate that will closely approximate the inside diameter of the container and be such a distance from the top of the container that when it is two thirds filled with water, the attached plate will be positioned 1 inch under the water.

Procedure

The test sample(s) shall be submerged in water contained in the vessel within the vacuum chamber. The uppermost surface of the sample shall be covered by not less than 1 inch of water.

Set the cover on the jar, close the outlet valve, and turn on the vacuum so that the gage rises slowly to a given gage reading, that is, 10, 15, or more depending on expected elevation above sea level for shipment of products.

During the rise in vacuum, observe the submerged sample for leakage in the form of a steady progression of bubbles from the flexible container.

Hold the vacuum for 30 seconds. Release the vacuum, remove the lid, and examine the sample for the presence of water inside the package.

Interpretation Of Results

If there are bubbles definitely attributable to leaks in a sample during the rise in vacuum, or during the hold at full vacuum, the package fails the test.

If water is inside a package, the sample fails the test.

If there are no bubbles observed, attributable to leaks, and if no water is inside a sample, it passes the test.

Reference

American Society of Testing Materials D 3078-84

CHAPTER 16

Sugar, Salt and Seasonings

One of the most useful factors in producing uniform quality in processed foods is the measurement and control of the density of the brines and sirups and the specific gravity or concentration of the product. The intent of this chapter is to explain the use of some of the basic equipment and to acquaint the quality control technologist with instruments and techniques that are generally available for control and measurement of this attribute of product quality.

Many vegetables as they mature increase in their density or specific gravity content. Examples are peas, corn, lima beans and potatoes (see Fig. 16.1). This increase in maturity can be measured by determining the specific gravity content of the product. Also, in the manufacture of these products the separation of overmature, mature or immature products can be accomplished by specific gravity separation.

Many fruits are packed in different concentrations of sugar solutions. The accuracy of the concentrations of these sugar solutions is important from a manufacturer's standpoint and for retaining the quality of the fruit as well as informative labeling of the finished product.

Salt is added to many products to improve the flavor, the amount varies depending on the commodity and the process. Several alternative methods are given in Test 16.2.

Finally, the determination of the specific gravity content or the concentration of food products like tomato products, citrus concentrate, or baby foods is important for manufacture and proper labeling.

HYDROMETRIC METHOD

The hydrometer is the simplest to use of all specific gravity methods. The necessary equipment is inexpensive. It is a very reliable method and has wide application. A hydrometer is simply a weighted calibrated sealed glass tube which floats upright in the testing solution. Like any floating body, it sinks to a depth inversely proportional to the specific gravity of the sample in which it is placed. (Fig. 16.2). When purchasing hydrometers, two factors should be kept in mind: (1) the stem should be approximately

162   TOTAL QUALITY ASSURANCE

12 in. long and (2) the hydrometer should be calibrated over a narrow range so that it can be read accurately. There are many scales in present day use. Hydrometers can be classified into five main groups.
(1) Specific gravity hydrometers, indicating the specific gravity of a liquid at a specified temperature in terms of the density of water at a specified temperature as unity.

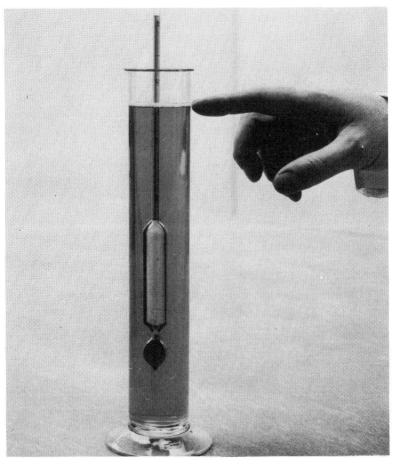

FIGURE 16.1 — Brix Hydrometer

Determining the concentration of syrup from a can of pears. The syrup is drained from the fruit into the cylinder. Then a Brix hydrometer is placed in the syrup and the concentration (% sugar in the syrup) is determined by reading the hydrometer at the bottom of the meniscus (syrup level). Photo made in the Food Technology Laboratory, Ohio State University.

SUGAR, SALT AND SEASONINGS

(2) Percentage hydrometers, indicating at a specified temperature the percentage of a substance dissolved in water; for example, the percentage of salts in a sample of brine, or the percentage of sugar in an aqueous solution.

(3) Arbitrary scale hydrometers, indicating concentration or strength of a specified liquid, or its density, referring to an arbitrarily defined scale at a specified temperature.

(4) Brix scale hydrometers, which are the most universal scale encountered for sugar solutions. A degree Brix equals the percentage (by weight) of sugar. It is the same as the Balling Scale. In Table 16.1 the relationships of Brix, Baume, specific gravity and other data are presented for sugar solutions.

(5) Hydrometers have also been developed to read the concentration of strength of other solutions or density of food products (see Fig 16.2). Table 16.2 shows the relationship of some of the different scales in use for salt solutions.

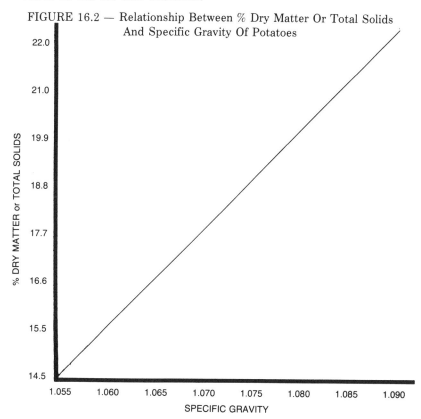

FIGURE 16.2 — Relationship Between % Dry Matter Or Total Solids And Specific Gravity Of Potatoes

Regardless of the type of scale for any hydrometer, the technologist must know the temperature of the solution in question when taking a reading. All hydrometers are calibrated for use at a given temperature.

TABLE 16.1 — Composition Of Sucrose Sugar Solutions At 20°C (68°F)

Refractive Index	% Sucrose by Wt Degrees Brix	Degrees Baume Modulus 145	Specific Gravity in Air at 20°C (68°F)	Wt. per 1 US Gal. of Syrup in Air at 20°C (68°F) lb	Wt of Sugar in 1 U.S. Gal. of Syrup, lb
	1	2	3	4	5
1.333	0.0	0.00	1.00000	8.322	0.000
1.3403	5.0	2.79	1.01968	8.485	0.424
1.3478	10.0	5.57	1.04003	8.655	0.866
1.3557	15.0	8.34	1.06111	8.830	1.325
1.3638	20.0	11.10	1.08297	9.012	1.803
1.3723	25.0	13.84	1.10564	9.201	2.301
1.3811	30.0	16.57	1.12913	9.396	2.820
1.3902	35.0	19.28	1.15350	9.599	3.361
1.3997	40.0	21.97	1.17874	9.809	3.925
1.4096	45.0	24.63	1.20491	10.027	4.513
1.4200	50.0	27.28	1.23202	10.252	5.127
1.4307	55.0	29.90	1.26007	10.486	5.768
1.4418	60.0	32.49	1.28908	10.727	6.438
1.4532	65.0	35.04	1.31905	10.977	7.136
1.4651	70.0	37.56	1.34997	11.234	7.865
1.4774	75.0	40.03	1.38187	11.499	8.626
1.4901	80.0	42.47	1.41471	11.773	9.418
1.5003	85.0	44.86	1.44848	12.054	10.244

This is usually stated on the stem of the hydrometer. The measurement of the temperature of the solution is above the given temperature used in standardizing the hydrometer—the observed reading will be below the actual value for the sample. Conversely, for solutions colder than the standard temperature, the readings will be above the actual values. Of course, the solutions can be cooled to the temperature as used for the calibration of the hydrometer and no temperature correction would be necessary.

Hydrometers are used extensively in the processing industries. Applications range from control of densities of brines and sirups used in packing the various products to use in determining the quality of the finished product, i.e., grading of peas (frozen and canned) and the "cut out" concentrations of many fruit products.

USE OF REFRACTOMETER FOR DENSITY MEASUREMENTS

The refractometer, which operates on a different principle than the hydrometer, is more easily employed as a production line tool. It is used

TABLE 16.2 — Baume, Specific Gravity, Percent Salt,
Degree Salometer And Ounces Of Salt Relationships
Per Gallon For Levels Of Concentration

Baumé Reading 60°F	Sp. Gr. 60°F	% of Sodium Chloride	Salometer Reading (0–100°)	Oz Salt per Gallon
0.8	1.0053	1	3.8	1.35
1.8	1.0125	2	7.6	2.72
2.8	1.0197	3	11.4	4.13
3.8	1.0268	4	15.1	5.56
4.8	1.0340	5	18.5	7.03
5.8	1.0413	6	22.6	8.52
6.3	1.0486	7	26.4	10.05
7.7	1.0559	8	30.2	11.61
8.6	1.0632	9	34.2	13.20
9.6	1.0707	10	37.7	14.83
10.6	1.0782	11	40.5	16.50
11.5	1.0857	12	45.3	18.21
12.4	1.0933	13	49.1	19.95
13.3	1.1009	14	52.8	21.73
14.2	1.1086	15	56.6	23.56
15.1	1.1162	16	60.4	25.43
16.0	1.1241	17	64.2	27.35
16.9	1.1319	18	68.0	29.31
17.8	1.1399	19	71.7	31.32
18.7	1.1478	20	75.5	33.38
19.5	1.1559	21	79.2	35.49
20.4	1.1640	22	83.0	37.66
22.2	1.1804	24	90.6	42.16
33.9	1.1972	26.5	100.0	48.14

as a direct line quality control instrument and is particularly valuable for controlling products which must be concentrated, like jams, jellies, preserves, tomatoes and citrus. The refractometer has a number of advantages in measuring solids or specific gravity, including:

(1) Results are quick.
(2) Operators can be easily trained.
(3) Quantity of samples needed is small.
(4) Precise values are obtained.

Principle of Operation

The refractometer is an instrument used to determine the refractive index. Some instruments, also, read in percent soluble solids or degrees Brix. The refractive index is a quantity which is constant for a pure substance under standard conditions of temperature and pressure. The

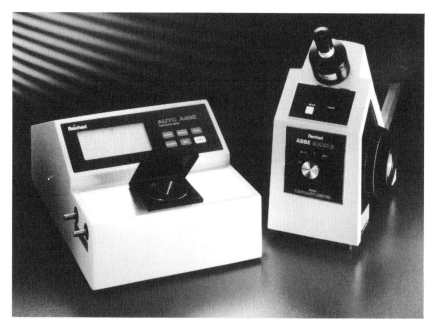

FIGURE 16.3 — Reichart-Jung Auto Abbe Refractometer and
Reichert-Jung Abbe Mark II
Courtesy of Leica Inc.

refractive index is the ratio of the sine of the angle of incidence of a light ray to the sine of the angle of refraction. This is expressed numerically as:

$$\frac{\text{Sine incidence }(i)}{\text{Sine refraction }(r)} = \text{Refractive Index }(n)$$

The index of refraction for water, as an example, is 1.33299 at 20°C (68°F).

The refractive index is determined with the refractometer, which compares the speed of light passing through a substance to the speed at which it passes through air. If water has a refractive index of 1.3 this means that light travels through water 1.3 times slower than through air. For example, if tomato juice has a refractive index of 1.3446 it means that light is slowed by a factor of 1.3446 when it passes from air to tomato juice. Thus, there is a direct relationship between the total soluble solids content and the refractive index of the solution.

Models

At present many types of refractometers are on the market. The Abbe 56 covers a wide index range (1.30 to 1.71—full scale for most items) with reading accuracy of two units to the fourth decimal place. The instrument has a water cooling system and for highest accuracy can be connected to an automatic temperature controller. The instrument is rugged and with reasonable care it should have very low maintenance costs.

The Spencer hand refractometer is an excellent instrument (range 1.30–1.38 or 0–30% solids). The scale is calibrated in increments of 0.2%. In addition, the accuracy is 0.1%. This instrument has a built-in illuminator which eliminates the need for an external light source.

The Goldberg "in-steam" or vacuum pan refractometer offers the industry an instrument directly applicable for line quality control for many products during the process of concentration.

Details of operation and calibration of each instrument should be followed carefully for good results. When making a reading, division between the light and dark portion of the field is most important. If this division is not distinct because of a color band, adjustment of the prisms may be necessary.

When taking a reading with most refractometers, the technologist should record: (1) index, (2) solids or Brix value, and (3) prism temperature. Since the refractive index is a measure of the speed of light, it is affected by temperature. If a material expands when heated, it will become less dense. Light will then be slowed down less and the refractive index will decrease. For a sugar solution the change amounts to 0.5% sugar for every 10°F. That is, if a solution measures 20% at 70°F and if it is heated to 80°F the same sample would measure only 19.5%. Thus, the temperature of the sample must be known exactly and the results interpreted accordingly. Most refractometers provide temperature compensation or correction values. Table 16.3 is a correction table for determining the Brix or solids by means of a refractometer when the readings are made at temperatures other than 20°C (68°F).

Furthermore, when using a refractometer on samples containing insoluble solids, it may be necessary to eliminate the insoluble solids portion before making a refractive index (that is, by filtering the sample or centrifugation with an ultracentrifuge and then reading the solids from the filtrate).

Particular care should be exercised in cleaning the instrument in order to avoid scratching the surface of the prisms. One should always use lens paper or absorbent cotton to clean the product from the prisms and to wipe dry after washing before reading the next sample. Furthermore, the technologist should note the manufacturer's warnings concerning the effects of acids, alkalies, etc. on the cement around the prisms. Many manufacturers have developed the newer models to withstand the normal acids and alkalies to be found around a food plant or laboratory.

With the greater demand today for products of uniform quality, the accurate control of such variables as solids or specific gravity is a must for the modern processor. Results for production personnel must be quick and accurate. The use of applicable instruments in the hands of competent technologists will provide management with the desired control for packing uniform quality products. A good refractometer is a "must" in most food operations or laboratories.

CONTROLLING THE QUALITY OF BRINE

The clearness of liquor in a can of vegetables is an important quality characteristic. Several vegetables are packed in brine, as the brine adds flavor to the product and aids in reducing process times.

Normal brines are made with water, salt and/or sugar. All brines should be added to the product as near the boiling point as possible. Salt and sugar may be added as a concentrate through the use of dispensing instruments such as the Flocron or Acuvol.

TABLE 16.3 — Correction Table For Determining The Brix Or % Solids By Means Of The Refractometer When The Readings Are Made At Temperatures Other Than 20 Degrees C (68 Degrees F)

Temp.	\multicolumn{8}{c}{Brix or Solids}							
	0	10	20	30	40	50	60	70
			SUBTRACT	FROM THE	% SOLIDS			
15	0.27	0.31	0.34	0.35	0.37	0.38	0.39	0.40
16	0.22	0.25	0.27	0.28	0.30	0.30	0.31	0.32
17	0.17	0.19	0.21	0.21	0.22	0.23	0.23	0.24
18	0.12	0.13	0.14	0.14	0.15	0.15	0.16	0.16
19	0.06	0.06	0.07	0.07	0.08	0.08	0.08	0.08
20	0	0	0	0	0	0	0	0
			ADD TO	THE %	SOLIDS			
21	0.06	0.07	0.07	0.08	0.08	0.08	0.08	0.08
22	0.13	0.14	0.15	0.15	0.15	0.16	0.16	0.16
23	0.19	0.21	0.22	0.23	0.23	0.24	0.24	0.24
24	0.26	0.28	0.30	0.31	0.31	0.31	0.32	0.32
25	0.33	0.36	0.38	0.39	0.40	0.40	0.40	0.40
26	0.40	0.43	0.45	0.47	0.48	0.48	0.48	0.48
27	0.48	0.52	0.54	0.55	0.56	0.56	0.56	0.56
28	0.56	0.60	0.62	0.63	0.64	0.64	0.64	0.64
29	0.64	0.68	0.71	0.72	0.73	0.73	0.73	0.73
30	0.72	0.77	0.79	0.80	0.81	0.81	0.81	0.81

Several factors may affect the color and clarity of the liquor. The color of the liquor should be clear, although it may be colored by soluble pigments of the products. Off-colored brine is usually caused by over-mature vegetables, minerals in the water and/or contamination from metals or other sources.

Suspended materials may appear in some brines. These are usually caused either by over-mature vegetables or quite immature vegetables that "cook-up" during the normal heating process. These suspended materials come from the actual breakdown of the product. Examples of these may be found with asparagus or green beans.

Sediment occurs in certain liquors. These are normally caused by the same factors as cited for suspended materials, except that they settle out. In addition, the sediment may result from poor washing of the product prior to processing. This again is particularly noticeable in items like asparagus.

Cloudy or cream-like viscous brines may be found in products such as lima beans or peas, particularly if they are over-mature or cooked for excessive periods. These "heavy" looking brines are caused by the dissolving out of the starch from the product during the cooking cycle. This is a frequent occurrence when packing field run produce where some of the product is more mature than other portions. Generally, the mature portions break down and these "starchy" materials dissolve out into the brine with the ultimate result of a thick, viscous or "heavy" looking brine.

There are several factors that the quality control technologist should consider to obtain a clear brine. Most important of these are: (1) choosing the right cultivar (using snap beans as an example, most dark seeded varieties such as regular Tendergreen will develop a dark off-colored brine); (2) segregation of qualities and packing like qualities together; and (3) using water free of minerals and organic matter for brine makeup. If a processor uses municipal water it is usually free of organic matter (however, this is not always true). Hardness in water (calcium and magnesium salts) may dissolve out and form precipitates when used in the brine, as the sodium in the salt will replace the calcium by the ion exchange principle and these will precipitate or settle out and form clouds or deposits in the brine.

Control Factors

Hard water will toughen certain vegetables, such as peas, lima beans, dried beans, etc. Water hardness between 85 and 170 ppm or 5–10 grains per U.S. gallon is desirable for peas, lima beans, field peas or most dry beans. For dried lima beans 170–225 ppm (10–15 grains per U.S. gallon) is preferred.

Too soft a water may cause noticeable "sloughing" of products like snap beans or potatoes. In the case of the latter, the Food & Drug Administration has approved the use of calcium salts such as calcium chloride, calcium sulfate, calcium citrate, monocalcium phosphate or any mixture of two or more of these salts. However, in no case should the calcium content exceed 0.051% of the net weight of the product. Recommended hardness values for several products are listed in Table 16.4.

Other minerals in the water such as iron and manganese may cause severe discoloration of the brine or product. Iron is particularly noteworthy for discoloration of red beets as it will cause them to turn black. Copper, in the presence of dissolved iron, may produce discoloration in many green products. All of these contaminants should be carefully controlled, and if any are present, major efforts should be undertaken to eliminate their source or one may expect discolored brines or products. Thus, water for brine should be of the finest quality and the quality control technologist should examine the water regularly, both chemically and physically, for quality.

Somers (1951) published a study on the effects of chlorine treatment on the flavor of canned foods (Table 16.5). Of the products he evaluated, apples, pears, cling peaches, figs, strawberries and yams are the most

TABLE 16.4 — Brine Table For Vegetables For Canning

Product	Recommended Hardness of Water (Grains/Gal.)	Lb Salt To Add Per 100 Gal. Water	Pounds Sugar To Add Per 100 Gal. Water
Asparagus	7–12	20–30	—
Snap Beans	7–12	12–20	0–15
Kidney Beans	5–8	15–20	30–45
Lima Beans	5–10	10–20	—
Dried Lima Beans	10–15	15–20	4–5
Beets	5–15	10–15	8–20
Carrots	8–15	15–20	—
Whole Kernel Corn	3–15	14–20	30–50
Peas	5–10	10–20	20–50
Dried Peas	5–10	15–20	12–20
Field Peas	5–10	15–20	—
Potatoes—less than			
1.075 specific gravity	5–10	12–20	—
over 1.075 sp gr	5–10[1]	12–20	—
Spinach	4–8	20–30	—

[1]Add 10.5 grains of calcium per 303 can or 1.22 lb of calcium per 100 gal. of brine.

susceptible to chlorine flavor at 5 ppm of chlorine. With green beans, lima beans, beets, carrots and spinach the flavors were affected at a concentration of 10 ppm chlorine in the water whether the water was used for brine or other purposes. Most municipal waters do not have this high a chlorine content, nevertheless, the quality control technologist

TABLE 16.5 — Effect of Chlorine Treatment On Flavor Of Canned Foods

Product	Lowest Concentration which Produced Off-Flavor when 2, 5, 10 and 50 ppm of Chlorine were Added	
	Partial Treatment Chlorination of All Water except Brines and Sirups	**Complete Treatment** Chlorination of All Water Including Brines and Sirups
	Chlorine, ppm	Chlorine, ppm
Applesauce, Rome Beauty[1]	10	5
Applesauce, Gravenstein[1]	(None at 50)	10
Apricots, halves unpeeled	(None at 50)	50
Apricots, whole peeled	(None at 50)	50
Asparagus, all green	50	50
Beans, green cut	50	10
Beans, green limas	50	10
Beans, with pork (recanned)[1]	—	50
Beets, red sliced	50	10
Carrots, sliced	(None at 50)	10
Carrots, pureed[1]	(None at 50)	50
Cherries, Royal Anne	(None at 50)	50
Corn	—	(None with 15)
Figs, whole Kadota	50	5
Grapefruit juice (recanned)[1]	—	50
Orange juice (recanned)[1]	—	50
Peaches, clingstone halves	(None at 50)	5
Peaches, Elberta halves	(None at 50)	10
Peas	—	(None with 10)
Pears	50	2 to 5
Pineapple juice (recanned)[1]	—	10
Potatoes, sweet, solid pack	(None at 50)	50
Pumpkin, solid pack[1]	(None at 50)	50
Prunes, Italian	(None at 50)	10
Spinach	50	10
Strawberries, whole	(None at 50)	10
Tomato Juice[1]	—	10
Vegetable Juice Cocktail (recanned)[1]	—	5
Yams, sirup pack	—	5

Source: Somers (1951).
[1]Chlorine added directly to the product.

should regularly evaluate the chlorine content of the water to make certain that it does not exceed these values.

Careful control of all washing, blanching, filling and processing operations will also help to prevent sediment, cloudiness and suspended matter in the brine. Each operation should be regularly checked and rigid specifications set for each operation. Too many times, simple changes or inefficient operations may lead to unfavorable qualities in the finished product.

At the present time, clearness of liquor and its evaluation is a very subjective factor. The usual methods of evaluation involve pouring the brine into a cylinder and observing its clarity by placing the cylinder against a background having standard lines, and observing the sharpness of these lines when looking through the cylinder containing the brine. Another method is to use two fingers in place of the background with standard lines. Transmission instruments could be used, but the need is not normally justified. Sediment can easily be measured by pouring the brine into a graduated cylinder and allowing it to settle for a given period of time and then determining the percentage of sediment. This, too, should not be necessary except for low grades of vegetables as the USDA Standards for Grades for vegetables do not highly emphasize the clarity of the liquor.

SUGARS AND SIRUPS

Sugar is an essential ingredient in the manufacture and preservation of jams, jellies, marmalades, apple butter, pickles, candy, most canned fruits, many fruit drinks and beverages, several vegetables and many specialty products. Sugar is added to these processed products to:
  (1) Improve flavor, color, texture and appearance
  (2) Exclude air
  (3) Provide energy or nutritive value to the product
  (4) Control microbial growth for highly concentrated products such as jams, jellies, and preserves

The most common types of sweeteners are: refined granulated sugar (can or beet sugar), sucrose, invert sugar, liquid sugar, dextrose, or corn sirup and high fructose corn sirup.

The data in Table 16.6 presents the relationship of Put-In (P-I) to Cut-Out (C-O) for the declared label statement for various Food and Drug Administration Standardized products. The P-I values are relative values since fruits vary in their natural or inherent sugar content according to the stages of maturity, variety and/or areas of production. One should sample and analyze raw fruits to determine the precise P-I sugar values needed. Sumner (1948) suggested the following method which v· have used in our laboratory quite successfully:
  (1) Take a sample after the sirup and fruit have been filled into the container.

(2) Remove sample from the container and place in a Waring Blendor and thoroughly mix for 15 seconds to 1 min.
(3) Pour into a strainer with cheesecloth as a filter and strain sufficient quantity to measure the sugar content.

TABLE 16.6 — Relationship Of Put-In (P-I) Syrup Versus Cut-Out (C-O) Syrup By Label Requirements For Standardized Fruit Products
(ALL DATA IN % OR BRIX VALUES)

Product	Extra Heavy P-I	C-O	Heavy P-I	C-O	Light P-I	C-O	Slightly Sweetened P-I	C-O
Apricots	55 or more	25–40	35–55	21–25	15–30	16–21	10–15	less than 16
Berries								
Black	40 or more	24–35	30–40	19–24	20–30	14–19	10–20	less than 14
Blue	40 or more	25–35	30–40	20–25	20–30	15–20	10–20	less than 15
Boysen	40 or more	24–35	30–40	19–24	20–30	14–19	10–20	less than 14
Dew	40 or more	24–35	30–40	19–24	20–30	14–19	10–20	less than 14
Goose	40 or more	26–35	30–40	20–26	20–30	14–20	10–20	less than 14
Huckle	40 or more	25–35	30–40	20–25	20–30	15–20	10–20	less than 15
Logan	40 or more	24–35	30–40	19–24	20–30	14–19	10–20	less than 14
Black Rasp	40 or more	27–35	30–40	20–27	20–30	14–20	10–20	less than 14
Red Rasp	40 or more	28–35	30–40	22–28	20–30	14–22	10–20	less than 14
Straw	40 or more	27–35	30–40	19–27	20–30	14–19	10–20	less than 14
Young	40 or more	24–35	30–40	19–24	20–30	14–19	10–20	less than 14
RSP Cherries	57 or more	28–45	34–56	22–28	25–33	18–22	10–25	less than 18
Sweet Cherries	45 or more	25–35	30–45	20–25	15–25	16–20	10–15	less than 16
Figs	40 or more	26–35	30–40	21–26	20–30	16–21		
Fruit Cocktail	40 or more	22–35	36–38	18–22	30–34	14–18		
Peaches	55 or more	24–35	40–55	19–24	15–25	14–19	10–15	less than 14
Pears	40 or more	22–35	25–40	18–22	15–25	14–18	10–15	less than 14
Pineapple	40 or more	22–35	30–40	18–22	20–30	14–18		
Plums								
Purple	40 or more	26–35	30–40	21–26	20–30	18–21	10–20	less than 18
Others	40 or more	24–35	30–40	19–24	20–30	16–19	10–20	less than 16
Prunes	50 or more	30–45	40–50	24–30	30–40	20–24		
Seedless								
Grapes	40 or more	22–35	30–40	18–22	20–30	14–18	10–20	less than 14

[1]Put-In concentration must be varied depending on variety, maturity or area of production to obtain desired Cut-Out sirup concentration.

(4) If using a hydrometer to measure the sugar content, allow the juice to stand for 10 min. and either dip off the scum of air bubbles or siphon off the clear liquor and measure. If using a refractometer, take a sample from down in the center of the sirup, so as to avoid any of the scum particles. This blended brix value will be equivalent to the equalized value after periods of 2-4 weeks.

In addition to the Put-In data in Table 16.4, for canned products, standards of identity have been established for jellies, preserves and jams which require not less than 45% fruit to 55% sugar. Fruit butters require five parts of fruit to two parts of sugar. The Federal minimum sugar content standards for jams, jellies and fruit butters are:

Jams—not less than 65% soluble solids

Jellies—not less than 65% soluble solids

Fruit Butter—not less than 43% soluble solids

For frozen fruit, there are no standards of identity, but most fruits are packed at 2+1 to 5+1. A 2+1 means two parts of fruit to one part of sugar, 3+1 means three parts of fruit to one part of sugar, etc. For some frozen fruit, dry sugar packs are recommended while for others, a sirup of comparable solids content is used.

TEST 16.1 — Determination Of Soluble Solids Refractometric Method

Equipment

(1) Refractometer—Abbe 56
(2) Lens paper
(3) Rubber policeman
(4) Distilled water

Procedure

(1) Place a few drops of the solution on the carefully cleaned fixed prism and slowly bring the two prisms together and clamp them.
(2) Turn on the inline switch and adjust the light arm for proper illumination.
(3) Bring the borderline into view with the coarse adjustment (lower right hand side). The color of the borderline may have to be compensated for by adjusting the compensator dial (directly under eyepiece). The borderline should be faintly red on the other.
(4) Observe the crosshairs sharply focusing the eye piece if necessary and bring the borderline up to the intersection by means of the coarse or fine hand control.
(5) Read the index by depressing the momentary contact switch (lower left hand side), estimating to the fourth place.
(6) Open prisms and clean very carefully with a piece of lens paper. The prisms are made of very soft glass and all possible precautions should be used to prevent damage to them.

Notes

(1) When working with materials with a high concentration of insoluble solids present such as the pulp in tomato paste, it is advisable to

filter out the solids from the serum. This can be accomplished quite easily by folding either filter paper or paper towels into a small triangle and placing the sample inside. Thus, by squeezing the filter paper the serum will diffuse through it and the solution can be applied directly to the prisms. Removal of the solids material does not affect the results, for refractive index is based only on the soluble solids. If the pulp is not removed a hazy image will be formed which is hard to center.
(2) The refractometer should be standardized periodically according to the amount of use it receives. A rapid and fairly accurate method of standardization is to use distilled water, which has a refractive index of 1.3330 at 20°C.
(3) In cleaning the prisms after the reading is taken, care should be taken to wash the surfaces of the prisms with water and a soft linen cloth so as to remove all soluble solids as well as moisture without scratching the surfaces of the prisms. If the funnel opening is used, as suggested above, the outer portions of the prism case must also be thoroughly cleansed. The glass surfaces of the prisms are easily scratched and all possible precaution should be taken to preserve the original smooth surface of the prism faces.

TEST 16.2 — Hydrometer Method

(1) Set cylinder on a tray or in sink and fill cylinder full with sirup.
(2) Lower hydrometer into cylinder and read scale at liquid level, not at top of meniscus.
(3) If sirup is not at same temperature as indicated on hydrometer (usually 20°C or 68°F), adjust sirup to temperature or make approximate corrections as indicated for hydrometer.
(4) Using the data in Table 16.2 convert Brix to Baume or specific gravity or vice versa.

TEST 16.3 — Glucose Determination By the YSI #27 Analyzer

General:

Potatoes containing high levels of glucose (greater than 0.25 mg/gm) and sucrose (greater than 2.0 mg/gm) have been shown to fry dark, absorb more oil and have off flavors.

Equipment:
1. YSI #27 analyzer and manuals.
2. Blank membranes.

3. Glucose-dextrose membranes.
4. Blender, appropriate jar and top neither of which leak.
5. Standards as purchased from the company.
   a. Use ranges that will allow most specimens to be in the middle. Thus, 500; 1000; 200 mg/100 ml can be diluted to appropriate amounts. (Parts of standard divided by parts of standard plus # of parts of H_2O equals dilution factor. Dilution factor times standard value equals new volume's value) ($V_{1A}V_{10} = V_{2A}V_{20}$).

   b. Make up your own standards by purchasing D glucose and diluting to desired volume (# milligram glucose per number of 100 ml).
6. Scales to weigh in tenths of grams.
7. Filter paper— size dependent upon #8.
8. Filter funnel.
9. Apparatus to hold funnel during filtration.
10. Knife to cut potato to desired weight.
11. Distilled water.
12. Containers to receive filtrate.

Procedure:
1. Standardize YSI 27— Follow procedure as outlined by 'Instruction Manual' and/or 'Service Manual'. Check membrane.
2. Choose tubers and remove about twenty grams from the center. Do not use apical or basal end. Research has shown apex to be low and basal high with the center to be the average and most reliable concentration of sugars.
3. Weigh slice and add four times this weight of water and the slice to the blender. Mix for two minutes.
4. Filter.
5. Inject filtrate into previously standardized instrument. Record result and push clear button.
6. Inject second filtrate to confirm #5. Record result and clear.
7. Go on to the next specimen, etc.
8. The results of five plus result of six divided by two, times your dilution factor, five, is the answer in whichever units you standardized the instrument to (usually milligrams per 100 ml).

$$\frac{(90 \text{ plus } 95)}{2} \text{ equals } 92.5 \text{ units of dextrose}$$

TEST 16.4 — Sowokinos Testing For Sucrose-Rating (SR)

General:

Low-sucrose potatoes demonstrated superior storage-life and processing characteristics. It has been known for many years that sucrose (12 carbon sugar) is the major free sugar found in immature potatoes. Since sucrose can be hydrolized to two 6-carbon reducing sugars in storage, the level of this sugar at harvest may be the critical factor in explaining varying rates of reducing sugar accumulation between varieties.

VAN HANDEL DETERMINATION METHOD
Equipment:
1. Triple beam balance.
2. Spectronic 20 (Bausch & Lomb).
3. Cuvettes one-half inch test tubes for spectronic 20.
4. Marbles to cover cuvettes.
5. Parafilm.
6. Refrigerator or ice bath at 4 degrees C.
7. 500 ml graduated cylinder.
8. 600 ml beaker.
9. 5 ml pipettes.
10. Rubber pipettor.
11. Boiling water bath.
12. Incubator (40 degrees C, 104 degrees F).

Reagents:
1. Sucrose standard (1 g sucrose/1000 ml water).
2. 30% KOH store in refrigerator (35.29 g solid 85% KOH, M, wt. 56.11 in 100 ml H_2O).
3. Anthrone reagent:
    a. Prepare a diluted sulfuric acid solution by adding 76 ml H_2SO_4 (analytical grade) slowly to 30 ml water while stirring.
    b. To the diluted H_2SO_4 solution from a., add 0.15 g anthrone. Mix well until all dissolved (around two hours). Store in dark bottle in refrigerator.
    Caution: Do not breathe anthrone power.
    Make sure the temperature of H_2SO_4 solution is below 60 degrees C before adding anthrone.

    Notes: (1) Make in well-ventilated area (under a hood).
    (2) Should **not** pipet anthrone reagent by mouth.

Procedure:
(Beaker, graduated cylinder and water should be pre-chilled in refrigerator at 4 degrees C.)

1. Weight 200 g of cut pieces from four to five washed-peeled tubers.
2. Juicerate (using Acme Juicerator) and collect juice in a 600 ml beaker. Wash juicerator with 100 ml water three times. Wait two to three minutes between washings.
3. Transfer to 500 ml graduate cylinder— take volume to 430 ml with water.
4. Cover mix, cool (4 degrees C) and let settle for one hour.
5. Take portion and dilute one part extract with four parts water.
6. Prepare clean and dry tubes and fill them as follows:
   a. To #1 and 2 add 0.1 ml water (reagent blank, duplicate).
   b. To #3, 4, and 5 add 0.1 ml of standard sucrose solution in each (0.1 ml equals 0.1 mg sucrose and should reach X equals $O.D._{620}$ gives 0.97 to 1.00).
   c. To 6,7, and 8 add 0.1 ml/tube of sucrose extract dilution (obtained from step (5) above).
7. Add 0.1 ml of 30% aqueous KOH reagent to each of all the tubes.
8. Mix, cover tubes with marbles, and heat at 100 degrees C for fifteen minutes (to destroy reducing sugars).
9. Cool to room temperature, and add 3 ml anthrone reagent.
10. Cover tubes with parafilm and mix.
11. Incubate at 40 degrees C (104 degrees F) for thirty minutes if one week old anthrone, or for sixty minutes if one month old anthrone.
12. Set colorimeter to zero with lowest reagant blank at 620 nm.
13. Read $O.D._{620}$ of the stable yellowish-green color developed in the standard sucrose and unknown solutions.

Calculation:

$$\frac{O.D. \text{ (unknown)} \times 0.1 \times 107.5 \text{ (factor)}}{O.D. \text{ (standard)}} \text{ equals mg sucrose/g tuber}$$

Tuber SR (Sucrose Rating)

A correlation was obtained as varieties with SR's 2.5 or less (1.0 to 2.5 mg sucrose/g tuber) at harvest, will produce acceptable chips from long-term storage (eight to eleven months) at 53 degrees F.

SUGAR, SALT AND SEASONINGS

TEST 16.5 — Measurement Of Sucrose By The Yellow Springs Instrument (YSI #27 Analyzer)

General:
Sucrose is becoming more important in determining storage and chip color of finished products.

Equipment:
1. Sucrose membrane.
2. Sucrose standards as purchased from the company of 1,000 and 2,000 mg/100 ml. See 4B Glucose Determination, for dilutions.

Procedure:
1. See Glucose Determination— do the same.
2. Install sucrose membrance.
3. Allow for stabilization.
4. Standardize using standards— diluted to 100 and 200 mg/100 ml.
5. Inject test specimen twice and average results. Apparent Sucrose reading.
6. Check standardization every ten test solutions.
7. Refer to No. 1 and find highest solution concentration of dextrose before dilution; i.e. highest dextrose test result divided by dilution factor. Ex: solution having highest dextrose is $\frac{340}{5}$ mg/100 ml equals 68 mg/100 ml.
8. Inject glucose standard in 50 mg/100 ml increment through the highest dextrose; rounded up to the next 50 mg/100 ml; Ex.; 0, 50, 100, 150 + (highest test = 198), 200, stop. If highest test was 210, then inject a 250 standard as well.
9. Record standard concentrations of dextrose (x) and corresponding "dextrose sensitivity reading (y)."
10. Calculation of true sucrose is as follows:

T.S. mg/100 ml equals D(A.S. mg/100 ml-D.S.)
    T.S. equals true sucrose of test.
A.S. equals apparent sucrose reading from 4 for test.
D. equals dilution factor from 1.
D.S. equals dextrose sensitivity coordinate (y) for corresponding (x) dextrose that has been divided by the dilution factor.

TABLE 16.7 — Percent Of Salt

Salt in Seasoning		Salt on Product		Salt on Product		Salt on Product	
MLS	% Salt	MLS	% Salt	MLS	% Salt	MLS	% Salt
4.50	26.2	1.80	1.05	4.35	2.54	6.90	4.02
4.55	26.5	1.85	1.08	4.40	2.56	6.95	4.05
4.60	26.8	1.90	1.11	4.45	2.59	7.00	4.08
4.65	27.1	1.95	1.14	4.50	2.62	7.05	4.11
4.70	27.4	2.00	1.17	4.55	2.65	7.10	4.14
4.75	27.7	2.05	1.19	4.60	2.68	7.15	4.17
4.80	28.0	2.10	1.22	4.65	2.71	7.20	4.20
4.85	28.3	2.15	1.25	4.70	2.74	7.25	4.23
4.90	28.6	2.20	1.28	4.75	2.77	7.30	4.25
4.95	28.8	2.25	1.31	4.80	2.80	7.35	4.28
5.00	29.1	2.30	1.34	4.85	2.83	7.40	4.31
5.05	29.4	2.35	1.37	4.90	2.86	7.45	4.34
5.10	29.7	2.40	1.40	4.95	2.88	7.50	4.37
5.15	30.0	2.45	1.43	5.00	2.91	7.55	4.40
5.20	30.3	2.50	1.46	5.05	2.94	7.60	4.43
5.25	30.6	2.55	1.49	5.10	2.97	7.65	4.46
5.30	30.9	2.60	1.52	5.15	3.00	7.70	4.49
5.35	31.2	2.65	1.54	5.20	3.03	7.75	4.52
5.40	31.5	2.70	1.57	5.25	3.06	7.80	4.55
5.45	31.8	2.75	1.60	5.30	3.09	7.85	4.57
5.50	32.1	2.80	1.63	5.35	3.12	7.90	4.60
5.55	32.3	2.85	1.66	5.40	3.15	7.95	4.63
5.60	32.6	2.90	1.69	5.45	3.18	8.00	4.66
5.65	32.9	2.95	1.72	5.50	3.21		
5.70	33.2	3.00	1.75	5.55	3.23		
5.75	33.5	3.05	1.78	5.60	3.26		
5.80	33.8	3.10	1.81	5.65	3.29		
5.85	34.1	3.15	1.84	5.70	3.32		
5.90	34.4	3.20	1.86	5.75	3.35		
5.95	34.7	3.25	1.89	5.80	3.38		
6.00	35.0	3.30	1.92	5.85	3.41		
6.05	35.3	3.35	1.95	5.90	3.44		
6.10	35.6	3.40	1.98	5.95	3.47		
6.15	35.8	3.45	2.01	6.00	3.50		
6.20	36.1	3.50	2.04	6.05	3.53		
6.25	36.4	3.55	2.07	6.10	3.56		
6.30	36.7	3.60	2.10	6.15	3.58		
6.35	37.0	3.65	2.13	6.20	3.61		
6.40	37.3	3.70	2.16	6.25	3.64		
6.45	37.6	3.75	2.19	6.30	3.67		
6.50	37.9	3.80	2.21	6.35	3.70		
6.55	38.2	3.85	2.24	6.40	3.73		
6.60	38.5	3.90	2.27	6.45	3.76		
6.65	38.8	3.95	2.30	6.50	3.79		
6.70	39.0	4.00	2.33	6.55	3.82		
6.75	39.3	4.05	2.36	6.60	3.85		
6.80	39.6	4.10	2.39	6.65	3.88		
6.85	39.9	4.15	2.42	6.70	3.90		
6.90	40.2	4.20	2.45	6.75	3.93		
6.95	40.5	4.25	2.48	6.80	3.96		
		4.30	2.51				

TEST 16.6 — Titrimetric Determination Of Salt Content In Food Products

Equipment and Reagents

(1) Centrifuge
(2) 100-ml Volumetric flasks
(3) 250-ml Erlenmeyer flasks
(4) Pipettes, 1, 2, 25 and 50 ml
(5) Centrifuge tubes
(6) Burette — 50 ml
(7) Ethyl alcohol 80%
(8) Saturated ferric ammonium sulfate indicator ($FeNH_4(SO_4)_2$)
(9) Concentrated nitric acid
(10) Ammonium thiocyanate $0.1N$ (NH_4CNS)
(11) Silver nitrate $0.1N$ ($AgNO_3$)

Procedure

(1) Remove 5 ml or 5 gm of brine or material on which the salt content is to be determined and transfer it to the volumetric flask (100 ml).
(2) Add 80% alcohol to about 50 ml. Shake well to get all insoluble material in suspension.
(3) Add 1 ml of HNO_3 and 25 ml of $0.1N$ $AgNO_3$ into volumetric flask and make to 100 ml volume with alcohol.
(4) Mix well and transfer mixture to six centrifuge tubes. Centrifuge at speed 6 for 5 min.
(5) Pipette 50 ml of supernatant liquid into 250-ml Erlenmeyer flask and add 2 ml of ferric ammonium sulfate solution and 2 ml of nitric acid.
(6) Titrate to permanent light brown color with ammonium thiocyanate.

Calculations

Divide number of ml of silver nitrate used by 2 and subtract number of ml of ammonium thiocyanate solution used. Multiply the difference by the factor 0.005843 to obtain the grams of salt present.

To determine the percentage sodium chloride, multiply the grams NaCl by 100/5 (20). This is the percentage NaCl in the sample since a 5 gm sample was used. Products vary in salt content from 0–2%.

$$(\tfrac{1}{2} \text{ of } AgNO_3 - NH_4CNS) \times 0.005843 = \text{NaCl in grams.}$$

TEST 16.7 — DETERMINATION OF SALT — MOHR Titration Method

Reagents

(1) Potassium chromate, 5% solution
(2) Silver nitrate, $0.1N$
(3) Distilled water

Equipment

(1) 500-ml Beaker or Erlenmeyer flask
(2) Burette, 0.1 ml division
(3) Erlenmeyer flasks, 125 ml and 250 ml
(4) Funnel
(5) Filter paper, S. & S. No. 595
(6) Volumetric pipette, 10 ml
(7) Graduated cylinder, 500 ml
(8) Balance

Procedure

(1) Weigh 25 gm of product (chips); for seasoning use 2.5 gm.
(2) Transfer to 500 ml beaker or Erlenmeyer flask.
(3) Add 250 ml distilled water from graduated cylinder.
(4) Stir and let stand at least 5 min. (Make certain all chips are covered with distilled water.)
(5) Stir again and filter into beaker.
(6) Pipette 10 ml of the clear filtrate into a clean 250 ml flask.
(7) Add 25 ml distilled water and stir.
(8) Add 6–7 drops of potassium chromate indicator and stir.
(9) Fill burette with $0.1N$ silver nitrate solution and titrate the salt solution including the indicator until the solution changes to a faint brick red color.
(10) Read the burette level and refer to Table 16.7 for percent of salt in the sample.

TEST 16.8 — Dicromat Salt Analyzer Method

The following is the general procedure for determining salt content in product used by the Diamond Crystal Salt Company. (DiCromat requires a minimum of one half hour warm-up. It is suggested that it be left on continuously during the working day.)

Standardization Procedure:
1. Weigh out twenty-five grams of product.
2. Add exactly 250 ml of distilled water.
3. Process the sample using your normal procedure (electric blender, crushing, dissolving, etc.) so as to obtain a slurry in which the salt containing medium can be separated.
4. Filter the slurry so as to obtain about 100-120 ml of filtrate (C).
    a. The bulk of the solids are removed using a common household cone wire mesh screen. It may be necessary to work the solids to obtain enough filtrate.
    b. Any solids remaining in the filtrate (C) can be removed by means of a milk strainer pad (six or seven inch) or loose woven paper wiping towel ("Webril" manufactured by the Kendall Company).
5. Pipette a sample of the filtrate (c) (10 ml when using 0.1 $NAgNO_3$) and determine the percent salt content in the liquid by your standard silver nitrate procedure.
6. Standardize the DiCromat Salt Analyzer as follows (normally, standardize about once a day).
    a. Unlock Microdial (A) by moving dial lever (B) counterclockwise.
    b. Pour remainder (C), about 100 ml, into DiCromat reservoir (D). Catch liquid in any suitable container (E) under the cylinder (F) discharge (G).
    c. With the liquid flowing into container (E) adjust the digital readout (H) by means of the Microdial (A), to read the percent salt calculated from step 5. (Be sure cylinder (F) is full—showing liquid in reservoir (D)—when readings are read.)
    d. Lock Microdial (A) by moving dial level (B) clockwise.
    e. Record Microdial setting (I) if more than one product is to be processed.

Operation Procedure:
1. Repeat steps 1-4.
2. Pour the liquid into the cylinder reservoir.
3. The percent salt content will be the maximum reading read directly on the digital readout.
    a. Liquid can be poured back through cylinder as many times as desired.
    Note: When using the DiCromat for more than one product, merely return the Microdial to the setting at which the product was standardized.

Cleaning:
It is very important that the electrodes, located in the cylinder, be kept clean. Normally pouring a strong detergent solution, such as toilet bowl cleaner, through the cylinder will suffice. Some food mediums will coat the electrodes more readily than others. **Never run a brush or other means into the cylinder to clean,** because of the danger of damaging the electronic probes (J, K, L).

TEST 16.9 — Sodium Electrode Method
Equipment:
1. pH meter with M.V. (millivolt) scale readable to 3 significant figures.
2. Orion Single Junction Sodium Electrode (Model 94-11 or 96-11).
3. Magnetic Stirrer with Spin Bar.
4. Mortar and Pestle.
5. Graduated cylinder— 250 ml.
6. Triple-beam balance accurate to 0.1 g.
7. 2 ml pipet (1).
   4 ml pipet (1).
   10 ml pipet.
8. 2 cycle semi-log paper.
9. 1000 ml volumetric flasks (3-4).
   100 ml volumetric pipet or flask (1).
   400 ml beakers (2 per sample).
   1 wash bottle.
10. Filter papers and funnels.
11. 1000 ml capacity plastic bottles with caps.

Reagents:
1. 5 Molar Na^+ electrode storage solution (11.0g NaCl/100 ml double distilled water).
2. Electrode Rinse Solution (10 ml Ammonium hydroxide/100 ml Double distilled water-ddw).
3. Lithium Trichloroacetate filling solution (LiCl3Ac) Orion Cat. No. 90-00-19.
4. Standards:
   a. Add 25.4 g. NaCl to a 1000 ml volumetric flask, fill to mark with ddw and stir, this is equal to 10% Na^+ standard since a 1:10 dilution of product is used.

b. Add 100.0 ml of the 10% Na^+ solution to a 1000 ml volumetric flask, fill to mark with ddw and stir; this is equal to 1% Na using a 1:10 dilution of product.

c. Add 10.0 ml of solution A (10% Na) to a 1000 ml volumetric flask fill to mark with ddw and stir. This is equal to a 0.1% Na using a 1.10 dilution of product.

d. Add 1.0 ml of solution A (10% Na) to a 1000 ml volumetric flask, fill to the mark with ddw and stir; this is equal to 0.01% Na using a 1:10 product dilution. **Note:** Solution D is only necessary for low sodium product.

e. Store the standard solutions in plastic bottles with caps.

5. Ionic Strength Adjuster (ISA). Add 20g ammonium chloride and 5 ml ammonium hydroxide to a 100 ml volumetric flask and bring the volume up to 100 ml and mix. Stir and store in a 100 ml plastic bottle.

6. Double Distilled Water (ddw) to prepare all solutions.

Standardization:

1. Change electrode filling solution. Take old solution out with an eye dropper and fill again with Orion Lithium Trichloroacetate.
2. Pour 100 ml of each standard solution into separate 250 ml beakers.
3. Pipet 2 ml of ISA into each beaker.
4. Standardize the pH meter to zero with 0.1% Na^+ standard solution. (Allow 15 second response time.)
5. Record M.V. reading for the 0.1%, 1.0% and 10% Na^+ standard solutions.
6. Plot M.V. vs. concentrations of standard (on log scale) on 2 cycle semi-long paper.

Procedure:

1. Weigh 20g sample into a 400 ml beaker.
2. Add 200 ml ddw (soak samples).
3. Let stand for 2 minutes; stir gently.
4. Filter if necessary.
5. Add 4 ml ISA.
6. Record M.V. for the unknown sample.
7. Determine % NA^+ from the standard curve by comparing MV of the sample against the standard solutions.

% NaCl equals % sodium (NA^+) x 2.54.

TEST 16.10 — Determination Of Percent Seasonings In Snack Foods

Reagents and Supplies
1. Solvents.
   a. Toluene (for barbecue flavoring).
   b. Methanol (for smoked flavor).
   c. Glacial acetic acid (for sour cream and onion).
2. Flavorings (same lot as on chips).
3. Chips (flavored and unflavored).

Equipment:
1. Mortar and pestle.
2. Glass-stoppered Erlenmeyer flasks, (three) 125 ml and (three) 250 ml.
3. Volumetric flask, 50 ml for solvent measurement.
4. Analytical balance.
5. Spectronic 20, selected half inch Q.D., I.D. 11.67 mm or equivalent; matched cuvettes.
6. Filter paper, Whatman No. 1.
7. Glass funnels, long stemmed, 250 ml.

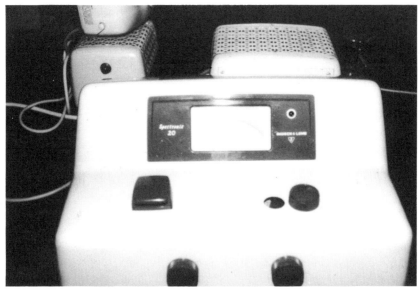

FIGURE 16.4 — Spectronic 20 Colorimeter

Procedure

1. Crush enough (one quarter inch pieces) to prepare the following mixtures in 250 ml Erlenmeyer flasks.
   a. 5.0 gram unseasoned chips, 50 ml solvent (blank solution).
   b. 4.70 gram unseasoned chips, 0.3 gram flavoring, 50 ml solvent (standard solution).
   c. 5.0 gram seasoned chips, 50 ml solvent (sample solution).
2. Shake each mixture for two minutes.
3. Filter each mixture into a 125 ml Erlenmeyer flask.
4. Set Spectronic 20 at wavelength that coincides with absorption maximum of the seasoning being analyzed. (440 mm for bacon, 475 for barbecue).
5. Set absorbance at zero by using blank solution.
6. Use matched cuvettes (1 cm cell path) half full of solution, making sure that outside of cuvette is clean before inserting into chamber and that marks on the cuvette face forward while in the chamber.
7. Using the blank, reset absorbance at zero after each reading, to prevent error from indicator drift.

Determination:

$$\frac{\text{Absorbance of sample}}{\text{Absorbance of standard}} = \frac{X}{6}$$

BIBLIOGRAPHY

ANON. 1955. This is Liquid Sugar. Refined Syrup and Sugars, Yonkers, N.Y.

GEISMAN, J.R., VERMA, S.S., and GOULD, W.A. 1962. New flavors developed for sauerkraut. Ohio Farm Home Res. 47, No. 1, 11-12.

GOULD, W.A. 1956A. Density can be a "tool" for control of quality. Food Packer 38, No. 1, 16, 30-32.

GOULD, W.A. 1957B. Refractometer is short-cut in density measurement. Food Packer 38, No. 3, 26, 49.

GOULD, W.A. 1957C Specific gravity tests can measure vegetable maturity. Food Packer 38, No. 2, 27, 76.

GOULD, W.A. 1962, Quality control techniques for checking syrup and sugar levels in foods. Canner/Packer 131, No. 10, 38-39.

GOULD, W.A. 1986. Quality Control Procedures for The Manufacture of Potato Chips and Snack Foods. Snack Food Association.

MURRAY, R.V., and PEDERSON, G.T. 1951. Water for canning. Continental Can Co. Bull. 22.

SOMERS, I. 1951. Studies on in-plant contamination. Food Technol. 5, 46-51.

TOWNSEND, C.T. SOMERS, I.I., LAMB, F.C. and ALAM, N.A. A Laboratory Manual for the Canning Industry. National Canners Assoc., Washington, D.C.

WECKELL, K.G. 1961-62. How sugar improved canned peas and corn. The Sugar Molecule XI, No. 3, 1-4.

CHAPTER 17

Flavor

The flavor of processed foods is probably the most important single quality factor of concern to the food technologist. Not only is the flavor factor important from the chemical additive standpoint, but it is also important from the standpoint of determining suitable cultivars of fruits or vegetables for processing, maturity levels of the crops, product formulation, processing methods, marketing, grading, and the actual storage stability of the processed products. Thus, a highly desirable flavor must be formulated and maintained in the production process.

EVALUATION OF FLAVOR

There are many techniques in use today in an attempt to evaluate the flavor of processed foods. The gas chromatograph is one of the newer instruments used to provide an objective measurement for specific chemical components of flavor. Unfortunately, chromatography is both expensive and somewhat sophisticated. In addition, other factors comprising the human judgment need to be considered, and these cannot be matched by evaluation with an instrument.

The subjective, or human evaluation approach, is still one of the best methods of determining the acceptability of a product. There are two broad types of flavor evaluation methods in present day use (Gould 1953). First, there is the consumer acceptance (preference) testing method and secondly, there is the panel difference method.

Consumer Acceptance Method

The consumer acceptance method is used when a new product is being developed, a change in the manufacturing procedure of the standard product is considered advisable, or for the constant quality checking of the manufactured product versus competitors' products. This type of flavor evaluation is ideally conducted with a true cross section of the population market where the manufacturer wishes to promote his product. The only disadvantage to this method is the large number of subjects to whom samples must be submitted to obtain reliable and meaningful results.

Panel Difference Method

The second flavor evaluation method utilizes a panel of judges and is easily adapted to small operations or laboratory testing. This type of tasting has two fundamental requirements for success:
(1) The selection of a panel that can detect the four senses of taste:
   (a) Sweet   (c) Bitter
   (b) Sour   (d) Salt
(2) The ability of panel members to repeat or duplicate themselves at any particular threshold level.

The particular method of taste evaluation depends upon the product and its inherent characteristics, probable market, and exactly what flavor components need investigation or evaluation. Regardless of the method used, the results are of little value unless they are taken carefully and unless the panel can repeat a test on the same samples at any given time. It is necessary, therefore, to measure threshold acuity levels for the taste panel members to ascertain whether the individual is capable of detecting differences in flavors or concentrations of specific components of flavor for the given commodity. In sensory work, the sense organs become easily fatigued. Usually there are 4-10 samples per session and at least 30 min between each session for flavor evaluation. For color evaluation there are up to 20 samples per session and for texture up to 12 samples per session.

Determining Product Differences.—Flavor panels for determining product differences can be classed according to the following methods.

(1) Paired Comparison Test.—Two unknown samples are submitted to the judges. The judges are then asked to select the sample having the most preferred flavor. A warm-up sample can be presented to the judges prior to the tasting of the two unknown samples. In this case the judges are asked to indicate which of the two unknown samples is the same as the warm-up sample.

(2) Triangle Taste Panel.—Three samples are presented to the judges, two alike and one different. The judges are asked to differentiate among the samples and to identify the two identical samples. This method is utilized when evaluating a small number of samples or when attempting to determine small differences in flavor between the samples.

(3) Triangular-scoring Flavor Panel.—This panel is conducted in the same manner as the triangle taste panel, however, the judges are asked to indicate their preferences for the samples on a numerical scoring

scale. Only those preference scores where the judges could identify the like samples are analyzed for significance.

(4) Dilution Test.—This panel is used to determine differences among homogeneous materials. It is important to know the threshold or acuity level of the taste panel members prior to conducting the taste evaluation. Samples which are presented to the judges will have differences in product formulation or in the amount of additives used in the product in uniform increments of decreasing intensity.

(5) Ranking Test.—Judges are asked to rank the samples in decreasing or increasing order of flavor intensity. This method has excellent applications for routine quality control of ingredient mixing, production changes, or formula alteration. With this method, samples from a day's run can be ranked in order of flavor differences. A triangle taste panel can then be conducted to find out if the differences between the best and the poorest samples are significant.

(6) Numerical Scoring.—Judges are asked to score the samples for preference on a numerical scale. The simplest scale for statistical analysis and for routine evaluation is from 1 to 10, with 10 being excellent or perfect; 9, 8, 7—good; 6, 5, 4—fair; 3, 2—poor; and 1—off or unacceptable. A scoring sheet is made up with the code for the samples on the left side of the sheet and the score evaluations on the top of the sheet. Several factors can be scored on the same sheet by using a code such as "C" for color, "F" for flavor, "T" for tenderness, and "S" for consistency. However, it is much better if sufficient samples are available for the panel to score each factor at one setting and on separate sheets as there is then less chance for error and greater ease in calculations after the scoring is completed.

For best success with this method of flavor evaluation, the judge(s) should be provided with a standard or "bench mark" indicating minimum or acceptable levels or tolerances.

In the USDA Standard for Grades where flavor is a factor, the inspector scores the product according to its grade on a numerical scale. The food technologist should set up a panel of flavor judges and constantly submit samples to this panel throughout the processing season to make certain that the quality meets product specifications.

(7) Descriptive Terms.—Judges attempt to describe the flavor or give their impression of the flavor. The word descriptions are very helpful when attempting to locate off-flavors or when defining highly desirable flavors. Much training in consistent use of testing is important when using this method. It is a highly desirable method when looking for "left out" ingredients in formulated foods.

(8) Flavor Difference Method.—This method is easy to use; a minimum amount of time and samples are needed, an accurate measure of the degree of flavor difference is provided between samples, the data can be quickly and easily analyzed for statistical differences, and indications of flavor acceptance can be obtained. The flavor difference method was developed for the laboratory evaluation of pesticide-treated samples. It has a strong place, however, as a standardized method or technique for all flavor evaluation studies.

Since this type of flavor analysis is a difference evaluation, a control or check sample needs to be provided. It can be used three ways: as a reference sample; as a coded control sample in the presentation; or as a sample for the preliminary screening of the panel member. A preliminary screening can be conducted by presenting actual test samples to potential members for their evaluation. Four replications should be presented to each member of the panel. If the potential member is unable to detect any flavor differences between the samples, he should be eliminated as a panel member. This would eliminate the need to test the threshold acuity level of each panel member.

Panel Judges.—After identifying prospective taste panel members that indicate normal organoleptic sensory abilities, it is further advisable to consider the following.

(1) No one should be asked to serve on a taste panel who is not acquainted with the product.
(2) Members with a strong dislike for the product to be tested should not be asked to serve on the panel.
(3) No one should be asked to serve who is not interested in the work and willing to give a conscientious, unbiased judgment of flavor.
(4) The full cooperation of all members of the panel is necessary for the success of the panel. Thus, the purpose of the tests and the objectives of the work should be fully explained to prospective panel members.
(5) Members with colds or other indispositions which might impair their ability to taste or smell should not be used.

In addition, all flavor evaluations should be conducted in a quiet, clean, odor-free room at a temperature of approximately 72°F. This room should be equipped with separate booths for each panel member. The flavor evaluation sessions should be held between 9 and 11 am and 2 and 4 pm. There should be no time limit set for an individual's judgment. All directions to panel members should be in writing with specific instructions clearly spelled out.

Interpretation

The panel members should not only detect low level concentrations but also be able to consistently repeat themselves. As a general guide in evaluating each subject's ability to detect and differentiate between the four solutions, the following show the average person's ability to discriminate in relation to flavor:

(1) Sour: 0.0027–0.0081 Molar
(2) Salt: 0.0810–0.2430 Molar
(3) Sweet: 0.0320–0.0640 Molar
(4) Bitter: 0.0080–0.0032 Molar

Subjects who experience difficulty in identifying higher concentrations obviously should not be asked to serve on product taste panels where meaningful results are desired.

TEST 17.2 — Triangular Taste Evaluation

Procedure

(1) Prepare the samples with two duplicate samples and one of the other sample. There are six possible orders in which the samples may be presented: AAB, ABA, BAA, BAB, BBA and ABB. Obviously the samples are not coded in this way but by numbers 1 to 3.
(2) Set up each setting with the data sheet as shown in Fig. 17.2.
(3) Check each panel member to see that he works individually and that his data sheet is completely filled in.
(4) Total those not correctly identified, the duplicate samples, and interpret the results with the help of the data in Tables 17.3 and 17.4.

Lastly, in samples where there are differences due to color, consistency, texture, etc., it may be necessary to disguise the color with artificial coloring or change the check samples by dilutions, etc.

TEST 17.3 — Numerical Scoring Taste Evaluation

Procedure

(1) Prepare and code 10 samples for 10 panel members (1 complete set for each panel member) at every individual setting. (See Fig.17.4).
(2) Fill 2-oz souffle cups ½ to ⅔ full of the coded sample, place on tray and distribute to flavor booths with score card and an empty cup for mouth rinsing.
(3) Notify panel members that the panel is ready and instruct them to score the samples for flavor only using the 10 point scoring system.

QC FORM 17.2 — Flavor Difference Evaluation Judge's Score Card

Date _____
Project _____
Name of Judge _____

Product _____
Treatment or Can Code _____ Panel Replicate _____
Code No. or Letter for Judge _____

Sample	Check Like Sample*	Separate for Flavor Only Flavor Difference Between Odd and Like Samples	Did you check by guess?
(1)	(2)	(3)	(4)
		___ None ___ Slight	___ Yes
		___ Moderate ___ Much	___ No

*Check by guess if no difference is detectable

5. If you checked a *Moderate* or *Much* in the flavor difference column (No. 3), then indicate below whether you consider either the odd sample or like samples to have an undesirable flavor. Odd Sample: Yes ___ No ___ Like Samples: Yes ___ No ___
6. *Rate for Flavor Only*. Place a check mark above either a short or long line on the scale below to indicate how you rate the like samples and the odd sample.

Like Samples |———————|———————|———————|———————|———————|
Odd Samples |———————|———————|———————|———————|———————|
            very poor   poor   fair   good   excellent

7. (a) Did you detect any kind of difference, other than flavor, between the samples? Yes ___ No ___
   (b) If "Yes", what kind of difference? _____

FIGURE 17.3 — Set Up For A Triangle Taste Panel

(4) Allow each judge as much time as he requires.
(5) Check to see that each judge is marking the evaluation card correctly and that he has signed it when finished.
(6) Tabulate the data from the cards and calculate the "F" value and the LSD using the method below.

Generally, it is advisable to duplicate the taste panel session to test the members' ability to repeat themselves. When this is done duplicate samples should never be given the panel with the same codes. In other words, the panel should be conducted in a blind manner.

Interpretation

The data, to determine if there are real or imaginary differences, should be statistically analyzed. The most applicable method where there are more than two different samples, is the Analysis of Variance as discussed in Chapter 5.

TEST 17.4 — FLAVOR DIFFERENCE TASTE EVALUATION
Procedure

(1) Code all samples to be evaluated in accordance with the scheme of figures shown in Table 17.1. These numbers have been so selected that no treatment will have the same designation in successive replicates nor will the numerical order of treatments be the same. First digits are assigned to indicate the replicate and thus facilitate the supervisor's tabulation of the data.

TABLE 17.3 — For Deciding Whether Panel Has
Ability To Differentiate Between Samples
In Triangular Test

N = No. of sets of samples (no. of judgments)
X = No. of correct separations of identical samples
Say "No ability" if X < K
Say "Has ability" if X > K

Significance Level

N	$\alpha = 0.05$ K	$\alpha = 0.01$ K	$\alpha = 0.001$ K
7	5	6	7
8	6	7	8
9	6	7	8
10	7	8	9
11	7	8	9
12	8	9	10
13	8	9	10
14	9	10	11
15	9	10	12
16	10	11	12
17	10	11	13
18	10	12	13
19	11	12	14
20	11	13	14
21	12	13	15
22	12	14	15
23	13	14	16
24	13	14	16
25	13	15	17
26	14	15	17
27	14	16	18
28	15	16	18
29	15	17	19
30	16	17	19
31	16	18	19
32	16	18	20
33	17	19	20
34	17	19	21
35	18	19	21
36	18	20	22
37	18	20	22
38	19	21	23
39	19	21	23
40	20	22	24
50	24	26	28
60	28	30	33
70	32	34	37
80	35	38	41
100	43	46	49
500	188	194	202
1,000	363	372	383

Source: Roessler *et al.* (1948).

TABLE 17.4 — Rule For Deciding Whether Panel Has Ability
To Classify Correctly In Triangular Test

N = No. of sets of samples
X = No. of sets of samples classified perfectly
Say "No ability" if X < K
Say "Has ability" if X > K + 2
If X = K + 1,
  Say "Has ability" with probability Z
  Say "No ability" with probability 1 − Z

Significance Level	$\alpha = 0.05$		$\alpha = 0.01$		$\alpha = 0.001$	
N	K	Z	K	Z	K	Z
10	3	0.64	4	.58	5	0.27
11	3	0.36	4	.27	5	0.09
12	3	0.15	4	.07	6	0.74
13	4	0.96	5	.73	6	0.32
14	4	0.62	5	.39	6	0.09
15	4	0.36	5	.16	7	0.76
16	4	0.16	5	.03	7	0.36
17	5	0.99	6	.80	7	0.12
18	5	0.66	6	.30	8	0.86
19	5	0.40	6	.10	8	0.43
20	5	0.20	7	.85	8	0.18
21	5	0.03	7	.51	8	0.01
22	6	0.73	7	.26	9	0.57
23	6	0.47	7	.07	9	0.27
24	6	0.26	8	.79	9	0.07
25	6	0.08	8	.48	10	0.76
26	7	0.48	8	.25	10	0.41
27	7	0.57	8	.06	10	0.17
28	7	0.35	9	.78	10	0.00
29	7	0.16	9	.48	11	0.60
30	8	0.98	9	.25	11	0.31
31	8	0.69	9	.07	11	0.10
32	8	0.46	10	.80	12	0.85
33	8	0.25	10	.51	12	0.50
34	8	0.08	10	.27	12	0.24
35	9	0.84	10	.09	12	0.06
36	9	0.59	11	.85	13	0.75
37	9	0.37	11	.55	13	0.43
38	9	0.18	11	.31	13	0.20
39	9	0.01	11	.12	13	0.03
40	10	0.74	12	.91	14	0.69
50	12	0.63				

Source: Roessler et al. (1948).

QC FORM 17.4 FLAVOR — Numerical Scoring Quality Control Form

Sample _____ Date _____

Samples	Perfect	Good		Fair			Poor		Off	REMARKS
	10	9	8	7	6	5	4	3	2	1

SIGNATURE _____

NOTE: Make check mark in columns corresponding to your rating of sample when scoring one factor. However, when scoring 2 factors, write in the following letter in the corresponding column or columns: (C) Color (F) Flavor (T) Texture (S) Consistency

(2) Present samples of the same replicate to each individual judge at the same time. Randomly arrange the samples and all the judges to evaluate the samples in any order.
(3) Instruct the judges to taste each sample slowly and to retaste samples as often as necessary in any order.
(4) Instruct each judge to taste the reference sample as often as he deems it necessary to determine the degree of flavor difference.
(5) Each judge should determine by flavor comparisons with the reference sample the degree of flavor difference for each numbered sample and then place a check in one of the eight boxes opposite or between the term(s) which best describe the degree of flavor difference as illustrated by the form in Fig. 17.5. If no flavor differences are detected, a check should be placed in the box opposite the word "None."
(6) After judges have rated the flavor difference, they should place a check in one of the two boxes at the bottom of the column indicating whether the flavor of the numbered sample is acceptable or not acceptable.

Interpretation

After the test is completed, the supervisor can assign numerical scores 1 to 9 to the flavor difference ratings checked by the judges. That is, a check in box 1 at the top of the form opposite "None" is equal to 1 and a check opposite "extreme" is equal to 9. The data is then ready for statistical tabulation and interpretation.

The following method of analysis permits an evaluation of results by simply adding the appropriate total, the ranges (differences between highest and lowest levels), and multiplying the sum of the ranges by a factor obtained from a reference table. The summary tabulation form (Table 17.5) was developed specifically for this type of flavor evaluation.

Four steps are presented here. Table 17.5, the summary sheet made from the score sheets of a 15-member panel, shows the results of tests on tomato juice.
 (1) The first step eliminates from the summary sheet the judges who failed to find much difference between the three "treatments."
 (2) The scores of all remaining judges are tested to see if there was a measurable difference among the three juices. This is done by finding the overall significant differences.
 (3) The extent of the difference between any two of the juices is then determined by finding the least significant differences.

TABLE 17.5 — Flavor Difference Summary Sheet Made From Scores Of 15 Judges Testing Tomato Juice For Off-Flavor

Treatment	Replicate	Flavor Difference Scores for Indicated Judge																				Total Sums	Range of Judge Sums	No. Not Acceptable
		1	2	3[1]	4	5	6[1]	7	8[1]	9	10	11	12[1]	13	14[1]	15[1]	16	17	18	19	20			
A Code 19225	I	2	5*	2	3	7*	7*	3	1	1	4	5*	3	6*	5*	3								
	II	3	1	2	2	6*	3	3	3	3	3	3	1	9*	3	1								
	III	4	7	2	3	4*	5	7*	1	2	3	3*	2	7*	2	1								
	IV	3	6*	2	3	4*	7*	5*	6	8	2	1	1	1	2	3								
	Sum	12	19	8	11	21	22	18	6	8	12	12	7	23	12	8						136	15	13
	Range	2	6	0	1	3	4	4	2	2	2	4	2	8	3	2						Percent not acceptable = 36.1		
B Code 29225	I	2	5*	4	3	5*	5*	7*	1	6*	3	7*	3	6*	7*	1								
	II	5*	5*	5*	3	5*	3	7*	1	6*	7*	5*	3	6*	5*	3								
	III	6*	9*	1	5	6*	5*	5*	3	5*	5*	3*	1	7*	1	1								
	IV	5	9*	4*	3	6*	5*	7*	1	5*	2	3*	2	7*	2	5*								
	Sum	18	28	14	14	22	18	26	6	22	17	18	9	26	15	10						191	14	28
	Range	4	4	4	2	1	2	2	2	1	5	4	2	1	6	4						Percent not acceptable = 77.8		
C Control 09225	I	2	1	3	3	1	3	1	1	5	1	1	1	1	3	1								
	II	1	1	1	2	1	6*	1	1	3	1	1	1	1	3	1								
	III	2	1	2	1	1	1	1	3	2	1	1	1	1	1	1								
	IV	1	1	4*	1	1	5*	1	4	2	1	1	1	1	3	1								
	Sum	6	4	10	7	4	17	4	9	12	4	4	4	4	10	4						49	8	0
	Range	1	0	3	2	0	3	0	3	3	0	0	0	0	2	0						Percent not acceptable = 0		
Grand sum of ranges		7	10	7	5	4	9	6	7	6	7	8	4	9	11	6							37	Judge range Total
Grand range of sums		12	24	6	7	18	5	22	3	14	13	14	5	22	5	6						142		Grand range of totals
OSD (1.25)		8.8	12.5	8.8	6.3	5.0	11.3	7.5	8.8	7.5	8.8	10.0	5.0	11.3	13.8	7.5								

[1]These judges (3, 6, 8, 12, 14, 15) were unable to distinguish flavor differences and their data were eliminated from consideration.

Note: Place an asterisk next to sample score when sample flavor was judged not acceptable.

QC FORM 17.5 — Quality Control Score Sheet Is Used By Panel Members For Evaluating Flavor Differences

Test Code _____ Date _____

Judge's No. _____ Replicate _____

INSTRUCTIONS

(1) Enter at the head of each column the code number of each sample in the test.
(2) Determine by flavor comparisons with the reference sample the degree of flavor difference for each numbered sample.
    (a) If you do not detect any flavor difference, place a check in the box below opposite the word none.
    (b) If in your judgment any flavor difference exists, place a check in one of the other eight boxes opposite or between the term(s) which best describes the degree of flavor difference.
(3) After rating the flavor difference, place a check in one of the two boxes at the bottom of the column indicating whether the flavor of the numbered sample is acceptable or not acceptable to you.

N.B.—The reference sample should be re-tasted as often as necessary to determine the degree of flavor difference for each sample.

                                                                   Sample Number

Degree of Flavor Difference

None ....

Slight ....

Moderate ....

Large ....

Extreme ....
Acceptable ....
Not Acceptable ...

Judge's Initials _____

(4) Finally, the importance of the samples rejected is assessed. This is done by testing the percentage of replicate samples judged "not acceptable."

In the example, which is followed step-by-step through this section, the conventional terminology among food technologists is used. For those not familiar with these terms, "treatment" refers to the material under test. In the example two tomato juice samples with off-flavor, Treatments A and B, are used with a sample of juice with no off-flavors, Treatment C (Control).

"Replicates" could be interpreted as "rounds." Each judge gets three small cups containing a sample of each "treatment" (coded of course) in each replicate. The test panel shown on the summary sheet had the procedure repeated four times, thus four replicates.

The detailed steps below should be followed to determine whether or not there is a statistical difference between any two treatments.

(1) For Each Judge
   (a) The sum of the scores in all of the replicates for each treatment. For example, for Judge 1, the sum of the scores for Treatment A is 12, for Treatment B, 18, etc.
   (b) The range (difference between the highest and lowest score) within each treatment (i.e., Judge 1, Treatment A, the range is 2, for Treatment B, 4 etc.)
   (c) The grand sum of the ranges computed in (b) previously, i.e., for Judge 1 is $2 + 4 + 1 = 7$.
   (d) The grand range of the treatment sums computed in preceding "a", i.e., the highest treatment sum for Judge 1 occurred in Treatment B (18), the lowest sum in Treatment C (6). The difference between these two is the grand range of 12 for Judge 1.
   (e) Calculate the overall significant difference (OSD) between treatment sums as follows: Enter Table 17.6 and obtain the appropriate factor in the 5% column for the number of replicates and treatments used. For example, in the illustration (Table 17.5) the number of replicates being 4 and the number of treatments 3, the factor at the 5% level of significance is 1.25. Enter this factor on the summary form and multiply the grand sum of the treatment ranges for each judge by this factor. The product which is the OSD is then entered at the bottom of the column for each judge, i.e., for Judge 1 it is $1.25 \times 7 = 8.8$. A comparison of this product with the grand range of the treatment sums will indicate whether or not the judge rated the various treatments significantly different. The range between the highest and low-

TABLE 17.6 — Flavor Difference Multipliers Of The Range
For Computing Overall Significant Difference In
Evaluating Judge's Performance

Number of Judges or Replicates	Number of Treatments and Significance Level															
	2		3		4		5		6		7		8		9	
	5%	1%	5%	1%	5%	1%	5%	1%	5%	1%	5%	1%	5%	1%	5%	1%
							Multipliers									
2	3.43	7.92	2.37	4.42	1.78	2.96	1.40	2.06	1.16	1.69	1.00	1.39	0.87	1.20	0.78	1.03
3	1.91	3.14	1.44	2.14	1.13	1.57	0.94	1.25	0.80	1.04	0.70	0.89	0.62	0.78	0.56	0.69
4	1.63	2.47	1.25	1.74	1.01	1.33	0.84	1.08	0.72	0.91	0.63	0.78	0.57	0.69	0.51	0.62
5	1.53	2.24	1.19	1.60	0.96	1.24	0.81	1.02	0.70	0.86	0.61	0.75	0.55	0.66	0.50	0.59
6	1.50	2.14	1.18	1.55	0.95	1.21	0.80	0.99	0.69	0.85	0.61	0.74	0.55	0.65	0.49	0.59
7	1.49	2.10	1.17	1.53	0.95	1.21	0.80	0.99	0.69	0.84	0.61	0.74	0.55	0.65	0.50	0.59
8	1.49	2.08	1.17	1.52	0.96	1.21	0.81	0.99	0.70	0.85	0.62	0.74	0.55	0.66	0.50	0.59
9	1.50	2.09	1.18	1.53	0.97	1.22	0.82	1.00	0.71	0.85	0.62	0.75	0.56	0.66	0.51	0.60
10	1.52	2.10	1.20	1.55	0.98	1.23	0.83	1.01	0.72	0.86	0.63	0.75	0.57	0.67	0.52	0.61
11	1.54	2.11	1.21	1.56	0.99	1.24	0.84	1.02	0.73	0.88	0.64	0.77	0.58	0.68	0.52	0.61
12	1.56	2.13	1.23	1.58	1.00	1.25	0.85	1.03	0.74	0.89	0.65	0.78	0.59	0.69	0.53	0.62
13	1.58	2.15	1.25	1.60	1.02	1.27	0.86	1.05	0.75	0.90	0.66	0.79	0.59	0.70	0.54	0.63
14	1.60	2.18	1.26	1.62	1.03	1.28	0.87	1.06	0.76	0.91	0.67	0.80	0.60	0.71	0.55	0.64
15	1.62	2.20	1.28	1.64	1.05	1.30	0.89	1.08	0.77	0.92	0.68	0.81	0.61	0.72	0.56	0.65
16	1.64	2.22	1.30	1.65	1.06	1.31	0.90	1.09	0.78	0.93	0.69	0.82	0.62	0.73	0.56	0.66
17	1.66	2.24	1.31	1.67	1.08	1.33	0.91	1.11	0.79	0.95	0.70	0.83	0.63	0.74	0.57	0.67
18	1.68	2.27	1.33	1.69	1.09	1.34	0.92	1.12	0.80	0.96	0.71	0.84	0.64	0.75	0.58	0.68
19	1.70	2.30	1.34	1.71	1.10	1.36	0.93	1.14	0.81	0.97	0.72	0.85	0.65	0.76	0.59	0.68
20	1.72	2.32	1.36	1.73	1.11	1.38	0.93	1.15	0.82	0.98	0.73	0.86	0.65	0.77	0.59	0.69

This table, shortened from tables prepared by Thomas E. Kurtz, Richard F. Link, Joh W. Tukey, and David L. Wallace (Gould et al. 1951).

est treatment score should be greater than the OSD value, i.e., for Judge 1 one grand range of treatment sums is 12, which is higher than the OSD value of 8.8. Whenever the range between the highest and lowest treatment sum for a given judge is equal to or less than the OSD value, it indicates lack of ability to distinguish between any of the treatments.
(f) Evaluation of judge performance: It will be noted in Table 17.5 after completing the calculations for OSD values for the 15 judges, that judges 3, 6, 8, 12, 14 and 15 had grand range of treatment sums which were equal to or lower than their individual OSD values at the 5% level. These judges, therefore, were unable to distinguish flavor differences, and in the illustration (Table 17.5) their data were eliminated from further consideration.

(2) For Each Treatment:
   (a) Add the sums of replicate scores for each of the judges not eliminated. For example, the total of the sums for Treatment A for the nine remaining judges is 136, for B, 191, and for C, 49. These figures should be entered in the total column to the right side of the summary tabulation form.
   (b) Compute the range of the sums for the remaining judges in each treatment and enter this value in the column on the right side of the summary tabulation form headed Range of Judge Sums. In this example, for Treatment A the highest sum, 23, was given by Judge 13 and the lowest sum of 8 by Judge 9. The range is, therefore, 15 and is entered on the next column on the same line as the 136.
   (c) Count the number of asterisks in all of the replicates for a single treatment for the 9 remaining judges and record the number in the column at the extreme right of the sheet headed Number Not Acceptable. In the example given in Table 17.5, number of asterisks for Treatment A for 9 judges is 13. For later evaluation of the significances, this number should be converted to a percentage. In the example cited, 13 is 36.1% of the total 36 evaluations made for Treatment A.

(3) All Treatments, All Judges:
   (a) Determine the range of the total scores for each treatment that were computed for 2a previously. In Table 17.5 this value is 142 which is obtained by subtracting 49 (sum for Treatment C) from 191, the sum for Treatment B. Enter this figure on the summary tabulation form on the lower right hand at the base of the Total Column.
   (b) Obtain the judge total by adding the range of judge sums for

each treatment. In the example cited, the 3 ranges are 15 for A, 14 for B, and 8 for C, making a total of 37. This figure should be entered at the base of the column headed Range of Judge Sums.

(c) The next step is to determine the overall significant difference (OSD) values to determine whether any significant differences exist among treatments. To do this, obtain the appropriate factor from Table 17.6. These factors are found in Table 17.6 in the column for the number of treatments that were used and on the line for the number of judges. In the example used, the appropriate factors are found in column 3 "for number of treatments" and on line 9, "for number of judges." For significance at the 5% level, the figure is 1.18 and at the 1% level, 1.53. Multiply this factor by the judge range total, 37. This will give 43.7 at the 5% level and 56.6 at the 1% level. In the example, Table 17.5, the grand range of totals is 142 which greatly exceeds the OSD value of 56.6 and, therefore, a highly significant difference among treatments exists.

(d) The next step is to determine the least significant difference (LSD) values which will permit an evaluation of differences between any two treatments. It is necessary, however, that the OSD value be significant before LSD values are calculated and specific comparisons between treatments are made. (To be significant the OSD value at the 5% level of significance must be less than the grand range of totals.) The procedure for calculating LSD values is essentially identical to that used for OSD. In the example, the judge range total, 37, is multiplied by the LSD factor found in Table 17.7 in the column headed 3 (number of treatments) and on the line 9 (number of judges). The LSD value at the 5% level is $0.98 \times 37 = 36.3$; at the 1% level it is $1.34 \times 37 = 49.6$. Therefore, the differences between any two treatments' totals (A − C = 87, B − C = 142, and B − A = 55) are significant at the 1% level in this illustration.

Table 17.8 contains the minimum percentage not acceptable that is necessary for significance at the 1% level for the indicated number of flavor difference evaluations. For example, with the 36 flavor difference evaluations used in this illustration, a minimum percentage not acceptable in both Treatments A and B (36.1 and 77.8) are highly significant.

IFT GLOSSARY OF SOME TERMS USED IN THE SENSORY (PANEL) EVALUATION OF FOODS AND BEVERAGES[1]

(1) *Acceptance.* (a) An experience, or feature of experience characterized by a positive (approaching a pleasant) attitude. (b) Actual

[1]Anon. (1959).

TABLE 17.7 — Flavor Difference Multipliers For Computing The Least Significant Differences When Using The Simplified Flavor Difference Procedure

Number of judges or replicates	2 5%	2 1%	3 5%	3 1%	4 5%	4 1%	5 5%	5 1%	6 5%	6 1%	7 5%	7 1%	8 5%	8 1%	9 5%	9 1%
							Multipliers									
2	3.43	7.92	1.76	3.25	1.18	1.96	0.88	0.39	0.70	1.07	0.58	0.87	0.50	0.74	0.44	0.63
3	1.63	3.14	1.14	1.73	0.81	1.19	0.63	0.91	0.52	0.73	0.44	0.61	0.38	0.53	0.33	0.46
4	1.91	2.47	1.02	1.47	0.74	1.04	0.58	0.80	0.48	0.68	0.40	0.55	0.35	0.48	0.31	0.44
5	1.53	2.24	0.98	1.37	0.72	0.98	0.56	0.77	0.47	0.63	0.40	0.54	0.34	0.47	0.30	0.43
6	1.50	2.14	0.96	1.32	0.71	0.96	0.56	0.76	0.46	0.62	0.40	0.53	0.34	0.46	0.30	0.42
7	1.49	2.10	0.96	1.33	0.71	0.96	0.56	0.77	0.47	0.63	0.40	0.53	0.35	0.46	0.31	0.42
8	1.49	2.08	0.97	1.33	0.72	0.97	0.57	0.77	0.47	0.63	0.41	0.54	0.35	0.47	0.31	0.42
9	1.50	2.09	0.98	1.34	0.73	0.98	0.58	0.77	0.48	0.64	0.41	0.55	0.36	0.48	0.31	0.43
10	1.52	2.10	0.99	1.35	0.74	0.99	0.59	0.78	0.49	0.65	0.42	0.55	0.37	0.48	0.32	0.43
11	1.54	2.11	1.00	1.35	0.74	0.99	0.59	0.79	0.49	0.65	0.42	0.56	0.37	0.49	0.32	0.43
12	1.56	2.13	1.01	1.36	0.75	1.00	0.60	0.80	0.50	0.67	0.43	0.57	0.38	0.50	0.33	0.44
13	1.58	2.15	1.03	1.38	0.76	1.01	0.61	0.81	0.51	0.68	0.43	0.57	0.38	0.50	0.34	0.45
14	1.60	2.18	1.04	1.39	0.77	1.03	0.62	0.82	0.52	0.69	0.44	0.58	0.38	0.50	0.34	0.45
15	1.62	2.20	1.06	1.42	0.79	1.05	0.63	0.84	0.52	0.69	0.45	0.60	0.39	0.52	0.35	0.46
16	1.64	2.22	1.07	1.43	0.80	1.07	0.64	0.85	0.53	0.70	0.45	0.61	0.40	0.53	0.35	0.46
17	1.66	2.24	1.08	1.44	0.81	1.08	0.65	0.86	0.54	0.72	0.46	0.61	0.40	0.53	0.36	0.48
18	1.68	2.27	1.10	1.46	0.82	1.09	0.65	0.86	0.54	0.72	0.47	0.62	0.41	0.54	0.36	0.48
19	1.70	2.30	1.11	1.48	0.83	1.10	0.66	0.88	0.55	0.73	0.47	0.62	0.42	0.56	0.37	0.49
20	1.72	2.32	1.13	1.51	0.83	1.10	0.67	0.89	0.56	0.74	0.48	0.64	0.42	0.56	0.37	0.49

Extension of table prepared by J. W. Tukey (Peryam and Pilgrim 1957).
Factors for 11 to 20 replicates calculated by H. L. Stier.

TABLE 17.8 — Flavor Difference Minimum Percntage Of
"Not Acceptable" Judgements for Significance At The 1% Level

Number judgements	20	25	30	31	32	33	34	35	40
Minimum percent	31.8	26.5	23.5	23.0	22.0	21.5	21.0	20.5	18.5
Number judgements	45	50	55	60	65	70	75	80	
Minimum percent	17.0	15.5	14.5	13.5	12.5	11.5	11.0	10.5	

utilization (purchase, eating). May be measured by preference or liking for specific food item. The two are often highly correlated, but they are not the same.

(2) *Acrid.* Sharp and harsh or bitterly pungent.

(3) *Acuity.* Ability to discern or perceive stimuli, sharpness or acuteness.

(4) *Adaptation.* Loss of sensitivity to a given stimulus or as result of continuous exposure to that stimulus or a similar one.

(5) *After-taste.* The experience which under certain conditions follows the removal of a taste stimulus; it may be continuous with the primary experience or may follow as a different quality after a period of time during which swallowing, saliva, dilution and other influences may have affected the stimulus substance.

(6) *Ageustia.* Lack or impairment of sensitivity to taste stimuli.

(7) *Anosmia.* Lack or impairment of sensitivity to odor stimuli.

(8) *Appetite.* Lust or desire for anything, but more especially for food. (If hunger be regarded as the specific sensory experience localized in the region of the stomach and presumably aroused by slow rhythmical contractions of the stomach walls, then appetite for food can exist without hunger, and in civilized man ordinarily does occur, especially as habituated behavior dependent upon past experience with food.)

(9) *Aroma.* A distinctive characteristic or suggestive of fragrance or odor.

(10) *Astringent.* Quality perceived due to the complex or sensations caused by shrinking, drawing, or puckering of the skin surfaces of the mouth, accompanied by a driving of the blood from the tissues. Dry feeling in the mouth.

(11) *Bitter.* A quality of taste sensation, the taste of quinine sulphate

being a typical example. Perceived by the circumvallate papillae at the back of the tongue.
(12) *Body.* The quality of a food or beverage, relating variously to its consistency, compactness of texture, fullness, or richness.
(13) *Burning.* As related to taste is used most often to describe the sensation of heat caused by inadequate stimulation, e.g., that of pepper, mustard, or other strong spices. It arises from the skin senses, including that of pain as well as those of temperature. When the adequate stimulus (hot material) is applied, the term hot is more likely to be used.
(14) *Chewy.* Tending to remain in the mouth without readily breaking up or dissolving. Requiring mastication.
(15) *Compensation.* The result of interaction of the components in a mixture of stimuli in which each component is perceived as less intense than each would be alone.
(16) *Cooling.* A physical sensation in the mouth resulting from the presence of a cold liquid or solid. Also result of chemical action (menthol) sensed by the skin.
(17) *Fatigue.* Condition of organs or organisms which have undergone excessive activity with resulting loss of power or capacity to respond to stimulation (see *adaptation*).
(18) *Flavor.* (a) A mingled but unitary experience which includes sensations of taste, smell and pressure, and often cutaneous sensations such as warmth, cold, or mild pain. (b) An attribute of foods, beverages, and seasonings, resulting from the stimulation of those senses which are grouped together at the entrance to the alimentary and respiratory tracts—especially odor and taste.
(19) *Flavoring.* Any substance, such as essence or extract, employed to give a particular flavor.
(20) *Fragrant.* A pleasing olfactory quality; odors which are distinctly pleasant smelling.
(21) *Gritty.* A hard, stone-like sensation, usually caused by the presence of sand particles or stone cells. Not a sensation, but a judgment based on complex perception.
(22) *Gust.* A unit of gustatory intensity relating to the threshold of a given substance.
(23) *Gustation.* A sense (taste) whose receptors lie in the mucous membrane covering the tongue and whose stimuli consist of certain soluble chemicals; e.g., salts, acids, sugars, etc.
(24) *Mealy.* A quality of mouth feel denoting a starch-like sensation. Friable.
(25) *Mouthfeel.* The mingled experience deriving from the sensations of

the skin in the mouth after ingestion of a food or beverage. It relates to density, viscosity, surface tension and other physical properties of the material being sampled.

(26) *Objective.* (a) Capable of being recorded by physical instruments or as a consequence of a repeatable operation. (b) Not dependent upon the observations and reports of an individual and thus verifiable by others.

(27) *Odor.* Sensation due to stimulation of the olfactory receptors in the nasal cavity by gaseous material.

(28) *Odorant.* A chemical substance which stimulates the olfactory receptors.

(29) *Olfactometer.* An instrument for controlled presentation of odor stimuli, used for measuring thresholds and other quantitative values.

(30) *Organoleptic.* (a) Affecting or making an impression upon an organ or the whole organism; (b) capable of receiving an impression; (c) sometimes used as a synonym for sensory when referring to examination by taste and smell (obsolescent).

(31) *Palatable.* Agreeable to the "taste," savor; hence, acceptable or pleasing.

(32) *Primary Qualities.* Within a specific sense, those qualities which are considered basic and from which all other qualities can be compounded. There is general belief that salt, sweet, bitter and sour are the four primary taste qualities.

(33) *Rancid.* Having a rank odor or taste, as that of old oil.

(34) *Salty (Saline).* A quality of taste sensation of which the taste of sodium chloride is the typical example.

(35) *Sensory.* Pertaining to the action of the sense organs.

(36) *Sharp.* Characterizing an intense or painful, well localized reaction to a substance being eaten or smelled; e.g., various acids and alcohols.

(37) *Sour.* A quality of taste sensation of which the taste of acid is the typical example.

(38) *Stimulus.* A change of energy in the environment which affects a sense organ.

(39) *Sweet.* A quality of taste sensation of which the taste of sucrose is the typical example.

(40) *Taste.* (See *gustation*). One of the senses, the receptors for which are located in the mouth and are activated by a large variety of different compounds in solution. Most investigators usually limit gustatory qualities to four: saline, sweet, sour, bitter. Distinguish from Flavor, the experience to which taste contributes.

(41) *Texture.* (See *Mouthfeel*). Those properties of a foodstuff, apprehended by the eyes and by the skin and muscle senses of the mouth, including the roughness, smoothness, graininess, etc.

BIBLIOGRAPHY

ANON. 1959. Glossary of some terms used in the sensory (panel) evaluation of foods and beverages. Food Technol. *13*, No. 12, 733–736.

GOULD, W.A. 1953. Methods for evaluating flavor in processed foods. Food Packer *34*, No. 1, 36, 44–45.

GOULD, W.A., SLEESMAN, J.P., RINGS, R.W., LYNN, M., KRANTZ, F. JR., and BROWN, H.D. 1951. Flavor evaluations of canned fruits and vegetables treated with newer organic insecticides. Food Technol. *5*, No. 4, 129–133.

INSTITUTE OF FOOD TECHNOLOGISTS. Not Dated. Sensory Testing Guide for Panel Evaluation of Foods and Beverages. Institute of Food Technologists, Champaign, Ill.

MAHONEY, C.H., STIER, H.L., and CROSBY, E.A. 1957. Evaluating flavor differences in canned foods. I. Genesis of the simplified procedure for making flavor difference tests. Food Technology Symposium, September, p. 29–36.

MAHONEY, C.H., STIER, H.L., and CROSBY, E.A. 1957. Evaluating flavor differences. II. Fundamentals of the simplified procedure. Food Technology Symposium, September, p. 37–42.

PERYAM, D.R., and PILGRIM, F.J. 1957. Hedonic scale method of measuring food preferences. Food Technology Symposium, September, p. 9–14.

ROESSLER, E.B., WARREN, J., and GUYMON, J.F. 1948. Significance in triangle taste tests. Food Res. *13*, 503–505.

SCHWARTZ, N., and FOSTER, D. 1957. Methods for rating quality and intensity of the psychological properties of foods. Food Technology Symposium, September, p. 15–20.

SIMONE, M., and PANGBORN, R.M. 1957. Consumer acceptance methodology: one vs. two samples. Food Technology Symposium, September, p. 25–29.

SJOSTROM, L.B., CAIRNCROSS, S.E., and CAUL, J.F. 1957. Methodology of the Flavor Profile. Food Technology Symposium, September, p. 20–25.

CHAPTER 18

Odor

In many foods and beverages odor is considered to be the most important quality. Obtaining and maintaining a desirable odor is quite important through the addition or deletion of chemicals, the methods of processing, the packaging materials and the storage conditions.

Various methods have been used to determine odors. Some of them include electromagnetic radiation, spectrograph, hygrometry and the stinkometer. One of the most promising instruments for component measurement in food aromas is the gas-liquid chromatograph.

Use of the sense of smell is the most important odor detection method. Smelling is done best by inhaling strongly through the nose for a period of two to three seconds with both nostrils open. When smelling an unknown specimen, first wave a little of the aroma toward the nose and sniff continuously to avoid too strong an excitation or perhaps a temporary injury of the sense of smell.

Taste has been divided into four primary tastes and all tastes can be made artificially by mixing the four primary tastes. When considering odors, however, the position is more difficult. Odors cannot be readily classified, for the reactions to odorants are usually mild and undefined, a sniff, a drawing away, or perhaps in extreme cases, disgust and nausea.

CLASSIFICATION

Endeavors to classify the odors have not been very successful. The number of odors seems large, and different judges do not agree as to distinctions. One of the best known classifications is that of Henning.

(1) Spicy, e.g., cloves, fennel, anise
(2) Flowery, e.g., heliotrope, coumarin, geranium
(3) Fruity, e.g., oil of orange, oil of bergamot, citronellal
(4) Resinous or balsamie, e.g., turpentine eucalyptus oil, Canadian balsam
(5) Burnt, e.g., pyridine, tar
(6) Foul, e.g., sulphuretted hydrogen, carbon bisulfide (Dawson and Harris 1951)

ODOR TRANSITIONS

Apart from the aesthetic enjoyment derived from odor or aroma in food, the food technologist must be aware of the transitions that take place in foods to prevent the development of undesirable aromas. Some of the transitions are the following:

(1) transfer of odors from one food to another;
(2) absorption of odors from the package or environment;
(3) changes produced by processing into currently marketable items;
(4) alterations caused by the treatments used to prevent microbial decomposition;
(5) changes resulting from the inherent life processes of respiring fruits and vegetables;
(6) modifications caused by high and low temperature and high and low humidity;
(7) changes produced by the impact of visible and invisible radiant energy;
(8) alterations resulting from chemical or enzymatic reactions within food tissues;
(9) atmospheric oxidation; and
(10) the paramount mechanism of microbial decomposition, including bacteria, yeasts and molds (Crocker and Dillion 1949).

ODOR AND FLAVOR DEFECTS

Milk and Cheese

One of the best demonstrations of the complexity of off-odor sources is the tabulation by Crocker for flavor-odor defects in milk and cheddar cheese (Crocker 1946).

	Milk
Barny, cowy	Absorbed from the surroundings in the process of production
Feed, ensilage	Feeding immediately prior to milking, also certain weeds and feeds that the cows may eat in pasturage (garlic, onions, turnips)
Metallic, cardboard oxidized, tallowy, oily, burnt feathers	Four principle causes: (1) Natural conditions affecting the cows (2) Copper and iron catalysts (3) Action sunlight (4) Irradiation by ultra violet light
High acidity, sour	Improper handling—inadequate refrigeration
Salty	Diseased udder
Bitter	Bitter flavored weeds in feed, lipase
Disinfectant	Sometimes the early stages of oxidized flavor
Neutralizer	Excessive addition of neutralizer
Rancid	Enzyme action and bacterial toxins
Additions	
Off-odors	Plastic and cardboard containers

	Cheese (Cheddar Type)
Musty	Old culture, or contamination by wild yeasts and molds
Old	Indication of runaway or other faulty curing
Metallic	Metals dissolving from equipment used in manufacturing
Salty	Too much salt, or abnormal milk
Flat	Not enough acidity or salt; improper manufacture or curing; or defective culture
Bitter	Uncontrolled enzymatic action, caused by improper processing or curing

Other Examples

The testing of tea, coffee, wine, and beer for aroma and flavor is commonly an appraisal step in the buying and selling of these commodities. For example: Lots of tea are first classified by appearance into types; then these rough gradings are further classified into price grades within the types by brewing a sample and then judging for aroma, appearance, color, and clarity.

The control of aromatic compositions is the actual function of the aging of liquors. Raw spirits contain a number of alcohols that have disagreeable odors, and long periods of storage give them time to convert into pleasantly odorous esters. Moreover, by storage in wooden containers, the flavors and colors are strengthened by absorption and extraction of woody constituents (Crocker and Dillion 1949).

Breweries and bottling plants for carbonated beverages require a "highly polished" water supply to avoid unusual odors in their products. Even the high quality potable supply delivered by the water treatment plant to the community may have sufficient residual chlorine or other barely perceptible substances that do not mix well with the flavors of beverages. Ice plants, bakeries, confectioneries, creameries, and similar food-processing industries likewise need an odor- and taste-free supply of water and frequently resort to the use of privately operated granular activated carbon filters of absorbers (Crocker and Dillion 1949).

The Crocker-Henderson Odor Standards designed an eight standard odor system as shown in Table 18.1. According to Crocker these standards were selected for premanency and unchangeableness of odor and comparative harmlessness. Further, they are relatively nontiring to the nose. To find the odor number of any substance, it is compared with a set of standards, first to find the fragrance figure, then for acidity, next for burntness and finally for caprylicness. By this means a five digit number can be obtained for any odor that is at convenient strength and in convenient condition for analysis. According to Crocker every odor appears to have some of each of the four components. The strength must be low enough that pungency is avoided and there is no stinging produced in the nose. Diluting, if necessary, with water, benzyl benzoate, diethyl phthalate, mineral oil, or other low odor substances is common.

The Crocker-Dillion Odor Directory for Caprylic (standards 1-8) are reproduced showing numerical sequence for many materials (Table 18.2).

TABLE 18.1 — Standards For Odor Analysis

Fragrant	Acid
1112 n-Butyl phthalate	7122 Vanillin
2424 Toluene	7213 Cinnamic acid
3336 α-Chloronaphthalene	5335 Resorcinol dimethyl ether
4344 α-Naphthyl methyl ether	2424 Toluene
5645 p-Cymene	5523 Isobutyl phenylacetate
6645 Citral	5626 Methyl phenylacetate
7343 Safrole	5726 Cineole (Eucalyptol)
8453 Methyl salicylate	3803 Acetic acid (20% solution)
Burnt	Caprylic
5414 Ethyl alcohol (very pure)	.... No suitable standard found
7423 Phenylethyl alcohol	7122 Vanillin
5335 Resorcinol dimethyl ether	7343 Safrole
4344 α-Naphthyl methyl ether	5624 Phenylacetic acid
4355 Veratrole	5645 Cymene
6665 Thujone	3336 α-Chloronaphthalene
4376 Paracresyl acetate	2577 Anisole
7584 Guaiacol	3518 2, 7-Dimethyl octane

TEST 18.1 — Evaluating Undesirable Odors In Packaging Materials

(1) Sample containers. Vaporproof containers may be necessary for sampling, storing and holding specimens for the development of maximum intensity of odor. Friction-top tin cans, Mason jars and laboratory glassware are suitable. Containers shall be clean, dry and free from odor. Aluminum foil or cellophane shall be substituted for any rubber gaskets or stoppers. Samples may also be wrapped directly in foil, cellophane or vegetable parchment.

(2) Testing room. Detection of low levels of odor requires working space in which individual members of the panel can concentrate. The room shall be free from plant or industrial odors and to avoid this it may be necessary to take samples home at night. Extremes of temperature should be avoided, except in special cases, and all distracting influences should be minimized.

FIGURE 18.1 — Crocker-Henderson Odor Standards

Materials

(1) Standard samples. In some instances, standard samples, representing satisfactory and maximum permissible levels of odor, may be set up. Maintenance of the samples is a serious problem, since age and other factors may alter odors drastically. In all cases, it is desirable to have some type of reference sample available.

(2) Water (odor-free). Moisture is commonly used to intensify odors or develop potential odors. Normally, tap water is satisfactory, except where residual odor or a chlorine smell is present. Distilled water is not always odorless. All water shall be smelled and tasted before acceptance.

(3) Fatty materials. A range of substances containing oil or fat may be used to pick up certain types of odors, with identification and degree of odor determined both by smell and taste. Common materials are mineral oil, milk chocolate, cream and fresh unsalted butter.

TABLE 18.2 — American Perfumer
And Essential Oil Odor Directory

Part I—Numerical Sequence

Caprylic I

Odor Number	Material	Application
3111	Benzyl Benzoate	Almost odorless; solvent and fixative, especially for artificial musk
3211	Diethyl Phthalate	Almost odorless; solvent and fixative, especially for artificial musk
4211	Dimethyl Phthalate	Almost odorless; solvent and fixative, especially for artificial musk

Caprylic II

Odor Number	Material	Application
3112	Anisic Acid	Mild, slightly rosey; sweetener and blender, especially for lily and lilac
4112	Phenylethyl Phenylacetate	Sweet mild rosey; a floral blender
3212	Anisyl Formate	Mild floral; used in heliotrope, tube rose, etc.
4212	Anisyl Alcohol	Mild floral; used as a blender for floral types
6212	Hydroxycitronellal Dimethyl Acetal	Slightly musty, fruity, floral; used as a base for lily, lilac, etc.
5312	Phenylethyl Propionate	Slightly fruity; used as a modifier for rose and other floral types
4412	Farnesol	Slightly fruity, rosey; used as a blender for rose and other floral types
5322	Phenylethyl Acetate	Fruity, peach-like; used in rose, jasmin, hyacinth, etc.
6322	Cyclamen Alcohol	Resembles lily, violet and hyacinth; used in perfumes of these and other floral types
6422	Cyclamen Propionate	Slightly more fruity than the alcohol; used in floral compounds such as rose, lily, lilac, violet, etc.
5522	Phenylethyl Isobutyrate	Fruity, rose-like; used in floral compounds
6522	Phenylethyl Butyrate	Fruity, rose-like; used in floral compounds
6622	Cyclamen Butyrate	Fruity, rose-like; used in floral compounds
7232	Methyl Inone	Mild, sweet, floral; used in violet and other floral bouquets
4332	Santalyl Acetate	Rosey; floral blender for rose, and other floral types
6332	Rhodinyl Phenylacetate	Rose de Mai type; used in floral compounds
7332	Rhodinyl Butyrate	Moss rose type; used in floral compounds
5432	Aldyhyde C-16	Fruity, floral; used as a flavor and in floral compounds
6432	Citronellyl Acetate	Fruity; used in rose, carnation, and other floral compounds
7432	Rhodinyl Isobutyrate	Fruity, rosey; used in rose and other floral compounds
5532	Neryl Acetate	Floral, fruity; used in rose, jasmin, etc.
6532	Citronellyl Butyrate	Fruity, rosy; used in rose, jasmin, etc.
7532	Citronellyl Propionate	Fruity, rosy; used in rose, jasmin, etc.

TABLE 18.2 — (Continued)

Odor Number	Material	Application
5632	Geramyl Propionate	Fruity; used in rose, jasmin, etc.
6342	Linalyl Isobutyrate	Fruity, woody floral; used with lavender, and in jasmin, lilac, etc.
	Caprylic III	
6113	Vanillin	Slightly musty, fragrant; used as a sweetner and blender for flavors and for perfumes
6123	Ethyl Vanillin	Slightly musty, fragrant; used as a sweetner and blender for perfumes and for flavors
7123	Musk Ketone	Smooth musky; the nicest of the artificial musks, used as a sweetener, blender and fixative in perfumes
2223	Acetate C-12	Mild fruity; blends with most floral notes
4223	Benzyl Alcohol	Mild fruity; used in jasmin, and other floral compounds
7223	Oil Peppermint, distilled	Fragrant, minty, slightly fruity; used mainly in flavors; in perfumes for its cooling effect and its fragrance
8223	Methyl Salicylate	Fragrant, minty, fruity; used mainly in flavors. In perfumes, used for its fragrant top notes, in cassie, tube rose, chypre, etc.
3323	Nerolidol	Mild floral, lily-like; used as a blender for floral compounds
4323	Anisyl Acetate	Mild floral; used as a blender especially for cassie
5323	Anisic Aldehyde (Aubepine) Ethyl Aubepine	Sweet odor of hawthorn; used in lilac, cassie, heliotrope, etc.
7323	Oil Anise, Russian	Heavy, fruity, floral, anethol-like; used mainly in flavors. Somewhat in perfumes as a sweetener
3423	Phenyl Cresyl Oxide	Suggestive of narcissus and rose; used in floral compounds
4423	Cinnamic Alcohol	Hyacinth-like; used as a fixative for hyacinth, lilac, lily, rose, jasmin, etc.
5423	Geranyl Acetate	Fragrant, suggestive of rose and lavender; used in rose, jasmin, lavender, etc.
6423	Cinnamyl Acetate	Soft, sweet, rosey; used in rose, jasmin, and other floral compounds
8423	Oil Ylang Ylang, Bourbon	Powerful, slightly fruity, floral; used in lilac, violet, oriental, and many floral compounds
4523	Anisyl Propionate	Fruity, floral; used in jasmin and other floral compounds
5523	Bensyl Acetate	Fruity, jasmin-like; used in jasmin, tuberose, etc.
6523	Isobutyl Penylacetate	Fruity, floral; used as a modifier for tuberose, rose, carnation, etc.
3623	Hexyl Butyrate	Fruity; modifier for floral compounds
5623	Benzyl Propionate	Fruity; used in jasmin, etc.
6623	Geranyl Butyrate	Fruity, rosey; used in rose, jasmin, etc.
7623	Benzyl Butyrate	Heavy, fruity; modifier for jasmin
6723	Hexyl Cinnamic Aldehyde	Fruity, jasmin-like; used in the same way as amyl cinnamic aldehyde
7723	Amyl Crotonyl Acetate	Fruity, jasmin-like; used in jasmin compounds

TABLE 18.2 — (Continued)

Odor Number	Material	Application
7823	Allyl Caproate	Fruity, pineapple-like; used mainly in flavors
4333	Phenylethyl Dimethyl Carbinol (Centifol)	Floral, citrusy; used in jasmin compounds
5333	Oil Grapefruit	Floral, citrusy; used mainly in flavors Blends with verbena, lemon, gardenia, chypre, etc.
6333	Oil Orange, Sweet California	Floral, citrusy; used mainly in flavors, and in perfumes for its orangey notes
7333	Oil Lime, distilled	Floral, citrusy; used mainly in flavors. In perfumes for its citrus character
4433	Oil Curacao Peel	A variety of orange; used mainly in flavors
6433	Oil Nergamot	Floral, orange; used in almost all floral types and in many oriental blends
5533	Oil Lemon, Messina	Italian lemon oil, used mainly in flavors. A smoother odor than the California lemon
2633	Hexyl Caproate	Sharp citrus, but weak and not very useful
3633	Hexyl Propionate	Sharp citrus, but weak and not very useful
5633	Hexyl Acetate	Sharp citrus, but not very useful
6633	Tolyl Acetate	Sharp, fruity; used in jasmin, lilac and tuberose
7633	m-Tolyl Carbinyl Acetate	About like tolyl acetate in odor and use
8633	Oil verbena	Heavy citrus, very fragrant and powerful; used in many floral bouquets
7733	Citral	Sharp, lemony; used in most citrus blends
7833	Amyl Cinnamic Aldehyde	Powerful, heavy, sharp; used in jasmin, lilac, etc.
4343	Phenylethyl Dimethyl Carbinyl Acetate	Floral, rosey; used in rose, etc.
6343	Linalyl Acetate	Fruity, floral; woody; used in jasmin, oriental, gardenia, etc.
7343	Terpinyl Propionate	Woody, fruity, floral; used in lavender compounds
4443	Oil Bitter Orange	Slightly more woody than the sweet orange; used in floral and oriental types, but mainly in flavors
5443	Oil Mandarin	Woody, citrus, tangerine-like; used mainly in flavors; in perfumery for its citrus notes
6443	Citronellyl Phenylacetate	Floral, citrus, woody; used in jasmin, rose, etc.
7443	Geraniol Palmarosa	Rosey, citrus, woody; used in floral bouquets
8443	Nerol	Sweet, rose, neroli-like; used in orange blossom, rose, etc.
5543	Rhodinyl Acetate	Red rose type; used in all rose bouquets
6543	Linalyl Benzoate	Heavy, resembling broom and tuberose; used in these and in oriental types
7543	Cyclogeraniol	Heavy, rosey; used in floral compounds
6643	Oil Lemon, California	Heavy, citrus; used in flavors and in some perfumes
7643	Jasmin absolute	Sharp, citrus, floral, woody; used in almost all types of perfumes
7743	Rhodinyl Formate	Powerful red rose type; used in all types of perfumes, especially carnation
7253	α-Ionone	Woody, orris, violet-like; used in all violet perfumes
5353	Terpinyl Acetate	Bergamot and lavender-like; used in lavender and cologne types
7353	β-Ionone	Slightly more fruity than the α-form; used in the same way
5653	Dimethyl Benzyl Carbinyl Acetate	Woody, hyacinth-like; used in hyacinth and lilac types

TABLE 18.2 — (Continued)

Odor Number	Material	Application
6263	Isocitronellyl Acetate	Fruity, woody; somewhat like ionone
7563	Oil Clove	Spicy, fruity, woody; used in flavors; and in oriental and spicy perfumes
7473	Oil Nutmeg	Spicy, fruity, woody; used in flavors and in oriental perfumes
	Caprylic IV	
5114	Coumarin	Resembles new mown hay; used in this and in lavender, fougere, chypre, etc., as a sweetener and intensifier; also used in flavors
6114	3-Methyl Coumarin	Slightly heavier odor than coumarin; used about the same way
6214	Methyl Naphthyl Ketone	Musty, orangey; used as a fixative for orange flower types
7214	Furanacrolein	Musty, orangey; used in rose and orange flower types
4314	Phenylethyl Salicylate	Faint, but lasting rose-hyacinth type; used in floral compounds as a fixative
6314	Benzyl Isoeugenol	Faint, carnation-like; used as a fixative, especially for violet compounds
3124	Phenyl Benzoate	Mild, musky; used as a fixative
7124	Musk Ambrette	Musky, aromatic; the most powerful of the musks, used as an intensifier and fixative in many fine perfumes
8124	Heliotropin	Heliotrope-like; used in lilac, carnation, sweet pea, etc., for lasting sweetness; a fixative
2224	Cyclohexyl Cinnamate	Mild, slightly balsamic; used somewhat in floral and oriental types
5224	Phenylethyl Alcohol	Honey-rose; used in rose, neroli, orange-blossom, jasmin, etc.
6224	Cinnamic Acid	Heavy, balsamic, vanilla-like; used as a fixative for oriental types
7224	Benzoin Siam, resin	Heavy, sweet, somewhat balsamic and vanilla-like; used as a fixative for oriental types
8224	Oil Sweet Birch	Slightly more musty and woody than methyl salicylate; used in about the same way
4324	Santalyl Penylacetate	Honey, floral; used as a fixative in floral and oriental types
5324	Hydroxycitronellyl	Floral, slightly musty; used in muguet, lilac, hyacinth, etc.
6324	Anethol	Heavy, musty, fruity, floral; used mainly in flavors
7324	Acetanisol	Sweet, musty, floral; used in fougere, trefle, mimosa, etc.
8324	Cinnamyl Cinnamate	Sweet, balsamic; used in heavy and oriental perfumes
4424	Geranyl Benzoate	Sweet, soft-rose; useful in rose compounds
6424	Cyclamen Aldehyde	Floral; used in muguet and cyclamen compounds
7424	Oil Lavender, 38–40%	A slightly low ester content not quite as powerful a lavender
8424	Oil Lavender, 50–52%	Woody, floral, minty; used in many floral compounds
6524	Acetyl Isoeugenol	Spicey, sharp; used in new mown hay, carnation, and other spicey compounds

TABLE 18.2 — (Continued)

Odor Number	Material	Application
3524	Oil Cananga	Similar to Ylang Ylang, but much cruder, used in soaps, etc.
6624	Linalyl Cinnamate	Spicey, heavy, lily, jasmin-like; used in jasmin, tuberose and rose compounds
7624	Isoeugenol	Spicey, sharp, heavy; used in carnation and oriental compounds
8624	Cyclohexyl Butyrate	Heavy, jasmin-like, spicey; used in jasmin, rose, oriental, etc.
5234	Isobutyl Cinnamate	Soft amber type; used in modern and oriental bouquets
6234	Balsam, Peru	Resin Honeysweet, balsamic, used in floral and oriental types
7234	Bois de Rose, Brazilian, South	Reminiscent of rose, orange and mignonette; used in floral perfumes
6334	Isobornyl Propionate	Woody, piney; used in lavender and woody compounds
7234	Balsam Tolu, resin	Woody, balsamic, piney; used as a fixative for floral compounds, especially lilac
8334	Tuberyl Acetate	Heavy, resembles tuberose; used in heavy, floral and oriental types
6434	Oil Cabreuva	Woody, nutty, with a slight violet character
7434	Aldehyde C-18	Coconut-like; used in flavors; and gardenia and tuberose perfumes
8434	Tuberic Alcohol	Heavy, floral, nutty; used in tuberose compounds
6534	Linalyl Propionate	Fruity, floral, woody; used in lavender, rose and lilac
5634	Geranyl Phenylacetate	Sharp, fruity, rosey, woody; used in floral bouquets
6634	Terpinyl Butyrate	Sharp, fruity, rosey, woody; used in floral bouquets
7634	Linalyl Butyrate	Sharp, fruity, floral, woody, rosey; used in floral bouquets
5734	Isopulegyl Acetate	Sweet, sharp, slightly minty; used in lavender compounds
6734	Geranyl Formate	Sharp, rose leaf type; used in rose and orange blossom compounds
7734	Amyl Cinnamic Aldehyde Dimethyl Acetal	Jasmin-like; a smoother note than amyl cinnamic aldehyde for jasmin compounds
3244	Cedrene	Mild cedarwood; used in woody compounds
4244	Cedrol	Mild cedarwood; used in woody compounds
5244	Bornyl Acetate Isobornyl Acetate	Pine needle type; used in woody compounds, disinfectants, and flavors
7244	Oil Serpolet	Smooth, piney; used in woody or piney compounds
4344	Santalol	Sweet, mild, sandalwood odor, used in rose, violet, and oriental compounds
6344	Oil Copaiba	Musty, woody; used in heavy and oriental compounds
5444	Oil Cedarwood	Mild, woody, spicey; used as a fixative for woody, heavy compounds
6444	Oil Sandalwood	Sweet, musty, woody; used in all heavy types, especially in oriental compounds
8444	Rose Otto, Kazanlik	True rose otto; used for finishing touches in many of the better floral and oriental compounds
8544	Rose absolute	Woody, rose; used in many fine perfumes for finishing touches
8644	Oil Cinnamon, Ceylon	Heavy, spicey, very smooth; used mainly in flavors; and for the better spicey perfumes
5254	Vetiveryl Acetate	Resembles vetiver but is milder, pronounced woody character; used in modern bouquets

TABLE 18.2 — (Continued)

Odor Number	Material	Application
8254	Oil Amyris Balsamifera	Source of West Indian Sandalwood; stronger, cruder; used about the same way
6354	Oil Pine Needle	Musty, woody, piney; used in woody, piney types and especially in disinfectants
7454	Oak Moss, resin	Heavy, woody, musty; used as a fixative in many modern and oriental types
7654	Cinnamic Aldehyde	The main constituent of oil cassia and cinnamon, somewhat cruder; used in flavors and perfumes where smoothness is not required
8654	Oil Cassia	A crude cinnamon; used in flavors and perfumes as is cinnamic aldehyde
8754	Oil Camomile	Sharp, citrus odor with smooth, woody tones; used mainly as a flavor but imparts interesting effects to the heavier type perfumes
5264	Myrrh, resin	Heavy, balsamic, somewhat like incense; used as a fixative in modern, French, and oriental types, also as an incense
5464	Olibanum, resin	Soft, incense-like; used in oriental and French types, as a fixative; and in incense
6464	Labdanum, resin	Balsamic, heavy, woody, resinous; used in many fancy perfumes of the heavier types, as a fixative
5564	Oil Vetiver, Haiti	A somewhat less powerful vetiver than the Bourbon variety
7564	Oil Vetiver, Bourbon	Heavy, resinous, woody; used as a fixative in many perfumes, especially modern and oriental types
4174	Oil Myrrh	Resinous, aromatic, balsamic, incense-like; used in heavy, oriental and modern types, as a fixative and in incense
	Caprylic V	
7115	Musk Xylene	The cheapest of the musks. It has a fatty note in addition to the musky character and is used mainly in soap
3215	Alcohol C-12	Fatty, fruity; used in traces to create fresh effects in perfumes
2315	Acetate C-10	Fatty, fruity; used in geranium, rose, orange flower, etc., also used as a flavor
5315	Beta Naphthyl Ethyl Ether (Nerolin)	Similar to the methyl ether but finer
4415	Butyl Anthranilate	A mild, fruity anthranilate, slightly fatty, grape-like
5415	Cinnamyl Propionate	Reminiscent of grapes; used in artificial fruit essences
6515	Veratraldehyde	Fruity, slightly musty, with a fatty note; used in flavors
2225	Acetate C-8	Mild, fruity; used in peach and other fruity flavors, and in jasmin and orange blossom perfumes
3225	Isobutyl Anthranilate	Orange Blossom-like; used in orange blossom perfumes
6474	Oil Cascarilla	Aromatic, cinnamon, peppery; used in incense and tobacco types, and as a flavor
8674	Oil Labdanum	Heavy, powerful, resinous, woody; used for leather-like and ambergris effects
6384	Oil Cade	Empyreumatic; used mainly in medicines; in perfumery, to give Russian leather effect
4225	Octyl Butyrate	Heavy, fruity, orangey; used as a floral modifier

TABLE 18.2 — (Continued)

Odor Number	Material	Application
5225	Benzyl Salicylate	Musty, somewhat balsamic, orangey; used as a solvent for artificial musks, and in carnation, jasmin, lilac, lily, etc.
6225	Isobutyl Salicylate	Resembles amyl salicylate, heavy, musty, somewhat orchid-like; used in trefle, orchidee, etc.
4325	Diphenyl Methane	Mild orange-geranium like; often replaces geranium in soap perfumes
5325	Furfural Acetone	Musty, fatty, slightly fruity
7325	Phenylethyl Anthranilate	Winey, grape-like; used in gardenia and broom as a fixative
2425	Acetate C-9	Flower, fruity; used in orange-flower, rose, etc.
5425	Terpineol	Lilac-like, musty, heavy, somewhat floral, slightly fruity; used in lilac and other floral compounds and in flavors
6425	Terpinyl Anthranilate	Resembles muguet and orange blossoms; used in these and lilac and other floral compounds
7425	Oil Sassafras	Musty, herby, minty, somewhat fruity, floral; used mainly in flavors
4525	Lauryl Acetate	Musty, fruity
5525	Phenylethyl Formate	Odor of eglantine; used in white rose blends
7525	Oil Star Anise	A cruder anise; used in cheap flavors and in soap perfumes
6625	Isoamyl Benzyl	Gardenia-like; used mainly in soap perfumes
4725	Lauryl Formate	Mild, but sharp, estery, musty
5725	Octyl Crotonyl Acetate	Sharp, fruity, musty
6725	Isopulegol	Sharp, minty, slightly like tuberose
7725	Amyl Butyrate	Sharp, estery, fruity; used in fruity flavors such as pineapple, raspberry, apricot, etc.
4825	Ethyl Formate	Fruity essence; used in artificial fruity flavors
3235	Ethyl Anthranilate	Orange flower type; used in synthetic neroli and jasmin compounds
4235	Dimethyl Anthranilate	Orange flower type; used like the ethyl anthranilate
5235	Methyl Anthranilate	Orange flower-like; used in orange, neroli jasmin, etc., and in flavors such as grape
6235	Isosafrol	Fragrant, musty; used in soap perfumes, and oriental types
4335	Amyl Cinnamate	Musty, amber-like; used as a fixative especially in oriental compounds
5335	Hydrocinnamic Alcohol (Phenylpropyl Alcohol)	Musty, slightly resinous, reminiscent of hyacinth, mignonette, and styrax; used as a modifier in floral compounds, especially in lilac, reseda and hyacinth
6335	Styrax, resin	Naphthalin-like, resinous, balsamic, musty, used in hyacinth, jonquille, tuberose, etc.
7335	Geraniol	Musty, rosey; used in many floral compounds
4435	Benzyl Cinnamate	Heavy, slightly musty; used in oriental compounds as a fixative
5435	Amyl Benzoate	Musty, amber-like; used as a fixative in oriental types
6435	Citronellol	Isomeric with rhodinol; sweet, rosey; used in rose, lily, etc.
7435	Rhodinol	Principle constituent of geranium oil; used in red rose and floral bouquets
5535	Linalyl Anthranilate	Orange blossom type; used in neroli, orange blossom and jasmin
7535	Linalool	Isomeric with geraniol; odor very much the same; used in lily, sweet pea, rose, neroli, etc.

TABLE 18.2 — (Continued)

Odor Number	Material	Application
5635	Hydrocinnamic Acetate (Phenylpropyl Acetate)	Sharper odor than the alcohol, but otherwise much the same; used in lilac, muguet, etc.
5735	Amyl Phenylacetate	Sharp, musty, fruity; used in rose, jasmin, etc.
6735	Isobutyl Furyl Propionate	Sharp, musty, fruity
8735	Amyl Propionate	Fruity, musty; used mainly in flavor for adding fruity touches
6835	Ethyl Amyl Ketone	Pungent, heavy, fruity; used in some fancy bouquets, but mainly in flavors
5245	Methyl Phenyl Carbinol	Gardenia and hyacinth-like; used in floral compounds
6245	Isobutyl Benzoate	Odor of eglantine; used in orange blossom, carnation, sweet pea, etc.
7245	Ethyl Salicylate	Similar to methyl salicylate but cruder; used in synthetic cassie
5345	Amyl Salicylate (Orchidee)	Suggestive of orchids; used in synthetic orchid, trefle, carnation, new mown hay, etc.
6345	Phenylacetaldehyde Dimethyl Acetal	Green leaf type; used in lilac, lily, rose, etc.
7345	Tetrahydrogeraniol (Dimethyl Octanol)	Resembles geraniol but with more of a green note; used in many floral compounds
8345	Neroli, absolute	Fresh, orange blossom-like, much concentrated so that the leaf notes are strongly evident; used in floral compounds
4445	Linalyl Methyl Anthranilate	Orange blossom-like, with mandarin notes; used in neroli, orange blossom, etc.
6445	Oil Spearmint	Heavy, fatty, herby, slightly anisey; used mainly in flavors
7445	Oil Fennel	Heavy, fatty, anisey, with green notes; used mainly as a flavor
8445	Tetrahydro Linalool	Heavy, fatty, linalool-like; used in lily, sweet pea, rose, neroli, etc.
5545	Dimethyl Benzyl Carbinol	Green leaf type; used in lilac, lily, hyacinth, etc.
6545	Methyl Isoeugenol	Sharp, spicey, green odor; used in carnation, lilac, etc.
5745	Citronellyl Formate	Sharp, bergamot-cucumber-rosey; used in rose, muguet, etc.
5255	Methyl Eugenol	Fatty, musty, spicey, but not as powerful or as spicey as eugenol, with pimenta-like notes; used in carnation, lilac, etc.
6255	Aldehyde C-14	Fatty, musty odor, resembling peach; used in flavors and in some perfumes
4355	Hyacinth, absolute	Pungent, slightly woody, green odor; resembles hyacinth only on dilution
5355	Oil Atlas Cedarwood, Moroccan	Green, cedary, of the mimosa type; used somewhat in floral bouquets
5455	Oil Carrot Seed	Musty orris-like; used in violet
6455	Oil Myrtle	Musty, piney; used in oriental compounds
7455	Eugenol	Heavy, musty, spicey; used in carnation and other spicey perfumes
5555	Phenyl Isobutyl Ketone	Heavy, musty, spicey, with a trace of green-wood
6555	Acetophenone	Sharp, musty, spicey; used in hawthorn, mimosa, foin coupe, etc.
5655	Methyl Phenyl Carbinyl Acetate	Gardenia-like; used in floral compounds
7655	Phenyl Acetaldehyde (50 per cent in Benzyl Alcohol)	Sharp, woody, green, floral; used in hyacinth, sweet pea, lilac, lily, jonquille, etc.
7265	Oil Cubeb	Odor resembles pepper; used in much the same way

TABLE 18.2 — (Continued)

Odor Number	Material	Application
6365	Oil Petitgrain	Resembles neroli, but is cruder and more woody; used in floral, oriental and soap compounds
7365	Oil Abies, Siberica	Pine needle-like; used in soaps, and insecticides
5465	Oil Juniper	Piney, spicey; used mainly in flavors
6465	Oil Guaiac Wood	Spicey tea-rose-like; used as a fixative in rose and violet types
7465	Oil Spruce	Piney, spicey; used in pine compounds
6565	Oil Mountain Laurel	Pungent with slightly woody top notes; little used
7565	Oil Coriander	Spicey, woody, pungent; used mainly in flavors. In perfumes such as carnation, lily, etc.
7665	Resin, Elemi	Woody, spicey, somewhat musty; used as a fixative for heather and verbena types
6375	Oil Parsley Seed	Woody, herby, musty; used mainly as a flavor
7375	Oil Cardamom	Woody, musty, herby; used in oriental, chypre, etc., and especially as a flavor
8375	Oil Angelica Root	Musky, resinous, spicey, peppery; used in flavors and in some perfumes of the oriental and chypre types
5475	Oil Canada Snakeroot	Musty, resinous, resembling patchouli and ginger; used in flavors and in the heavier type perfumes
6475	Oil Black Pepper	Musty, woody, resinous, not very pungent; used somewhat as a flavor
7475	Oil Celery Seed	Musty, resinous, herby; used in sweet pea, tuberose, etc.
5575	Resin, Galbanum	Musky, musty, resinous; used as a fixative for mignonette and fern in soap perfumes
7575	Oil Ginger	Musty, resinous, spicey; used in oriental types. Not used much as a flavor since it lacks pungency
8575	Oil Olibanum	Musty, resinous, incense like; used as a fixative for heavy, oriental type and champaca
7675	Oil Patchouli	Heavy, musty, resinous, woody; used in oriental, woody, and most heavy amber types as a fixative
8775	Oil Cedarleaf (Oil of Thuja)	Heavy, musty, resinous, spicey, pungent; used in pine and cedar compounds
4285	Hexyl Salicylate	Phenolic, slightly thyme-like
5285	Carvacrol	Phenolic, minty, somewhat thyme-like; used as an antiseptic and a flavor
6385	Oil Clary Sage	Musty, resinous, burnt, with amber-like notes used in chypre and heavy perfumes
6485	Salicylic Aldehyde Caprylic VI	Heavy, phenolic; used somewhat in flavors
5226	Oil Peppermint, natural	Crude mint; used for flavors
6226	Eucalyptol (Cineol)	Odor resembles eucalyptus and spike lavender; used somewhat as a flavor
7226	Oil Eucalyptus	Mild, lavender, camphor-like, with a piney note; used mainly as a flavor
5326	Camphor	Heavy, somewhat pungent and cooling, not much used in perfumery
6326	Oil Lavender, 30–32%	A low ester lavender, crude; used in cheaper perfumes
6326	Oil Lavendin	Heavy, burnt odor resembling benzaldehyde; used as a modifier for bitter almond flavors
7326	Oil Spike Lavender	Camphoraceous, somewhat like rosemary, more powerful lavender than the one of low ester content; used in soap perfumes

ODOR

TABLE 18.2 — (Continued)

Odor Number	Material	Application
6286	Tolyl Aldehyde	Heavy, burnt odor resembling bensaldehyde; used as a modifier for bitter almond flavors
7286	Oil Bitter Almond	Heavy, burnt, pungent odor; used mainly as a flavor
4236	Benzophenone	Musty, rosey; a fixative for rose and geranium perfumes
6236	Beta Naphthol Methyl Ether (Yara Yara)	Heavy, musty, orange flower type; used in orange blossom, etc.
4336	Diphenyl Oxide	Musty, somewhat harsh; used in rose and geranium as a fixative.
6336	Phenylacetic Acid	Honey-like; somewhat horsey, reminiscent of civet; used in floral compounds and as a flavor
4436	α-Methyl β-Furyl-acrolein	Mild, musty, honey-like; used in floral compounds and as a flavor
6436	Methyl Acetophenone	Aromatic, slightly musty; used in new mown hay, mimosa, cassie, lilac, hawthorn, etc.
7436	Ethyl Cinnamate	Penetrating, slightly musty, spicey; used in floral and oriental types as a fixative
5536	Paramethoxy Acetophenone (Epenone)	Musty, sharp, heavy; used as a fixative for lilac, tuberose, mimosa, etc.
6536	Benzylidene Acetone	Heavy, pungent, tenacious; used in lilac, sweet pea, heliotrope, etc.
6636	Ethyl Phenylacetate	Sharp, sweet, suggestive of honey; used as a fixative in rose, orange blossom, etc.
7636	Methyl Phenylacetate	Sharp, sweet, suggestive of honey and rose; used as a modifier in rose, etc.
6246	Menthol	Strong cooling effect, used mainly for flavoring in the same way as peppermint
4346	Borneol	Camphoraceous, piney; used in pine compounds
5346	Isoborneol	Camphoraceous, piney; used in pine compounds
7346	Methyl Cinnamate	Heavy, musty, spicey, slightly fruity; used as a floral fixative
8346	Oil Cajeput	Odor somewhat like Eucalyptus; used slightly as a flavor
6446	Cyclonol (1-Methyl 3-Dimethyl) Cyclohexanol-5	A menthol substitute, but cruder
7446	Oil Rosemary	Minty, herby, somewhat camphoraceous; used as a flavor and in perfumes such as toilet waters
8546	Oil Geranium	Slightly musty, heavy, floral; used in geranium, rose, etc.
7256	Oil Pennyroyal, Imported	Somewhat harsh and penetrating, mentholic, herby; used with origanum, thyme, etc.
6356	Ethyl Benzoate	Harsh, musty; used mostly in soaps
7356	Methyl Benzoate	Harsh, musty; used mostly in soaps
7456	Cedrenol	Harsh, musty, woody; one of the main constituents of cedarwood oil
8556	Oil Bay	Heavy, musty, citrus, somewhat metholic; used in bay rum compounds and somewhat in carnation
7656	Oil Lemongrass	Sharp, citral notes with a greasy heavy character; used in soaps
6966	Oil Mint, Timidja	A crude, woody mint
7266	Oil Marjoram, sweet	Minty, herby, woody; used mainly in flavors
6366	Oil Origanum	Minty, herby, woody; used mainly in flavors
7366	Oil Estragon (Tarragon)	Minty, herby, somewhat anise-like; used mainly in flavors, but in perfumes of the chypre, fougere types

TABLE 18.2 — (Continued)

Odor Number	Material	Application
8366	Oil Pimenta Berries	Minty, musty, all-spice; used in carnation perfumes and in flavors
6466	Oil Thyme, white	A purified thyme with a smoother odor than the red; minty, herby, woody; used in flavors and in some perfumes
8466	Oil Basil, sweet	Minty, herby, woody; used in chypre, mignonette, jonquille, etc., also in flavors
6566	Oil Citronella, Ceylon	Lemon, herby, somewhat minty; used in fly spray compounds
6276	Oil Artemisia Morrocan (Absinthe)	Minty, woody, herby, somewhat like rosemary; used mainly in flavors
7576	Oil Savory	Minty, woody, herby; used mainly for flavors
8576	Oil Calamus	Spicy, herby, woody; used mainly for flavors
7676	Oil Sage, Dalmation	Heavy, woody, herby; used mainly for flavors
7386	Benzaldehyde	Heavy, burnt, pungent odor of bitter almonds; used in flavors for bitter almond effect and in a few perfumes of the violet type

Caprylic VII

4527	Delta Hydroxy Valeraldehyde	Sharp, musty, fatty, aldehyde odor
5527	Methyl 1-3 Pentanediol	Sharp, musty, fatty, aldehyde odor
6527	α-Ethyl β-urylacrolein	Sharp, musty, fatty
3337	Alcohol C-10	Musty, fatty; used in small amounts in floral bouquets, especially orange blossom
4337	Alcohol C-9	Musty, fatty; used in small amounts in floral bouquets, especially in orange blossom
3437	Alcohol C-11	Musty, fatty; a modifier for floral bouquets, especially rose
5437	Methyl Tuberate	Heavy, musty, fatty; used in rose, tuberose, etc.
5637	Octyl Methyl Ketone	Heavy, sharp, musty, fatty; used in traces in gardenia and sweet pea
4737	Hexyl Formate	Heavy, pungent, musty, fatty
5737	Methanyl n-Butyrate	Heavy, pungent, musty, fatty
6737	Amyl Acetate	Heavy, fatty, fruity; used mainly in flavors for fruity effects
7737	Isobutyl Acetate	Heavy, fatty, fruity; used mainly in flavors for fruity effects
5837	Linalyl Formate	Heavy, fruity, fatty; used for red rose effects
6837	Ethyl Propionate	Heavy, fruity, fatty; used in flavors almost exclusively
3447	Aldehyde C-9	Heavy, fatty, musty; used in rose and orange blossom bouquets
4447	Alcohol C-8	Heavy, fatty, musty; used in floral bouquets, especially rose, jasmin and modern French as a modifier
5447	Aldehyde C-10	Heavy, fatty, musty; occurs in neroli and sweet orange; used in floral bouquets
4547	Aldehyde C-11	Heavy, fruity, fatty, musty; used in rose bouquets
5547	Aldehyde C-12	Heavy, fruity, fatty, musty; used in floral bouquets
5647	Ethyl Oenanthate	Heavy, fruity, musty, wine-like; used mainly in flavors
6647	Aldehyde C-8	Heavy, fruity, musty; used in jasmin, rose and violet
6747	Ethyl Butyrate	Heavy, fruity, musty; used mainly in flavors
7747	Amyl Formate	Heavy, fruity, musty; used mainly in flavors
7847	Ethyl Acetate	Heavy, fruity, musty; used mainly in flavors
5557	Methyl Heptine Carbonate	Green violet-leaf odor; used very sparingly in floral bouquets, especially violet and sweet pea

Odor Number	Material	Application
	TABLE 18.2 — (Continued)	
6557	Bromstyrol	Strong, pungent, hyacinth-like; used in hyacinth, lilac, etc., especially in soaps
5757	Hydrotropic Aldehyde (Phenyl-propionaldehyde)	Harsh, hyacinth-like odor; used in hyacinth, lilac, rose, especially in soaps
6757	Hydrocinnamic Aldehyde (Phenylpropyl Aldehyde)	Powerful, harsh, hyacinth-like; used in floral, oriental, etc.
5857	Methyl Heptenone	Heavy, harsh, fruity, odor like isobutyl acetate; used in cheap soap perfumes
7367	Oil Manevoro	Much like costus, heavy, pungent, burnt
6467	Citronellal	Heavy, musty, fruity, much like citronella but cruder; used in cheap compounds
5567	Oil Niaouli (Gomenal)	Heavy, musty, burnt, fruity; used with eucalyptus in medical preparations
6567	Oil Thyme, red	Heavy, musty, herby, burnt, fruity; used mainly in flavors; in traces in some perfumes
5667	Oil Opoponax	Heavy, musty, balsamic, sharp; used in some modern types
7667	Oil Tansy	Heavy, musty, burnt; used mainly in flavors; occasionally in some perfumes
5377	Thymol	Heavy, burnt, minty; used mainly as an antiseptic
5477	Oil Galbanum	Balsamic, green, slightly flowery; used occasionally in fancy bouquets
6477	Oil Wormseed, American	Heavy, balsamic, green odor; used mainly in medicines
6577	Oil Rue	Heavy, pungent, burnt, herby; used in some floral bouquets such as sweet pea
7577	Fig Leaf Resin	Heavy, pungent, green herby
6667	Oil Wormwood	A somewhat cruder Artemisia (absinthe); used mainly as a flavor
7677	Oil Angelica Seed	Much like Angelica root, but cruder; used in flavors and in some modern perfumes
	Caprylic VIII	
3328	Tincture Ambergris	Mild, animal odor, the least animal-like of the four animal extracts. The concentrate material is much more burnt and resinous; used as a fixative for most of the finer perfumes
5428	Tincture Castoreum	Heavy, musty, animal-like; used as a fixative in many perfumes
6238	Indol	Powerful, somewhat like civet; used in traces in some floral bouquets as jasmin, orange flower, gardenia, etc.
4338	Tetraquinone	Odor resembles civet; used as a fixative in small amounts
5438	Tincture Tonquin Musk	Heavy, animal-like; used as a fixative in many perfumes
7638	Ethyl Valerianate	Heavy, musty, fruity, estery; used in flavors
5448	Tincture Civet	Heavy, musty, animal-like; used as a fixative in many perfumes
6458	Cuminic Aldehyde	Heavy, buggy, musty; used sparingly in some floral types as lilac, cassie, orris, lily, mimosa, etc.
8558	Oil Cumin	Same as cuminic aldehyde but more powerful; used only in traces
6368	Oil Caraway	Heavy, herby, burnt, somewhat buggy; used very sparingly in perfumes as cassie; also used in flavors
7468	Oil Costus	Heavy, herby, burnt; used in small amounts in violet types
5578	Octyl Phenol	Heavy, phenolic, animal-like, almost nauseous

TABLE 18.2 — (Continued)

Odor Number	Material	Application
5778	Diacetyl	Heavy, burnt, very powerful; in great dilution it has a buttery note for which it is used in flavors

Test Specimen

The preparation of the test sample for presentation to the panel is very important. No one standard method of doing this is sufficient, since the type of materials under test and the possible odors present allow a wide range of possibilities. A few of the methods which have been used are given below as a guide; others may be devised as necessary. Each laboratory should develop and record the detailed preparation procedure found satisfactory for specific products.

(1) Direct testing. Cut or tear representative samples of the packaging materials to a convenient size. It is often desirable to open up fresh surfaces at the instant of smelling (tearing paper board, scraping wax with a knife, breaking open a glue joint, etc.)
(2) Moistening. Water brings out some types of odors. By various mechanisms, and especially when the product may be subjected to moisture, try one of the following.
   (a) Breathing on sample. The tester exhales on the specimen, then smells it.
   (b) Dampening. Sprinkle the sample lightly with water, then smell. Time is sometimes required to develop an odor. Normally, the specimen is dampened, stored at 70–80°F or at 100°F, in an odorless glass jar for 24 hr, then tested.
   (c) Soaking. Saturate or submerge the specimen in water for a period of time and at a temperature judged suitable for each case. Test by smelling both the water and the specimen.
(3) Transfer to oily product.
   (a) Place the specimen in a covered glass dish adjacent to but not in contact with mineral oil for 4–24 hr at room temperature. Oil soluble materials such as printing-ink odor, kerosene, etc., can be detected by smelling the oil.
   (b) Prepare three sandwiches with a pat of unsalted butter between two pieces of the specimen and overwrap with parchment or cellophane. Hold for one, two and three days, respectively, at room temperature; refrigerate for 30 min; then taste and smell the butter. In preparing the reference samples, use odor-free parchment.

(c) Place the folded sample in a water tight tray and add an inch depth of coffee cream (18% butterfat). Cover with aluminum foil, hold at 45°F for 24 hr, then taste the cream. This procedure has been used widely for printing-ink odors and is very sensitive. As an alternative, strips of material may be placed in a dish or flask and covered with cream. Run a blank test on the cream in a glass bottle with a foil cap.
(d) Place the specimen in a covered glass dish adjacent to a piece of plain milk chocolate and hold at room temperature for one to two days. Taste the chocolate and compare with a control.
(e) Prepare a package by wrapping some commercial product, known to be sensitive to odor, with the specimen material. Hold for the desired time and temperature, then smell and taste the product in comparison with a control.

Procedure

All samples shall be identified only by code numbers, including the control, and the panel director shall give the testers only the minimum amount of information prior to the test. The panel director shall give each member a set of specimens and a report form. Each tester shall work independently.

Report

(1) Unknown odor. Each individual panel member shall report the type of odor (oil, solvent, musty, etc.), its chemical nature and (if possible) origin. Complete agreement by the panel on type of odor cannot be expected, but with experience the group report will serve as valuable data.
(2) Known odor. When the odor is familiar or is established, each individual panel member shall rate the coded samples numerically as follows:
  Essentially none (pleasant)—1
  Slight, but not objectionable (neutral)—2
  Moderate objectionable—3
  Strong (offensive)—4
Intermediate ratings, e.g., 2.5, may be made where increased sensitivity is possible.
Note: The decision between "slight" or "not objectionable" and "moderate" or "objectionable" will obviously depend upon the product tested, the functional requirements and the experience of the test panel. The objectionable odor level must be remembered, with

some help from the standard samples. The individual ratings shall be tabulated and if possible averaged. The exact procedure followed shall be recorded.

(3) Taste as odor index. The same reporting system shall be used on samples under Procedure 3b to 3e, inclusive.

NOTE: After all the test results are in, it is usually desirable to review individual reports in a group meeting, for educational purposes.

Reproducibility

If the average rating is over the previously agreed rejection point (generally two), make a check test, preferably using different panel members.

BIBLIOGRAPHY

AMERICAN SOCIETY FOR TESTING AND MATERIALS. 1968. Correlation of subjective—objective methods in the study of odors and taste. A.S.T.M. Spec. Tech. Publ. *440*.

CROCKER, E. C. 1946. Comprehensive methods for the classifications of odors. Proceedings of the Scientific Section of the Toilet Goods Assoc. *6* (Dec. 5). Arthur D. Little Inc., Cambridge, Mass.

CROCKER, E.C. 1954. Series in Food Technology Flavor. McGraw Hill Book Co., New York.

CROCKER, E. C., and DILLION, F. N. 1949. Odor directory. Am. Perfumer Essential Oil Rev. *53*, No. 5, 396–400.

DAWSON, ELSIE H., and HARRIS, BETSY L. 1951. Sensory methods for measuring differences in food quality. U.S. Dept. Agr. Infor. Bull. *34*.

McCORD, P., and WITHERIDGE, W. N. 1949. Odors Physiology and Control. McGraw Hill Book Co., New York.

MONCRIEFF, R. W. 1967. F.R.I.C. The Chemical Senses. Chemical Rubber Co., Cleveland, Ohio.

SOUTHWICK, C.A., JR. 1953. Measuring package odors. Mod. Packaging *26*, No. 10, 149–150, 213–214.

CHAPTER 19

Physical Evaluation of Color

Color is a sensation experienced by an individual when energy in the form of radiation within the visible spectrum falls upon the retina of the eye. Without light, color does not exist. Color is in the mind and is not, strictly speaking, a property of the object. Thus, color belongs to visual experience. It is not an inherent characteristic of an object; the object merely emits, transmits, or reflects light of a certain spectral distribution which is translated by the eye, nerve, and brain complex into a color response.

The color perceived when the eye views an illuminated object depends on: (1) the spectral composition of the light source (standard light source), (2) the chemical and physical characteristics of the object or the colorant, and (3) the spectral sensitivity characteristics of the viewer's eye (standard observer).

Color evaluation and control of fresh and processed foods is becoming more important every day. The food processing industry places greater emphasis upon the color factor. In the case of most fruits and vegetables, fresh or processed, the color of the product is one of the most important attributes of quality.

The processor finds that it is desirable to measure and control color on virtually every product that he produces. The reason is relatively obvious—production of a standard product. Variations in color not only detract from the product's quality, but marked variances may materially affect the product's flavor or consistency since many products rely on a substantial amount of the constituent to provide color and body.

Electronic sorting is now in use for sorting off-colored products from acceptable products such as, tomatoes, citrus, cherries, potato chips, etc. Units have been developed for use directly on the field harvesting equipment or for in-line in the factory. The units sort for a wide range of color differences and for defects or other material that is not acceptable. Sorting is accomplished at rates of 3600 units per second or 25 tons per hour per sorting unit. The electronic sorter can be adjusted to remove a wide range of off-colored or any given levels of defective units depending on the demand by the user.

There are a number of methods available to measure color. These range from simple subjective visual product comparisons to sophisticated objective instruments specifically designed to measure color and color differences. The specific method that should be used for any given product is dependent upon a number of factors. Before discussing any of these methods and procedures, the subject of lights and lighting should be considered.

LIGHTS AND LIGHTING

Investigations on color evaluation show that during daylight hours natural daylight varies between wide limits (12 to over 350 ft candles and in color temperature from 5,000–40,000° Kelvin color temperature, and from yellow through white to deep blue). These variations are shown in Fig. 19.1. By examining the data in Fig. 19.1 it can readily be seen how very unstable daylight is as a source of light for color grading.

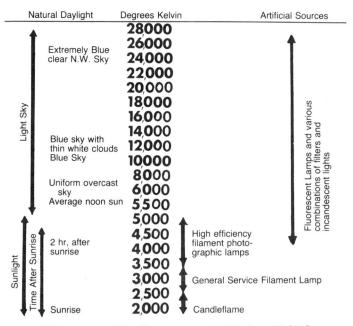

FIGURE 19.1 — Color Temperature Of Various Light Sources

Many food processing plants operate longer than daylight hours and normal daylight varies tremendously during different times of the day, with varying cloud conditions, with geographic locations and at different times of the year. For this reason, color evaluation, especially subjective evaluations, needs to be performed in a standardized environment.

MacBeth Examolites installed in color laboratories, over grading tables or in specific areas where color is to be evaluated, are one method to realize standardized illumination. These artificial lights are similar to those used in the cotton, tobacco, grain, printing, and other industries where color control and specifications, too, are factors in determining the product's value or worth.

Such artificial lighting produces a very close approximation to North sky natural daylight for a moderately overcast day. The Examolites emit 75–100 foot candles of light intensity and a color temperature of 7,500° Kelvin. In the color evaluation laboratory at the Ohio State University, the lighting produced is uniform throughout the room and illumination is closer to daylight than that obtained with ordinary fluorescent "daylight" sources.

The energy distribution curve for daylight fluorescent tubes shows the presence of violet, blue, green, and yellow spectrum lines as well as low energy in red and blue. (It is the deficiency in red and blue light or the inability to produce an efficient phosphor for these parts of the color spectrum.) Further, the presence of specific spectrum lines tends to minimize or accentuate color difference, depending on the color in question, and normally rules out fluorescent daylight as a source of light for accurate color matching.

The illumination for the Examolite is produced by two sources of light: (1) four 40 watt (6,500°K) "daylight" fluorescent lamps; and (2) four silver bowl reflectors on 25 watt incandescent lamps. The color temperature produced is 7,500°K, the same as that of the accurate artificial (filtered) daylight lamps and/or ideal natural daylight. The Examolite light energy distribution curve has been improved over straight fluorescence by the addition of red and blue light energy to the daylight fluorescent tubes, in addition to the elimination of some of the effects of the spectrum lines.

The MacBeth Examolite provides its own reflector of a permanent baked white finish, similar to a white refrigerator finish. This eliminates variation due to reflection from a ceiling, as used in open reflector type lighting, where different colors alter the illumination as the ceiling color varies. In addition, the unit is closed so that dust, dirt, and lint cannot collect on the tubes and bulbs and further alter the color.

The color of the walls is, also, a very important factor in the lighting of a color grading room. The walls of the room should be painted a light neutral gray (reflectance of approximately 70%) of a flat finish (mat 8). This has been found to be very satisfactory in reducing glare and it provides a comfortable background free from reflecting color influence for food grading or product evaluation.

For a more accurate duplication of natural daylight which is required for color standards and critical color evaluation, the MacBeth (Model BBX36) filtered incandescent lamp installation may be required or the use of the MacBeth Executive light.

Thus, a prerequisite for an accurate and proper scheme of color evaluation and measurement is, among other things, correct illumination and adequate lighting.

SUBJECTIVE COLOR EVALUATION

Subjective color evaluation is any process wherein the human element is involved as an eventual deciding factor in the determination of color quality for product acceptability.

Although the eye is extemely sensitive to small differences in color, its memory of colors is not very reliable in the absence of control

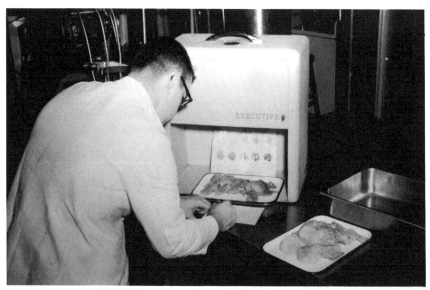

FIGURE 19.2. — MacBeth Executive Light

standards. All human biases, personality differences, and associated frailties subject the color judgment to various interpretations. Equally important in understanding the limitations of visual color measurements is the near impossibility of describing a color in subjective terms. For example, how red is tomato red?

Although the human eye has some definite weaknesses when used to evaluate color, subjective determinations can provide meaningful results. The eye is capable of distinguishing small color differences when it is able to compare to the sample or product color standards. Thus, a standard color chart, plate, or model for comparison is needed if meaningful subjective color evaluation is to be made.

Advantages of subjective methods include speed and the relative ease with which the measurement can be made. Disadvantages include human error, variance between individuals, and situations with less than standardized conditions. There are, however, reasons for devising methods to subjectively measure color along with the means to significantly reduce relative error. Before the technologist attempts to evaluate color subjectively, he should be evaluated for color assurance to determine that he is not color blind, using the AO H-R-R Pseudoisochromatic Plates test, and evaluated for color acuity, using the Inter-Society Color Council Color Aptitude test. On the latter test, he should score at least in the upper "fair" range, preferably 65 or over.

Color Comparison or Matching

Perhaps the simplest way to evaluate color of processed products is to compare the product with a recently processed product or standard product. The recently processed product should have been previously evaluated and considered to possess preferable color characteristics. A case or two of this standard product should be stored at low temperature for future reference. It is important that any color evaluation be made with the aid of a reference standard. The human eye can perceive and distinguish small color differences, but it is a known fact that difficulties occur when an evaluation is made without some type of reference or "bench mark" for comparison; i.e., the human eye needs help in remembering what "standard" color really is.

The following are a few systems available for color matching and comparison of food and other items.

(1) **Ridgway Charts.**—The Ridgway Charts were developed in 1886 and enlarged in 1912 by Robert Ridgway of the U.S. Biological Survey. The charts contain 1,113 colors consisting of 36 hues which are reduced by regular proportions of gray producing systematic groupings.

(2) **Maerz and Paul Color Dictionary.**—The dictionary was published in 1930 and contains 7,056 colors on 56 charts. The charts are divided into 7 main groups with the order of hue presentation following the spectrum.

USDA Color Standards

Color is evaluated in certain food products with the aid of a USDA permanent plastic color standard. Examples include tomatoes, apple butter, peaches, sauerkraut, and lima beans. Color quality is evaluated on these products according to an individual's perception of surface color in contrast to the preference standard. The plastic standards are made to simulate prescribed color grades for Grades A, B, C, etc., according to the product being evaluated. A product is considered to be equal to a specific color grade if it matches or exceeds the color shown on the plastic standard, or in certain instances, if a given percentage of the product matches the standard.

The procedure for differentiating color is a relatively accurate method, as usually only an "equal to" or "less than" decision concerning color must be made. The method is also suitable for those products that are somewhat less than homogeneous.

MacBeth-Munsell Disk Colorimeter

The MacBeth-Munsell Disk Colorimeter was developed especially for the color grading of tomato products such as tomato juice, tomato pulp (puree), tomato catsup, tomato paste, chili sauce, and tomato sauce. As such, the unit has been adopted by the U.S. Department of Agriculture for the grading of tomato products. The colorimeter could be considered to be a combination of subjective and objective evaluation methods since the unit along with an individual's perception of color is involved in determining the color grade. For close control or evaluation of color of any sample, disks can be established for specific score points of different grades. Table 19.1 gives Munsell color percentages for scoring tomato juice.

The Munsell system is based on the use of the three visual color attributes: hue, value, and chroma. The Munsell charts have 40 hues containing 982 colors. This system has the advantage of being devised on a psychological system of visually equal steps in order that the three color attributes become parameters by which color may be analyzed and described accurately under standard conditions.

TABLE 19.1 — Munsell Color Percentages for Scoring Tomato Juice

Score Points	Red (5R 2.6/13)	Yellow (2% YR 5/12)	Black[1]	Gray[1]
28[2]	73	16.34	5.33	5.33
26[3]	65	21.0	7.0	7.0
24.5[2]	59	24.5	8.25	8.25
23[3]	53	28.0	9.5	9.5

[1]Any combination of neutral black and gray is allowed. For "cold break" extracted and high temperature, short-time sterilized juice a greater percentage of gray may be necessary to match the sample; and for "hot break" extracted and conventional processed juice a greater percentage of black may be necessary to match the sample.
[2]Percentages determined by interpolation and research.
[3]Minimum standards established by the U.S. Department of Agriculture.

The chromatic colors of the Munsell System of Color Notation are divided into five principal hues, i.e., red, yellow, green, blue, and purple. Subdividing the primary hues yields five intermediate hues of yellow-red, green-yellow, blue-green, purple-blue, and red-purple. Capitalized letters such as "R" for red, or "YR" for yellow-red are used as symbols for the hue names. Similarly, further subdivision yields hue combinations such as red-yellow-red, which is symbolized by "RYR." For finer divisions, each hue is divided into ten equal steps such as 1R to 10R, or 6YR to 10YR.

The value notation indicates the degree of lightness or darkness of a color. In the Munsell system, a pure black is symbolized as 0/ and a pure white as 10/.

The chroma notations of a color indicate the degree of departure of a particular hue from a neutral gray of the same value. The scales of chroma extend from /0 for a neutral gray to /12,/14,/16 or farther depending on the saturation of a particular color.

Whenever a finer division is needed for any of the three attributes, decimals are used to indicate a particular color. Thus, the complete Munsell notation for any chromatic color is written hue value/chroma, such as 2.5R 4.5/2.4.

The definite evaluation of the color of food products according to the Munsell System consists of two parts: The first is the percentage of the different specific colors which when blended together give a composite color which exactly matches the sample. The percentage notations of a particular color must always add up to 100%. The second essential part of each color notation is the description of each of the color cards—hue (specific color), value (lightness) and chroma (saturation). Each class of agricultural products requires a particular set of color cards. For example, the color of canned tomatoes and tomato products may be matched by varying the proportions of particular colors, namely, definite shades of red, yellow, and combination of neutral black and neutral gray. The formula of the red color used in matching tomato products is 5R2.6/13. The designation 5R is a pure red free from purple or yellow. The second notation, "2.6," indicates the value or intensity of the black constituent of the color. A value of 2.6 indicates a considerable proportion of the black constituent. The final designation shows the chroma, which expresses the strength or degree of departure of a particular hue from a neutral gray of the same value. A chroma of 13 is very saturated.

The required color cards, which can be obtained only from the Munsell Color Co., Newburgh, New York, are cut in the form of Maxwell Disks, which are uniform circles, 3¾ in. in diameter. Each of the cards has a small hole (¼ in.) at the center and a single radial slit from the center to the edge so that the disks may be placed one in the other and so slipped together such that any desired proportion of one or more of the color disks may be exposed as a segment of the single circle that is visible. When these disks are held together with a suitable binding post at the center and spun at a speed great enough to eliminate flicker, the color seen by the eye is the sum of the different segments exposed. The disks should be slipped together in such a direction that the spinning will cause them to lie flat rather than to fly apart. The amount of each color card exposed is changed until the combined effect exactly matches the color of the sample. The amount of each color exposed is then measured by the percent of the circumference occupied by each segment.

The requirement of color of Grade A tomato pulp according to USDA is that it shall be "equal or better than that produced by spinning a combination of the following Munsell color disks: R65% (5R 2.6/13) glossy finish; Y21 (2.5YR 5/12) finish, glossy; black and gray 14% total or any combination of the two (Black N1 glossy finish and/or Gray N4 mat finish). For minimum Grade C, 53% Red, 28% Yellow, and 19% of the black and gray in any combination according to USDA and F&DA."

Similar disks could be established for other tomato products enabling the grader to score samples within a specific range if desired, in addition to comparing to the minimum USDA standards. The Magnuson Engineers of Reno, Nevada, have developed plastic disks for use with this colorimeter. These offer the advantage of ease of cleanliness and since the color disks are combined into one disk, there is greater accuracy when using.

MacBeth-Munsell Colorimeter

The MacBeth-Munsell Colorimeter unit consists of an arrangement of two spinning disks mounted directly beneath a color-correct light source with controlled viewing conditions. The light source is composed of two R40 300-watt reflector flood lamps used with two 7¼ in. MacBeth daylight filters. The light source and filter combination is mounted on a deck in the upper portion of the unit and is enclosed with a special nonselective diffusing glass. The light source-filter combination produces the closest duplication of North sky daylight (7500°) that is commercially available. The filters used are carefully graded and selected and a code is etched on the rim of the filter for identification. All Type No. 1 disk colorimeters are supplied with filters which produce a color temperature of 7500° unless otherwise specified.

The center section of the unit, enclosed by two hinged doors, constitutes the viewing mask support (horizontal bar) attached to a positioning guide for a constant angle of viewing. The interior of the viewing area is painted a light neutral (Munsell mat 8) in order to standardize the surrounding conditions when color judgments are made. The sample holder consists of two parts which include a tray and a 3½ in. diameter x ⅞ in. deep sample cup. The cup fits against a guide on the tray which, in turn, slides on to the viewing surface by means of ways or tracks. The sample holder is constructed in this manner so that any samples which spill over from the cup will be caught by the tray which is easily removable for cleaning. On both sides of the sample holder are mounted the spinning disk motors. The color disks are held in place on the spindle by a knob which is a friction fit for the spindle.

The conversion of Munsell disk mixtures to ICI tristimulus values facilitates the comparison of Munsell disk mixtures with other methods of evaluating color. The ICI color notation is a standard notation and data from a majority of the methods employed in evaluating color is convertible to the ICI system of color notation.

A computation form for converting disk mixtures to ICI tristimulus values and then to trilinear coordinates x, y, and z is shown in Table 19.2.

The C.I.E. Color System

The Commission Internationale de l'Eclavinge system is based on the effect any color may have on an agreed "standard observer." Three saturated spectral colors—red, green, and blue—are considered to be the vertices of an equilateral triangle. The assumption is made that at

244 TOTAL QUALITY ASSURANCE

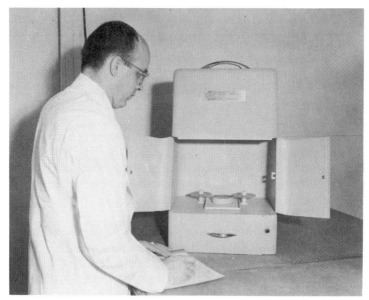

FIGURE 19.3 — Munsell Disk Colorimeter

the corners of the triangle the amount of the particular color is 100%. As light progresses farther away from the corner, the light becomes uniformly weaker until it reaches any point on the opposite side where its intensity is zero.

In 1931 at Cambridge, England, the International Commission on Illumination (ICI) met and adapted certain conventions for use in colorimetry which have been used throughout the world. As a result, nearly all color data can be translated to ICI standards which provide a common system for color comparisons. This system of color notation is based on observations made by an imaginary standard observer. The establishment and need for such an observer has been reviewed. The colors perceptible to the standard observer are expressed in equal progressions of coordinate systems. The data are expressed as the absolute (X,Y,Z) and fractional (x,y,z) amounts of three lights—red, blue, and green—necessary for a standard observer to match a given sample under a given set of conditions.

OBJECTIVE COLOR EVALUATION

Any method which, in the process of measuring or evaluating color, eliminates the human element as a decisive factor in determining color

PHYSICAL EVALUATION OF COLOR

TABLE 19.2 — Mathematical Relationships Between Color Scales for Illuminant C 1931 2" Standard Observer

TO CONVERT FROM ↓ / TO →	L, a, b	R_d, a_{Rd}, b_{Rd}	X_G, Y_G, Z_G	CIE X, Y, Z
L, a, b		$R_d = 0.01L^2$ $a_{Rd} = a\dfrac{f_Y L}{100}$ $b_{Rd} = b\dfrac{f_Y L}{100}$ Where $f_Y = 0.51\left(\dfrac{21 + 0.2Y}{1 + 0.2Y}\right)$	$Y = 0.01L^2$ $X_G = 0.01L^2 + \left(\dfrac{aL}{175}\right)$ $Z_G = 0.01L^2 - \left(\dfrac{bL}{70}\right)$	$Y = 0.01L^2$ $X = \left[0.01L^2 + \left(\dfrac{aL}{175}\right)\right]0.9804$ $Z = \left[0.01L^2 - \left(\dfrac{bL}{70}\right)\right]1.181$
R_d, a_{Rd}, b_{Rd}	$L = 10\sqrt{R_d}$ $a = \dfrac{a_{Rd}}{f_Y}\dfrac{100}{L}$ $b = \dfrac{b_{Rd}}{f_Y}\dfrac{100}{L}$ Where $f_Y = 0.51\left(\dfrac{21 + 0.2Y}{1 + 0.2Y}\right)$		$Y_G = R_d$ $X_G = R_d + \left(\dfrac{a_{Rd}}{1.75 f_Y}\right)$ $Z_G = R_d - \left(\dfrac{b_{Rd}}{0.70 f_Y}\right)$ Where $f_Y = 0.51\left(\dfrac{21 + 0.2Y}{1 + 0.2Y}\right)$	$Y = R_d$ $X = \left[R_d + \left(\dfrac{a_{Rd}}{1.75 f_Y}\right)\right]0.9804$ $Z = \left[R_d - \left(\dfrac{b_{Rd}}{0.70 f_Y}\right)\right]1.181$ Where $f_Y = 0.51\left(\dfrac{21 + 0.2Y}{1 + 0.2Y}\right)$
X_G, Y_G, Z_G	$L = 10\sqrt{Y}$ $a = \dfrac{17.5(X_G - Y)}{\sqrt{Y}}$ $b = \dfrac{7.0(Y - Z_G)}{\sqrt{Y}}$	$R_d = Y$ $a_{Rd} = 1.75 f_Y (X_G - Y)$ $b_{Rd} = 0.70 f_Y (Y - Z_G)$ Where $f_Y = 0.51\left(\dfrac{21 + 0.2Y}{1 + 0.2Y}\right)$		$Y = Y_G$ $X = 0.9804 X_G = \dfrac{X_G}{1.02}$ $Z = 1.181 Z_G = \dfrac{Z_G}{0.847}$
CIE X, Y, Z	$L = 10\sqrt{Y}$ $a = \dfrac{17.5(1.02X - Y)}{\sqrt{Y}}$ $b = \dfrac{7.0(Y - 0.847Z)}{\sqrt{Y}}$	$R_d = Y$ $a_{Rd} = 1.75 f_Y (1.02X - Y)$ $b_{Rd} = 0.70 f_Y (Y - 0.847Z)$ Where $f_Y = 0.51\left(\dfrac{21 + 0.2Y}{1 + 0.2Y}\right)$	$Y_G = Y$ $X_G = \dfrac{X}{.9804} = 1.02X$ $Z_G = \dfrac{Z}{1.1181} = 0.847Z$	Other Relationships: $x = \dfrac{X}{X + Y + Z}$ $y = \dfrac{Y}{X + Y + Z}$ Hunter Gardner Multipurpose Reflectometer: $G = Y$; $A = R = 1.25X - 0.25Z$ $B = Z$.

Source: Hunterlab Reflections and Transmissions, Hunter Assoc. Laboratory, Inc.

NOTES: The I.C.I. notation for each disk used is found in Table 7.2, pp. 368–373 and Table 7.1, pp. 377–380 of the July J. Opt. Soc. Am. The values are placed in columns 1, 2, 3 of the form. In column 4 is placed the percentage of each disk which is exposed. In columns 5, 6, 7 are placed the products of column 4 and columns 1, 2, 3, respectively, i.e., column 5 is column 4 times column 1, column 6 is column 4 times column 2, and the Total is the sum of values in columns 5, 6, 7. X is the sum of the values in column 5, Y the sum of values in column 6, and Z the sum of the values in column 7. The trilinear coordinates are found by the above formulae.

If desired the Munsell renotation for the disk mixture can be obtained. Convert Y to Value from Table 2, p. 406, July 1943, J. Opt. Soc. Am. Hue and Chroma are read from the large scale diagrams (Fig. 1–9, pp. 387–395, July 1943, J. Opt. Soc. Am.).

quality can be considered to be an objective measurement. Obvious advantages to objective measurement are the standardization of procedures and the reproduction of results. Disadvantages include varying degrees of error inherent in the instrument along with untimely equipment malfunctions. In many instances, however, the advantages heavily outweigh the disadvantages. Regardless, any mechanical device should serve to assist human evaluation rather than to completely replace it.

Measurement of tomato juice color by means of the Agtron F, a single wavelength, objective color instrument, has been suggested. A similar instrument, the Agtron E, was adapted in 1953 as the official grading instrument for tomato products in California. In 1957, the Agtron was accepted by the Ontario Department of Agriculture as the official grading instrument for tomato color in Ontario. Several models of the Agtron instruments are now available including the Agtron E5 which is used extensively to measure tomato color.

Agtron Model E4

The Agtron E4 is designed to measure the internal color of a tomato. The instrument is an abridged spectrophotometer which measures the spectral reflectance of tomatoes at two monochromatic wavelengths and provides a reading which is a dimensionless ratio of the two reflectances. The unit uses gas discharge tubes for illumination and incorporates selected glass filters to isolate individual spectrum lines. The spectrum lines use the 546 nanometer line of mercury and the 640 nanometer line of neon. These lines were selected because they are in spectral regions which are critical to tomato color. The ratio of reflectance at these two wavelengths correlates highly with visual evaluation of tomato color.

Key to Symbols:
G = AGTRON grade
X_G = Green reflectance (546 nm) $G = 276 \dfrac{(X_G - 0.7)}{(X_R - 0.7)}$
X_R = Red reflectance (640 nm)

Tomato Grading Formula
  Therefore:
    Tomato "A" $G = 276 \dfrac{(3.5 - 0.7)}{(29.1 - 0.7)} = 27.2$
    Tomato "B" $G = 276 \dfrac{(9.3 - 9.7)}{(36.3 - 0.7)} = 66.6$

In operating the Model E4 Agtron, the red and green zero levels are established by standardizing on a special black disk. The relative sensitivity of the two circuits is then standardized on a red plastic standard, similar to "tomato red." Tomato halves are placed in position and illuminated by a combination of the mercury vapor and neon gas discharge tubes. A red filter is used to isolate the 640 nanometer line of neon and the meter is adjusted to read 100. The filter is then changed to a green filter which isolates the 546 nanometer line of mercury, and the Agtron reading obtained is the color score assigned the product.

Agtron Model F

The Agtron Model F is also an abridged spectrophotometer in which a single filter is employed for isolating a selected monochromatic line of the light source. The Model F was originally developed for the USDA to measure the color of extracted pulp of raw tomatoes and establish a color grade based upon the USDA Strained Tomato Products Standards. The instrument can be obtained with any one of four light sources and four filters—red, green, blue, or yellow. A gas discharge tube is employed as the light source for illuminating the sample. For the red measurement, the 640 nanometer line of neon is used; for the green measurement the 546 nanometer line of mercury is used; for the blue measurement the 436 nanometer line of mercury is used; and, for the yellow measurement the 585 nanometer line of neon is used. The green illuminating light source and the green filter are used for the evaluation of liquid or pureed tomato products. Since tomato ripening is accompanied by a color change from green to red, the values obtained provide an indication of fruit maturity.

The Model F Agtron has a meter which is calibrated from 0 to 100. The amplifier circuit is designed to independently adjust the zero, or null point, and circuit sensitivity. Two different reference materials are used to standardize the instrument. One of these materials usually has a color slightly darker than the darkest sample to be measured and the other has a color which is slightly lighter than the lightest sample. Then, by means of separate controls, the instrument is standardized so that it reads zero on the dark standard and 100 on the light standard. In this way, the range of color between the two standards can be spread over the full 100 points on the meter scale. Thus, a manufacturer can establish close color tolerances.

The instrument is used primarily as a stationary, single sample reading instrument, i.e., the color measurement is made by placing a sample of the product in a standard sample cup. The cup is then placed in a reversed opening in the top of the instrument and its reflectance is measured. However, at the Ohio State University, the instrument has been utilized as a continuous color measuring instrument for tomato juice. As such, it is placed in the line and juice is continuously pumped through a cell. Under this method, color can be continuously measured and evaluated.

Data indicate that continuous color readings can be of significant value since they allow the manufacturer to know at all times the color of his product. With this knowledge, he can make any necessary adjustments of the raw product; sorting, trimming, blending, or other processing variables to keep the product within the desired color ranges.

The USDA, when testing the Agtron unit as a stationary single sample reading instrument on samples of raw tomato juice, reported a correlation coefficient of 0.92 with the Hunter Color and Color-Difference Meter (L,a,b). Similar results have been obtained at the Ohio State University on raw tomato juice with the Agtron Model F and the Hunter Color and Color-Difference Meter (L,a,b) giving a correlation coefficient of 0.94. Moreover, excellent relationships have been obtained between the Model E5 Agtron (official cut-surface instrument used in California for tomato grading) and the Model F Agtron (for liquid or puree grading).

Use for Other Products.—The Model F Agtron has been used on several other products with excellent results. Examples include toasted breakfast cereals, baby foods, spices, roasted ground coffee, starch, brewer's malt, citrus concentrates, and ground chocolate.

Agtron Model E5M

The Agtron Model E5M is a redesigned spectrophotometer capable of measuring tomato reflectance values on both cut and liquid surfaces. Thus, it is able to do the job of both the Agtron Models E and F.

The E5M relies on a system of improved optics and electronics. The unit consists of two mercury and one neon tube lights for illumination of the sample. The mercury tubes produce green light and the neon tube produces red light. Upon illumination of the sample, light is reflected and passes through a short wavepath filter which suppresses infrared light. At this point, light is divided into either green or red spectrums by means of a dichroic mirror. The mirror, an interference filter, has a high transmittance for red and a high reflectance for green. The split beams of color then pass through individual red and green narrow-pass filters (also interference filters) and reach photocell amplifier units. The three interference filters dramatically sharpen spectrophotometric values and precise readings are obtained for the ratio of green and red light intensities.

Unlike the Models E4 and F, standardizing disks and filter changes are not necessary. The unit incorporates simplicity and ease of operation. As such, the Agtron Model EM is being used officially in California at field inspection stations and other locations to determine the color grade for raw tomatoes.

A formula has been developed by Agtron, Reno, NV., for scoring tomato color using the readings of Agtron Model E5. The formula is:

$$Ye = 63.2036 - 0.4372(E5)$$

where: Ye = Tomato score
(E5) = Reading of Agtron Model E5

The Agtron Model E5M is similar to the Agtron Model E5F except that the E5M divides a green (546nm) signal by the red (640 nm) signal, whereas, the E5F divides a green (546nm) signal by an infrared (811nm) signal.

Recently, a new microprocessor-based, fully automatic, self calibrating colorimeter, Agtron Model E10, has been developed. This Model replaces the previous E5 Models and has a standard RS232 signal port for easy connection to a printer or a computer. The E10 has been designed to read and process any combination of light sources, optical filters, and arithemetic formulas. The E10 is able to duplicate the E5C for coffee, the E5F for potato chips and snack foods, and the E5M for tomato products. When used for snack foods and tomato color analysis the instrument ratios two signals. For snack foods the E10 divides a green (546 nm) signal by an infrared (811 nm) signal and for tomato products the E10 divides a green (546 nm) signal by the red (640 nm) signal.

Agtron M-400—M-30A—M 35 and 45

The Agtron Model M-400 is designed to measure the external color of food products. The instrument is an abridged spectrophotometer which measures the spectral reflectance at three monochromatic wavelengths and provides a reading at each wavelength.

The M-30 Wide Area Viewer is an optical viewing attachment designed to operate with the M-400 Agtron to measure the relative spectral characteristics of nonhomogeneous and partly colored products. The capability of the M-30 to provide accurate and repeatable measurements of such products is based on concentric diffuse illumination of a 30 sq. in. sample. The reflected light from the sample represents an average color character of the sample which is largely unaffected by particle size, particle geometry, irregularities, voids and shadows.

The unit uses light obtained from two concentric gaseous discharge tubes; one mercury vapor and one neon which radiates light at specific lines or wavelengths which have zero band width. The spectral lines are at 435, 546 and 640 nanometers. Interference filters are used to "isolate" these lines and a phototube, which receives the reflected light, produces an output signal proportional to the luminous intensity.

Use of monochromatic light also permits shades-of-gray calibration standards. The meter is affected only by intensity without regard to spectral wavelength. Standardization consists simply of selecting the desired gray calibration disks and using two vernier controls to set the meter zero and full scale span. Thus, standardization and sensitivity are set simultaneously without use of a colored reference and without complicated calibration procedures.

After making the two external connections between the M-30 and M-400, the black (M-00) disk and white (M-90) disk are used to establish meter zero and full scale. The ZERO control sets the zero reflectance level and the STANDARDIZE control sets the maximum reflectance level. The normal setup procedure would be to obtain the lightest and darkest acceptable products.

Read the reflectance levels in the red, green, and blue modes and record data. From this, determine which of the three spectral modes will be the most effective. With a given product, one, two, or all three modes may be needed for most effective control. There may be very little difference in reflectance between the two samples in one mode, with a larger difference in one or both of the other modes.

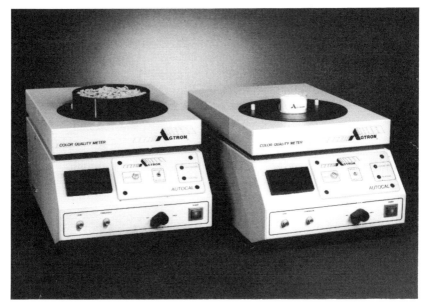

FIGURE 19.4 — Autocal Agtron M35 & M45

Assume the lightest and darkest samples read 36 and 52 on the red mode when standardized with the M-00 and M-97 disks. Select the M-33 and M-56 disks for resetting the zero and full scale meter points. This increases the sensitivity by a factor of five, and causes the two samples to produce nearly a full-scale meter deflection. It may be decided then that product samples reading above 80 are too light and samples reading below 20 are too dark.

Much more sensitive operation can be obtained depending upon the nature of the products produced, the type of information desired, and the requirements of the user. The sensitivity of the M-30 is sufficient in most cases to permit zeroing on the darkest sample and full-scale standardizing on the lightest sample of a given product.

Hunter Color and Color-Difference Meter (Gardner Model)

A number of instruments have been developed which are capable of translating the qualitative psychological aspects of color into a quantitative physical measurement using three numbers to define a color. Since the development of filters which approximate the I.C.I. tristimulus functions, simple objective, physical measurements are possible which predict the color response of a human observer.

One of these instruments is the Hunter Color and Color-Difference Meter. It is designed for rapid and precise measurement of color and small color differences by an objective tristimulus method of color measurement. It has been used extensively in the United States both for tomato research and for color control in the production of tomato products. The unit, designed by Gardner Laboratories, represents some of the earlier attempts to objectively measure color.

This instrument evaluates color by means of three photocells connected to a galvanometer which measures in numerical terms the hue, value, and chroma of a sample viewed in the reflected light of a tungsten lamp. One dial gives either Rd (45° luminous reflectance) or L (visual lightness), depending on the type of measuring circuit selected by the user. The other two scales of the instrument measure "a," which is redness when plus, gray when zero, and greenness when minus, and "b," which is yellowness when plus, gray when zero, and blueness when minus. The three potentiometer dials of the instrument give numbers which relate to the standard I.C.I. specifications of color by the following equation for the L scale:

$$L = 100\sqrt{Y}$$
$$a = 17.5 \, (1.02 \, X - Y)/\sqrt{Y}$$
$$b = 7.0 \, (Y - 0.847Z)/\sqrt{Y}$$

For the Rd scale two nomographs relating "a" and "b" to I.C.I. X,Y,Z are available on request from E. I. du Pont de Nemours & Company, Inc.

When used as a color difference meter the Hunter Instrument is usually standardized against ceramic tiles whose color characteristics are not far removed from the material under examination. The instrument is most accurate when the three circuits are balanced with a color standard similar to the material under investigation.

The instrument can be used to measure color and color difference on virtually all food products that are uniform or reasonably homogeneous in nature. A standardized procedure for measurement and preparation are necessary prerequisites.

Hunterlab D25 Color and Color-Difference Meter (Hunter Laboratories)

The Hunterlab D25 Color and Color-Difference Meter is a redesigned and improved version of the Hunter Color and Color-Difference Meter. Standardization and color measurement are made easier on this unit as a zero-centered potentiometer eliminates polarity switches and the adjustment knobs turn in the direction indicated by the meter needle. Color values are read directly from three scales in the form of L, a and b values. The theory involved relates the L, a and b values to a three-dimensional coordinate system for colors as shown in Figure 19.7. The vertical axis of the solid is white at the top to black at the bottom. Hue (varying from red through orange, yellow, green, blue, and purple back to red) is determined by direction from this axis. Thus, the readings of the D25 Color Difference Meter uniquely define any color on scales that are easy to relate to visual impressions.

A formula has been developed to score tomato color using the D25 Hunter Color and Color Difference Meter:

$$Ye = 121.9511 + 4.1386L - 0.0819L^2 + 130.5724 \cos 0$$

where: $\cos 0 = -\dfrac{a}{\sqrt{a^2 + b^2}}$

Ye = tomato color
L, a, b = Hunter color readings

The unit can be used on almost all foods having a homogeneous nature. Failure to reproduce results on the same sample often indicates that certain characteristics of the specimen prevent objective color evaluation.

Hunterlab Digital Color Difference Meter, D25D3A

The Hunterlab D25D3A provides for the color measurement of a wide range of samples and a digital readout of color scale values. Depending upon the particular color systems with which the instrument is equipped, values may be determined for several color measurement systems without recalibration. Tomato color (TC) can be computed automatically and displayed without further calculation (/D6 option).

The Hunterlab D25D3A is built in two sections—the optical unit and the measurement unit. In the optical unit, light from the quartz halogen cycle lamp is directed downward to the specimen port at an angle of 45°

FIGURE 19.5 — Three Dimensional Coordinate System For Colors

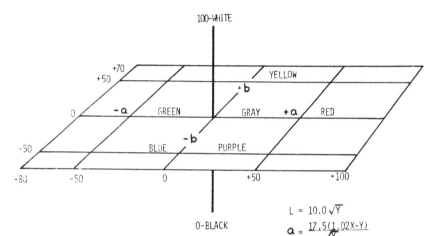

$$L = 10.0\sqrt{Y}$$
$$a = \frac{17.5(1.02X-Y)}{\sqrt{Y}}$$
$$b = \frac{7.0(Y-0.847Z)}{\sqrt{Y}}$$

from the perpendicular. A specimen lift is used to position the specimen to be measured in the port, and the light receptor is placed directly above the specimen area. Electrical signals are directed by cable to the measurement unit.

The optical unit can be mounted on the stand with the specimen port facing either upward or downward. It can also be removed entirely from the vertical supports and used in any position that may be desirable, the only limit being the length of the cables connecting it with the measurement unit.

The measurement unit converts the signals from the optical unit into color specifications which are displayed on the digital readout on the front control panel.

USDA Hunterlab D6 Tomato Colorimeter

The Hunterlab Model D6 Tomato Colorimeter developed by the U.S. Department of Agriculture and the Hunter Associates Laboratory, Inc. is used on tomato grading platforms for measuring the color of a composite raw tomato juice sample prepared from a representative sample of tomatoes obtained from the producer's loads. The color of the juice

FIGURE 19.6 — USDA-Hunter Tomato Colorimeter – D6

sample is expressed as a single index number of three digits from 00.0 to 99.9. The instrument is a modification of the general-purpose Hunterlab Color Difference Meter mentioned previously.

General Electric Recording Spectrophotometer

Objective detection of color and color differences in food products has also been attempted in an indirect manner which involved measurement of properties of extracted color. The use of an extracted color method to indicate possible color changes in the original product depends upon a close relationship between the extracted portion and the color characteristics of the original product. It has been pointed out that some fractions which influence color may be difficult to remove by the extraction method. It was indicated, however, that total color differences, measured directly in terms of what an observer saw depended upon the differences in the reflectance characteristics of the sample and could be measured by reflectance spectrophotometry or tristimulus colorimetry.

Since color is associated with wavelength, an obvious way of measuring it is by using the spectrophotometer. While very useful in color, matching the spectrophotometric method has several disadvantages. First of all, the color must be present in sufficient quantity. In addition, one cannot easily convey the description of a color by describing a reflectance curve.

The General Electric Recording Spectrophotometer is a precision laboratory instrument for automatically and accurately measuring and recording color. The spectrophotometer consists of a light source which radiates all wavelengths of the visible and near infrared spectrum, a monochromator for selecting monochromatic light as the spectrum is scanned, a filter system for shining the light alternately on the sample and on a standard, a phototube and associated amplifier circuits, and a recorder for making pen and ink record of the measurement. From the reflectance curve, an integrator can be used to obtain X, Y, and Z values of the specific product.

CHEMICAL EVALUATION OF COLOR

Color is one of the most important factors in the quality evaluation and grading of nearly all agricultural products. The intensity of color of these products depends primarily upon the quantity and quality of the colorant materials within the product. In the plant kingdom, such factors as "freshness," variety, maturity and process variables play an important role as to the actual color of the product. In subjective or physical (reflectance) color evaluation, the colorants (pigments) are found together in plant tissues. However, they may be extracted and quantitatively evaluated individually. Such information is of value to the plant breeder, food processor or technologist to determine precisely the "cause and effect" of pigments or pigment changes in a food sample.

There are four major pigments in plant chloroplasts. The green pigments are called chlorophylls. They consist of two components: chlorophyll a ($C_{55}H_{72}O_5N_4Mg$)—blue-black in the solid state and

PHYSICAL EVALUATION OF COLOR 257

FIGURE 19.7 — Hunterlab Digital Color Difference Meter, D25D3A

greenish-blue in solution; and chlorophyll b ($C_{55}H_{72}O_6N_4Mg$)— green-black in the solid state and in solution pure green. The term "chlorophyll" as it is commonly used refers to these two green pigments. Chlorophyll always contains 2.5% magnesium, which is the only metal in the ash. These pigments are insoluble in water and soluble in organic solvents, especially ethyl alcohol, ether, acetone, etc. Chlorophylls have the property of fluorescence; that is, they absorb special wavelengths of the incident light and transmit it in another wavelength which is longer than the absorbed wavelength. Chlorophyll a gives a deep red and chlorophyll b, a brownish-red fluorescence.

The other two pigments, carotenes and xanthophylls (carotenoids), are yellow pigments; xanthophylls are always associated with

TABLE 19.3 — Effect Of Cooking Conditions Upon Vegetable Pigments

Color of Vegetable	Type of Pigment	Solubility	Effect of				
			Acid	Alkalis	Salt	Hard Water (Ca and Mg Ions)	Heavy Metals (Iron and Tin)	Long Cooking

Color of Vegetable	Type of Pigment	Solubility	Acid	Alkalis	Salt	Hard Water (Ca and Mg Ions)	Heavy Metals (Iron and Tin)	Long Cooking
White Green	chlorophyll	— fat soluble	— darkens	yellows brightens	— brightens	yellows browns	browns browns	grays browns
Blue, red or purple	anthocyanin	water soluble	reddens	grays	—	darkens	dulls	darkens
Pale yellow	flavone	water soluble	brightens	browns	—	darkens	dulls	darkens
Orange, yellow, or orange red	carotene or lycopene	fat soluble	—	—	—	—	darkens	—

chlorophyll. The carotenes may also occur in fruits, roots, and other plant parts where chlorophyll is absent. The carotenes have the formula $C_{40}H_{56}$. There are three prevalent types: Alpha-, Beta-, and Gamma-carotenes. Lycopene, the red pigment of tomatoes, is a precursor of Beta-carotene.

The xanthophylls are oxygen derivatives of the carotenes. These pigments are yellow in color. Both carotene and xanthophyll pigments are insoluble in water but soluble in organic solvents, such as chloroform and alcohol.

In addition to the chlorophylls and the carotenoids, some other pigments found in plants are the anthocyanins. The anthocyanins, which change from red color in acid to medium blue in the alkaline range, are water soluble pigments found in the cell sap. They are glucosides, composed of a sugar and an anthocyanidin. Betanin, the red pigment in beets, is a beta cyanin similar to the anthocyanins except it turns yellow in alkaline solution. Nitrogen is a part of its structure. See Table 19.3.

General Statement on Spectrophotometry

The theory behind the use of the spectrophotometer for the measurement of the concentration of color pigments is the Beer-Lambert Law which states that absorption is proportional to the concentration of the solution. Absorption is a function of $A = E \times c \times l$, where E is wavelength, c is concentration and l is the length in centimeters of the cell path. Thus, using an adaptation of this formula, the concentration of the solution can be calculated. For determining the amount of color pigments present in a given sample the data is plotted with the concentration of a known sample against its optical density. The concentration can be determined by comparing the optical density obtained from the known to the standard curve.

Every chemical solution has a wavelength which is best suited for spectrophotometric analysis. This is known as its optimum wavelength and must be experimentally determined for an unknown sample. The optimum wavelength is found by plotting either the percent transmittance or optical density against a series of wavelengths. The data will peak out at the optimum wavelength.

The visible spectrum indicates the following color for each wavelength: 400–435 nm—violet; 435–480—blue; 480–490—green-blue; 490–500—blue-green; 500–560—green; 560–580—yellow-green; 580–595—yellow; 595–610—orange; and 610–750—red.

Transmittance, T, is a ratio of the energy transmitted by the sample, I, to the incident radiation, Io. Both I and Io must refer to the same wavelength; therefore,

$$T = I/Io \text{ (Bouguer's Law)}$$

Since transmittance is generally evaluated in percentage values,

$$I/Io \times 100 = \%T$$

The extent of absorption of light by a solution containing a colored solute is a function of concentration. Beer's Law states that the intensity of a beam of monochromatic light decreases exponentially as the concentration of the absorbing medium increases. It is analogous to Bouguer's Law in ascribing an exponential decrease in light intensity, which is accompanied by an arithmetical increase in concentration of the colored solute. The mathematical relationship is:

$$\log Io/I = k_2 c$$

where k_2 is a constant depending on the wavelength, the nature of the absorbing solution, and the sample thickness; c is the concentration of the solution. Beer's Law states the fundamental aspects of absorptiometry, but it is not complete unless combined with Bouguer's Law. No deviations from Bouguer's Law are known; however, those from Beer's Law are quite numerous.

TEST 19.1 — Tomato Juice Color Evaluation Using MacBeth-Munsell Disk Colorimeter

Equipment

(1) MacBeth-Munsell Disk Colorimeter

Procedure

(1) Place the colorimeter in a position so that the operator does not face into a strongly contrasting light, but not necessarily in a darkened room.

(2) Fill the sample cup level with the tomato product and place between the two spindles of the colorimeter.

(3) By using Grade A or Grade C disks, spinning on each side of the sample, determine which of the three combinations of black and gray (Disk 1, 2, or 3- Table 19.4) most nearly looks like the sample and use disks with this combination for both A and C for the grading comparison.

(4) After deciding the proper black and gray combination for the product, place the Grade A disk on the left hand spindle and the Grade C disk on the right spindle.

TABLE 19.4 — Tomato Color Disk Specification And Percentages
By Grade For Products

Color Notation		Disk No. 1 (%)	Disk No. 2 (%)	Disk No. 2 (%)
a. U.S. Grade A— U.S. Fancy Color Catsup, pulp, paste, and juice				
Red	(5R2.6/13)	65	65	65
Yellow	(2.5YR5/12)	21	21	21
Black	(N1)	7	14	0
Gray	(N4)	7	0	14
b. U.S. Grade C—U.S. Standard Color Catsup, pulp, paste, and tomato juice (23–24 pts)				
Red	(5R2.6/13)	53	53	53
Yellow	(2.5YR5/12)	28	28	28
Black	(N1)	9.5	19	0
Gray	(N4)	9.5	0	19
c. U.S. Grade C Tomato juice (25 pts)				
Red	(5R2.6/13)	59	59	59
Yellow	(2.5YR5/12)	24½	24½	24½
Black	(N1)	8¼	16½	0
Gray	(N4)	8¼	0	16½

(5) Turn on lamp and spinning motors and adjust the viewing mask at the 45° position and in such a manner that when the eye is about 18 in. from the mask, a section of the sample will be seen through one aperture of the mask and a section of spinning disk will be seen alongside of it in the adjacent aperture in the mask. Make all grading comparisons in this manner with the optional use of a hand mask. The hand mask is used to provide what is known as "aperture colors." This means that with the use of such a mask the outside factors which distract color judgments are removed.

(6) If there is a perfect color match, assign score points corresponding to the color disk matched, in accordance with the standard. Other score points are assigned by interpolation, using the two disks as reference points, or disks can be made up for score points as previously described.

(7) Do not adjust scores assigned even if they appear to be wrong when the sample is viewed in natural daylight. The color temperature and light energy distribution of the artificial light source in the col-

orimeter remains constant while that of natural daylight varies through a wide range, depending on time of day, time of year, latitude, atmospheric conditions, surrounding contrast and other factors. Extensive experimentation has proven that grading under these conditions with a constant light source of the proper color and intensity is comparable to that under the most ideal natural daylight conditions available and much more uniform than when care is not used to wait for the natural daylight conditions.

(8) Interpretation: (See Table 19.4 for color disks for Color Specification by Grade for Products.)

Product	Grade A	Grade B	Grade C	Grade D
Tomato juice	26–30		23–25[1]	0–22[2]
Tomato pulp and paste	45–54		40–44[2]	0–39[2]
Tomato catsup	21–25	21–25	17–20[2]	0–16[2]

[1]Partial Limiting Rule
[2]Limiting Rule

TEST 19.2 — Fresh Tomato Color Measurement With Model E Agtron

(1) Calibration

(a) Turn the meter switch ON and allow a 30 min warm-up period.

(b) Open the drawer and place both black calibration disks in the supports. The disks must be clean and free from scratches. Center them beneath the photo cells using the guide marks at the lower edge of the front panel.

(c) Put the Filter Lever in the RED position (right side), with the Red Zero control, adjust the meter needle on zero.

(d) Put the Filter Lever in the GREEN position (left side), with the Green Zero control, adjust the meter needle to zero again. This will not affect the previous adjustment of the Red Zero.

(e) Open the drawer, remove the black calibration disks, and put in the red calibration disks.

(f) Use the Standardize Control to set the meter on 46 (with Filter Lever still in the Green position).

(g) Move the Filter Lever to the RED position, and, if the meter needle goes to 100 ± 2, the instrument is properly calibrated.

(2) Tomato Color Measurement

(a) Cut tomato in half through a plane perpendicular to the stem-blossom axis. Avoid making a curved or rough cut surface; it is important that the surfaces be smooth and flat.

(b) Support the two halves in the drawer by means of the two horizontal wires and clips below the wires. Center them beneath the photo cells, using the guide marks at the lower edge of the front panel.
(c) Put the Filter Lever in the RED position and set the meter on 100 with the Standardize control.
(d) Move the Filter Lever to the GREEN position and read the meter. This indicates the color grade. Riper tomatoes register a low numerical reading, greener ones register a higher reading.

Interpretation

The Agtron Model E4 and a newer model, the E5, are used in California to score the inside color of tomatoes and to provide a basis for grade determination. The instrument is suited for use in California because the standards require that color be measured on the inside of the fruit.

Experiments conducted with the instrument at the Ohio State University have established lines for the categories of No. 1 and No. 2 quality for tomatoes as follows: (1) minimum No. 1—48.0, (2) minimum No. 2—84.0.

In addition, the following figures can be used as a general guide to evaluate tomato color: (1) ripe tomatoes, 10–35; (2) moderately ripe tomatoes, 40–60; (3) cull tomatoes, 70-90.

TEST 19.3 — Color Evaluation With Agtron Model F

Procedure

(1) Standardization
   (a) Switch the instrument on and allow a 45 min warm-up period.
   (b) Place the red standardizing cup in the recessed well and turn the standardizing knob (right side) until the meter reads 70.
   (c) Remove the red standardizing cup and replace it with the black standardizing disk. Turn the zero adjustment knob (left side) until the meter reads zero.
   (d) Repeat steps (1) (b) and (c) until no variance is noted from the initial setting.
   (e) Keep the recessed well covered when color readings are not being taken. Handle the disks in such a manner as to avoid scratching and periodically clean the disks in a mild detergent solution.

(2) Sample Preparation of Tomato Pulp
  (a) Remove 8½ lb of fruit from the inspection sample taken in conjunction with and in addition to each normal 50 lb inspection sample.
  (b) Wash the sample, if necessary, and dry.
  (c) Place into a gallon blender container.
  (d) Blend for 1 min.
  (e) Pour 200 ml of blended sample into a filtering flask through a strainer.
  (f) Seal top with rubber stopper.
  (g) Deaerate until all bubbles recede to the surface.
(3) (a) Fill a clean, dry, sample cup with extracted, deaerated tomato product.
  (b) Seat cup in the recessed well.
  (c) Read reflectance on meter and record.

Interpretation

The following figures can be used as a general guide in interpreting color scores: (1) good color, 20–40, (2) fair color, 41–60, (3) poor color, 61–80.

At O.S.U. the following minimum standards have been set for tomato color: (1) No. 1—42.0, No. 2—78.0.

TEST 19.4 — Color Measurement With Agtron E5

Procedure

(1) Calibration
  (a) Plug in the Agtron instrument and turn it on by pressing the on-off (red) button until it clicks and lights up.
  (b) Allow a warm-up period of 15–20 min.
  (c) Just before using, flip meter switch (above meter) to ON position (UP).
  (d) Open the drawer and standardize the unit by adjusting the meter with the calibration knob until the needle is over the calibration mark (48) on the 0–100 scale.
(2) Color measurement for tomato halves
  (a) Randomly select 20 tomatoes from each lot and prepare them by cutting the fruit in half midway between stem end and blossom end. Make a clean, even cut. Keep the two halves together.

PHYSICAL EVALUATION OF COLOR

    (b) Open the sample drawer by pulling down and out on the lever. Remove the black juice tray.
    (c) Place the tomato halves on the two circular supports in the drawer with the cut side up.
    (d) With the drawer open, readjust, if necessary, the CAL mark (48) with the calibration knob on the right.
    (e) Close the drawer, lift the lever up, and read the meter.
    (f) Record value on meter. Repeat steps (2)(c), (d), and (e) for all 20 tomatoes, and calculate the average value for the lot.

(3) Color measurement for tomato juice, pulp, and other tomato products.
    (a) Open the instrument drawer by pulling down and out on the lever. Install black juice tray in drawer.
    (b) Fill large petri dish (150 mm × 25 mm) with 175 ml of tomato product.
    (c) Place petri dish in center of black tray.
    (d) With drawer open, adjust meter to CAL mark (48), if necessary, with the calibration knob.
    (e) Slowly close the drawer, read the meter, and record the value indicated.
    (f) Wash the petri dish between samples, and dry it with cheesecloth.

Interpretation

See Test 19.2 and 19.3.

TEST 19.5 — Color Evaluation Using the Agtron M-400a And M-30

General

Use the Agtron M-30 unit for large particle samples and use the Agtron M-400A unit for small particle samples.

Equipment and Materials

(1) Zero calibration disk
(2) 90 Calibration disk
(3) Sample cups

Procedure

(1) Make sure M-30 is plugged into M-400A if you are using it.
(2) Turn power switch on to the M-30 or M-400A position and allow the instrument to warm up 1 hr for most stable operation.

(3) Obtain the desired "zero" and "90" calibration disks. Select the desired spectral mode and set the selector for that color on the unit.
(4) Place sample cup on unit and insert the "zero" calibration disk.
(5) On the M-400A, set the Gain Control at minimum (full counter-clockwise) and adjust the "zero" control for a meter reading of the same value as the "zero" disk.
(6) Place the "90" disk in the sample cup and adjust the Standardize Control for a meter reading of 90.
(7) Recheck settings by repeating steps (4), (5), and (6).
(8) Place sample in cup (approximately 100 gm or enough to fill cup), shake the sample slightly by hand, place on the viewer area, and record the meter readings.

TEST 19.6 — Color Evaluation Using The Hunter Color And Color Difference Meter

(1) Unit Preparation
   (a) Locate the instrument where there is medium or subdued illumination, no drafts and relatively dry air of constant temperature.
   (b) Situate the galvanometer such that the scale will be close to the height of the eyes of the instrument operator. Where there is marked vibration, one should choose a foundation-supported wall of the building rather than a partition to further minimize the galvanometer unsteadiness.
   (c) Attach the five cords from the measurement unit in the following manner: (1) Plug one cord into a 110 volt, 60 cycle alternating current supply. (2) Connect two cords to the galvanometer. (3) Connect the two cords with curve connectors to the exposure unit. (Each of these cords has a different terminal, therefore, there is no possibility of making wrong connections.)
   (d) Turn on the instrument lamp at least 20 min before use so that the photo cells in the apparatus may warm up and come to a fairly stable temperature.
(2) Sample Preparation
   (a) It is essential that the sample to be measured is a homogeneous mass (pulper, comminutor, etc.) and free from air bubbles (heating, deaeration, etc.).
   (b) Certain samples should be measured at standardized temperatures and others mixed well beforehand.

(3) Standardization
   (a) Start with "off-on" switch "up," galvanometer switch "off," and scale switch at circuit desired (Rd or L). Galvanometer spot should rest on "0" at the center of the scale.
   (b) Place the colored standard, selected to be nearest to color of specimens, in exposure position.
   (c) Set the three instrument dials and polarity switches to color values of the standard.
   (d) With the standard in position, and dials thus set, turn selector switch to Rd and pull galvanometer switch to "low." If deflection is not more than 2 or 3 divisions, pull switch to "high" and then turn adjusting knob above Rd dial until the galvanometer spot returns to "0." This adjustment is a screw in front of the exposure unit.
   (e) When balanced for Rd, pull galvanometer switch off, turn selector switch to "a," and again balance, using the knurled screw on the lower right side of the exposure unit.
   (f) When balanced for "a," repeat the operation for "b," this time using the screw on the left side.
(4) Measurement of Color
   (a) Balance and recheck standardization of the unit with the standard between sample measurements.
   (b) Place the sample in a special measuring flask and then seat the flask in measuring position.
   (c) Turn the selector switch to Rd.
   (d) Pull the switch to "high" and set the dial for zero deflection of the galvanometer.
   (e) Pull the galvanometer switch to "off" and turn the selector switch to "a." Repeat the same steps as for Rd above. Also for "b."
   (f) Read the Rd, a, and b values for the color from the dials to the first decimal place.
   (g) Turn the galvanometer switch to "off" before changing specimens and before every setting on a specimen appreciably different in color from the specimen previously exposed. If the switch is left on high, the galvanometer deflection may be so violent that the galvanometer suspension will be bent.
   (h) Record the data and plot it on "Data and Graph Sheet for Hunter Color and Color-Difference Meter" or on a chromaticity chart.

Interpretation

Interpret the data in terms of the standard and comparison of one sample to another. If desired, the data can be expressed in terms of the ICI specifications according to the formula given in Fig. 19.2 Some processors prefer to record the data in the form of an a/b ratio. This simplifies the amount of data that needs to be collected and facilitates quicker interpretation.

TEST 19.7 — Color Evaluation Using The Hunter Lab D25 Color And Color Difference Meter

Procedure for Operation

(1) Standardization
   (a) Turn instrument on at least 30 min prior to use.
   (b) Place the enameled color standard over the light source that is closest to the color of the specimen to be evaluated.
   (c) Zero adjust the galvanometer by holding down the push button switch on the left of the meter and turning the zero-adjust knob on the right until the meter shows no deflection to either the right or left, i.e., the needle is centered vertically on the null-balance bridge.
   (d) Adjust the digital dials (right side) to the "L," "a," and "b" values assigned the standard. These three values are located on the back of the color standard. Always adjust the "L" dial first before going to "a" and "b" dials. Check to make sure that the color standard is completely covering the light source such that no light is escaping.
   (e) Adjust each of the standardizing knobs (left side) until the galvanometer zeros by starting with "L" and going to "a" and "b" by switching the tristimulus selector at the bottom of the instrument.

(2) Color Measurement
   (a) Remove the color standard and seat the color cup containing the sample over the light source in such fashion that it is equidistantly centered around the light source.
   (b) Starting with "L," then progressing to "a" and "b," center the galvanometer by adjusting the appropriate digital dials. Frequently depress the zero adjustment button and check for null-deflection.
   (c) Read each scale to the nearest hundredth and record the data.

Interpretation

Interpret the data in terms of standards and differences between samples. An a/b ratio can also be used. A 1.90 reading, for example, represents an excellent tomato juice for color.

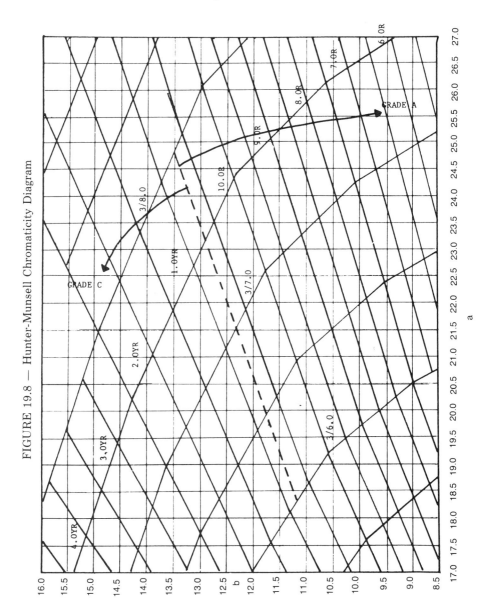

FIGURE 19.8 — Hunter-Munsell Chromaticity Diagram

TEST 19.8 — Color Evaluation Using The Hunter Lab Digital Color Difference Meter, D25D3A

General

The Hunterlab D25D3A provides for the color measurement of a wide range of samples and a digital readout of color scale values. Depending upon the particular color systems with which the instrument is equipped, values may be determined for several color measurement systems without recalibration. Tomato color (TC) can be computed automatically and displayed without further calculation (/D6 option).

Procedure

(1) Unit Operation
   (a) Source Voltage Check (perform daily)
      (1) Move standby switch to operation position, depress L button and allow display to stabilize (15–30 min).
      (2) Depress "S" button and note source voltage supplied to lamp displayed on the digital readout. If necessary adjust to correct value indicated on the side of the optical unit by turning knob on rear of measurement unit of instrument.
   (b) Standardization (perform daily)
      (1) Place the calibrated white tile on the specimen port and depress the Y button. Three values will be displayed on the digital readout and updated sequentially every 2.4 sec.
      (2) Depress the "1" display hold button to stop sequencing of digital readout. Adjust L/Y/0 standardizing control knob until the calibrated tile value is displayed on the digital readout.
      After setting L/Y/0 value, depress the "2" Display Hold button and adjust the a/X/0 control knob until the correct value is displayed.
      Do likewise for the "3" display hold button and b/Z/0 control knob to set correct value.
      (3) Depress the R button to release the sequencing hold.
      (4) After completion of the above steps, replace the calibrated white tile with the black glass provided with the set of standard tiles. This is the "Zero Reflectance Adjustment." *This can be performed on the Y,X,Z scale only.*
      (5) Depress the "1" display hold button. Using the small screwdriver supplied with the instrument, adjust the screw

accessible through the Y zero aid. hole directly below the calibration knob until the digital readout shows +00.07 ± 0.03. Perform the same steps for X and Z functions using "2" and "3" display hold buttons and corresponding adjustment screws.
  (6) Depress the R button to release the sequencing hold.
  (7) Repeat steps 1 through 5 until standard values of the white standard and the black glass remain stable without adjustment.
(c) Measuring a Wide Range of Colors Using White Standard as a Reference
  (1) Standardize the instrument on the L (or Y) scale using the white standard as described in steps (b) (1) and (2). Push appropriate button for scale desired and set corresponding scale values on instrument.
  (2) Place sample or specimen on specimen port, check for proper position through viewing aperture on the front of the optical unit. Depress L (or Y) button and record values displayed on the digital readout.
(d) Measuring Specimens or Samples of Similar Color
  (1) Choose a standard tile as close as possible in color to the samples to be measured and standardize the instrument to the calibrated values on the back of the tile as performed in steps (b) (1) and (2). Remove the standard tile and place the sample to be measured on the specimen port. Record color values displayed on the digital readout.
(e) Measuring Color Differences
  (1) Standardize the instrument on the L (or Y) scale using the white standard tile.
  (2) Place the standard sample or standard tile from which color differences of other samples are to be measured on the specimen port. Depress the L (or Y) button.
  (3) Enter the values displayed on the digital readout into the instrument memory by depressing the "EM" button. Wait until the light on the "EM" button goes out before proceeding.
  (4) Replace the standard sample or standard tile with the sample whose differences from the standard are to be measured.
  (5) Depress the "1" button and then the L (or Y) button. The display will show the L,a,b (or Y,X,Z) differences, with

proper signs, of the specimen values minus the standard sample or tile values.
- (f) Use of Instrument when Equipped with Tomato Color Option (TC)
  - (1) With the specimen port on top, install the specimen block with plexiglass insert and the stainless steel shield on the optical unit.
  - (2) Standardize the instrument on the L,a,b scale as described in steps (b) (1) and (2), but use the red TC standard tile.
  - (3) Depress the 4th function "4" button, then depress the "4" Display Hold button. The TC value displayed should be within ± 0.3 units of the calibrated TC value on the back of the red standard.
  - (4) Proceed with the measurement of TC of samples, i.e., depress 4th function button, and then depress "4" Display Hold button.

TEST 19.9 — Evaluation Of Product Color Using G.E. Recording Spectrophotometer

Procedure

(1) Color Measurement
  - (a) Attach graph paper to cylinder.
  - (b) Check to see if lenses are in or out.
  - (c) Insert black specular cup in sample side of sphere if it is desired to run test with "gloss out." For performing test with "gloss in," insert white specular cup.
  - (d) If test is run with "lenses in" it is necessary to insert a "mask" between the sample exit port and the sample.
  - (e) After the "mask" is in position, check to see that the light beam strikes in the center of the slit opening in the mask.
  - (f) Place sample at sample exit port on spectrophotometer sphere.
  - (g) Remove standard specular cup and observe sample. Use extreme care in handling specular cup which you remove from port. It is coated with MgO which is extremely fragile.
  - (h) If any other color except the one being measured shows at the sample exit port, adjust the position of the sample to eliminate the unwanted color.
  - (i) With only the power switch in the "on" position, adjust the wavelength indicator by hand to 400 nanometers and the reflectance dial to 50.0%.

- (j) Make sure paper is square on cylinder. Make sure paper is wrapped tightly about cylinder and place top edge of paper over bottom edge.
- (k) Secure center of paper with a 1 in. piece of ½ in. cellulose tape.
- (l) Adjust position of graph paper on cylinder so that when pen is lowered, it falls on the intersection of 400 nanometers and 50% on the graph paper.
- (m) Secure paper on cylinder by placing end clips down.
- (n) Lift the pen from the paper.
- (o) Place the "balance motor" switch in the "on" position.
- (p) Adjust the "powerstat" to a reading of 110.
- (q) When pen comes to a stop, lower it to the paper and place the "recorder" switch in the "on" (right) position.
- (r) When the curve is completed, lift pen from the paper.
- (s) Turn powerstat to "0."
- (t) Place "balance motor" switch in "off" position.
- (u) Place "recorder" switch in the "neutral" position.
- (v) Remove paper from cylinder and prepare to integrate.

Interpretation

Upon integration the X, Y, Z values can then be evaluated for comparison and calculating dominant wavelength, purity, etc.

TEST 19.10 — Determination Of Red Anthocyanin Color (Betanin) In Red Beets

Equipment and Materials
(1) 1-ml Pipette
(2) 100-ml Volumetric flask
(3) Spectrophotometer

Procedure

(1) Remove the liquid portion from a can of brine packed beets.
(2) Dilute 1:100 with distilled water.
(3) Determine % transmission—a spectrophotometer at wavelength of 530 nm zeroed with distilled water using a B & L Spectronic 20 as follows.
   (a) Turn amplifier control clockwise to turn on the colorimeter. Allow it to "warm up" 15 min.

(b) Adjust wavelength control to desired wavelength appearing opposite mark in wavelength scale window.
(c) Adjust amplifier control to bring dial indicator to 0% transmittance.
(d) Match three vials: insert test tube or cuvette containing distilled water into holder and close lid. Bring dial indicator to 90% transmittance with light control. Fill another tube with distilled water and check if transmittance is also 90%. Find three tubes with maximum difference in transmittance ±1.
(e) Insert test tube containing solvent, distilled water or blank sample into sample holder, close the lid and bring dial indicator to 100% transmittance with light control.
(f) Replace blank with unknown and close lid.
(g) Read percent transmittance of unknown under indicator on dial.
(h) Before changing to other wavelength, adjust light control counterclockwise to avoid damaging percent transmittance meter.
(i) Repeat steps (b) and (e) for each wavelength used in spectrophotometry.
NOTE: Always place the cuvette in the sample holder in exactly the same way. A line will be scratched on each cuvette and this line should be opposite the index line on the sample holder. The cover on the sample holder should be closed whenever a reading is made.

Interpretation

(1) Light Red—60% or more transmission.
(2) Bright Red—25–59% transmission.
(3) Deep Red—0–24% transmission.

TEST 19.11 — Qualitative Determination Of Lycopene Pigment In Tomatoes

Procedure

(1) Blend the sample in a Waring Blendor to a uniform pulp (5 min).
(2) Remove from the blender and weigh a 5-gm sample. Transfer to the cup of a clean blender, add 10 ml of benzol and blend exactly 5 min.
(3) Remove 4 ml of the benzol extract to a 15 ml centrifuge tube, make up to 20 ml with benzol and centrifuge at 100 rpm for 10 min. (If the sample is left for overnight to settle, this step can be omitted.)
(4) Pour off the centrifuged extract into the glass cell for the spectrophotometer and measure the percent transmittance at 485 nm.

(The spectrophotometer should be previously standardized using pure benzol as 100% transmittance.) See Test 19.10.

TEST 19.12 — Quantitative Determination Of Tomato Pigments (Lycopene And Beta Carotene)

Equipment

(1) Pan balance
(2) Waring Blendor or Osterizer
(3) 500-ml Separatory funnel
(4) 9-cm Wide mouth funnel
(5) Petroleum ether (65–110°C)
(6) Acetone
(7) Methanol 90%
(8) KOH (20%) in Methanol
(9) Glass wool
(10) 100-ml Volumetric flask
(11) Sodium sulfate (anhydrous)
(12) 100-ml Graduated cylinder

Procedure and Method

(1) Blend the sample in Waring Blendor (or Osterizer) to a uniform pulp—if whole tomatoes are being used, the seeds should be removed by forcing the sample through cheesecloth.
(2) Remove from the blender and accurately weigh a 5-gm sample, transfer to a clean blender cup, add 75 ml of acetone and 60 ml petroleum ether (65–100°C) and blend for exactly 5 min.
(3) Quantitatively transfer contents to a 500-ml separatory funnel. A 9-cm funnel loosely plugged with glass wool and a wash bottle containing acetone facilitate this transfer and prevent tomato pulp from entering the separatory funnel.
(4) Wash the extract three times with distilled water. Shake the separatory funnel gently in an inverted position for 0.5 min per wash. This step removes the acetone whose function is to remove the water of the sample, thus helping to prevent the formation of stable emulsions. Drain and discard the wash water—hypophase = lower phase.
(5) Add 20 ml of 90% methanol and mix 0.5 min. (Carotenols and chlorophylls are separated from the carotenes by immiscible solvent extraction.) The hypophase is discarded.

(6) Add 20 ml of 20% KOH in methanol and mix for 0.5 min. This saponification removes all but the nonsaponifiable fraction, which includes the carotenoids.
(7) Repeat step (5).
(8) Repeat step (4)—the separatory funnel stem is dried with absorbent cotton.
(9) Draw the petroleum ether (hyperphase) into a 100-ml volumetric flask through a 9-cm funnel, which has been loosely plugged with glass wool on which approximately 4 gm of sodium sulfate has been placed. This step removes any moisture which tends to cloud the extract.
(10) Bring to 100 ml volume with petroleum ether (65–110°C). (Part or all of which may be used to rinse the separatory funnel and 9-cm funnel containing sodium sulfate.)

NOTE: Samples may be analyzed immediately or refrigerated (in darkness) at 0°C for a maximum of 72 hr.

(11) Transfer 20 ml ± 2 ml of the cleared extract into a glass absorption cell of the spectrophotometer. The second cell is to be used for the solvent (petroleum ether).
(12) Measure the % transmittance and density of the extract over the visible spectrum from 400–550 nm.

NOTE: The absorption maxima of this extract mixture would be expected at approximately 445, 474 and 505 nm. Therefore, at these wavelengths it is desirable to take readings at every 2–5 nm. For the rest of the spectrum readings at every 10 nm are usually satisfactory.

TEST 19.13 — Qualitative Green Color Of Raw, Canned And Frozen Asparagus

Procedure

(1) Blend 100 gm of asparagus with 10 ml of water in Waring Blendor to a uniform pulp.
(2) Weigh 20 gm of the above mixture, and transfer to the cup of a Waring Blendor with about 70 ml acetone, and blend for exactly 5 min.
(3) Transfer to a 100-ml graduated cylinder, make up to 100 ml with acetone.
(4) Stir the mixture in the cylinder thoroughly and transfer part to a centrifuge tube and centrifuge for about 10 min at 2000 rpm.
(5) Measure transmittance of the clarified pigment solution in a suitable instrument at the spectral band centered about 665 nm.

(6) Convert percent transmittance to parts per million chlorophyll using Table 19.5. This method can also be used for raw, canned or frozen snap beans. The optimum wavelength is 575 nm. The color instrument can be standardized with a known solution of chlorophyll dissolved in acetone.

TABLE 19.5 — Comparison Of Transmittance And PPM Chlorophyll

Subjective		Objective	
		% Transmittance	PPM Chlorophyll
Greenest	1	30.5	144
	2	35.0	126
	3	45.1	95
	4	58.0	64
	5	60.4	60
	6	69.3	43
Light-green	7	73.6	36
Border	8	84.1	19
White	9	97.5	3

TEST 19.14 — Spectrometric Method For Total Chlorophyll And The a And b Components

Procedure
(1) Weigh 2–10 gm samples of plant material and macerate it in blender or in mortar with 85% acetone.
(2) Filter and wash the sample with acetone.
(3) Transfer filtrate to volumetric flask of appropriate size and dilute to volume with 85% acetone.
(4) Pipet aliquot of 25–50 ml into separator containing 50 ml ether.
(5) Add water carefully until it is apparent that all fat soluble pigments have entered the ether layer.
(6) Drain and discard water layer.
(7) Add 100 ml water to second separator placed in rack below first and let ether solution run through it to bottom of lower separator. When all solution has left upper separator, rinse it with ether.
(8) Drain and discard water, add a portion of fresh water. Continue washing process until all acetone is removed.
(9) Transfer ether solution to 100-ml flask, dilute to volume and mix.
(10) Remove a sample and determine the absorbance spectrometrically at 660 nm and at 642.5 nm.

Calculation of Chlorophyll Concentration

(1) Total chlorophyll (mg/1) =
$7.12 \log \frac{Io}{I}$ (at 660 nm) + $16.8 \log \frac{Io}{I}$ (at 642.5 nm).

(2) Chlorophyll a = $9.93 \log \frac{Io}{I}$ (at 660 nm) − $0.77 \log \frac{Io}{I}$ (at 642.5 nm).

(3) Chlorophyll b = $17.6 \log \frac{Io}{I}$ (at 660 nm) − $2.81 \log \frac{Io}{I}$ (at 642.5 nm).

BIBLIOGRAPHY

AMERICAN SOCIETY FOR TESTING MATERIALS. 1954. Symposium on color of transparent and translucent products. Am. Soc. Testing Mater. Bull. *201* and *202*.

BATE-SMITH, E.C. 1954. Flavonoid compounds in foods. *In* Advances in Food Research, Vol. 5, E.M. Mrak, and G.F. Stewart (Editors). Academic Press, New York.

BLANK, F. 1954. The anthocyanin pigment of plants. Botan. Rev. *13*, 241.

CHEMICAL RUBBER CO. 1966. Handbook of Chemistry and Physics. Chemical Rubber Publishing Co., Cleveland, Ohio.

CULPEPPER, C.E., and CALDWELL, J.S. 1927. The behavior of the anthocyan pigments in canning. J. Agr. Res. *35*, No. 2, 107–132.

DAVIS, R.B., and GOULD, W.A. 1955. A proposed method for converting Hunter color difference meter readings to Munsell hue, value and chroma renotations corrected for Munsell value. Food Technol. *9*, No. 11, 536–540.

GILREATH, S. 1969. Elementary Quantitative Chemistry. W.H. Freeman and Co., San Francisco.

HABIB, A.T., and BROWN, H.D. 1956. The effect of oxygen and hydrogen ion concentration of color changes in preserved beets, strawberries and raspberries. Proc. Am. Soc. Hort. Sci. *68*, 482.

HUNTER, R. S. Third Quarter, 1953 and Second Quarter, 1954. Optics and Appearance Instrumentation. Instrumentation. Minneapolis-Honeywell Regulator Co., Industrial Division, Philadelphia, Pa.

KRAMER, A., and SMITH, H.R. 1947. Electrophotometric methods for measuring ripeness and color of canned peaches and apricots. Food Technol. *1*, No. 4, 527–539.

LUCAS, E.W., RICE, A.C., and WECKEL, R.G. 1960. Changes in the color of canning beets. Univ. Wisc. Res. Bull. *218*.

YOUNKIN, S.G. 1950. Color measurement of tomato purees. Food Technol. *4*, No. 9, 350–354.

YOUNKIN, S.G. 1950. Measurement of small color differences in tomato purees. J. Opt. Soc. Am. *40*, No. 9, 596–599.

ZSCHEILE, F. P., and PROTER, J.W. 1917. Analytical methods for carotenes of Lycopersicon species and strains. Ind. Eng. Chem. Anal. Edition *19*, 47.

CHAPTER 20

Size, Shape, Symmetry and Style

SIZE

One of the chief factors to be considered in respect to the quality of fruits and vegetables is size. The size, and the knowledge of the percentage of the respective size, are important to the grower, processor and consumer. In some cases, large size may be preferred, while in others, the small size is paramount. In practically all cases the uniformity of size within a lot, whether this is the load as received at the factory, or the package as received by the consumer, is most desired.

The U. S. Department of Agriculture in the Standards for Grades has set up size classifications for raw and processed fruits and vegetables. Some of these are determined by actually counting the number in a given box, crate, barrel, hamper or container. With other fruits and vegetables the actual size classification may be measured with a standard gauge, with standard screens, or by determining the diameter, weight or length.

Fruits and vegetables as received at the processing plant are usually graded and sorted into various sizes. The equipment generally used for mechanically sizing fruits and vegetables in the production lines may consist of screens; shakers or vibrating belts; cups, rolls; rings; drums, with slatted openings or screens; or diverging belts. In addition to the above types, fruits and vegetables may be sized by weight. For efficient processing operations and for uniformity of the finished product, careful supervision and quality control checks must be made of the sizing operation. Some of the quality control checks and standards are given below for specific fruits and vegetables.

Apples

Size of apples is indicated by numerical count per box pack, as shown in Table 20.1. Apples also may be graded on the basis of diameter (take the greatest diameter at right angles to a line from stem to blossom end for minimum size). For maximum size take the shortest diameter at right angles to a line from stem to blossom end. The size of apple in the container is expressed in terms of inches, such as 2½ to 2¾ in., etc. There usually is not more than ¼ in. variation in the transverse diameter within a container of graded apples.

Apricots

The size of apricots is determined by counting the number per can, as shown in Table 20.2.

TABLE 20.1 — Size of Apples

Classification	Count
Very large	88 or less
Large	96–125
Medium	138–163
Small	175–200
Very small	216 or more

TABLE 20.2 — Size Of Apricots

Size	#2	Can Size—Count/Can #2½	#10
Small	25 or more	36	130
Medium	18–24	26–35	93–129
Large	12–17	18–25	66–92
Extra large	—	17 or less	65 or less

Asparagus

For the raw product using a vernier caliper, the diameter of the butt end of the stalk (not more than 5 in. from tip) is measured and the spears are classified as shown in Table 20.3. Also within each class the length of the spear may be established. The range will usually run from 5 in. to 11 in.

For the canned product, the number of spears per can is determined and the size classified in accordance with Table 20.4.

SIZE, SHAPE, SYMMETRY AND STYLE

Bananas
If packaged in singles, bananas shall be divided into five grades as shown in Table 20.5.

Lima Beans
Lima beans are sized mechanically with screens (28/64 up to 34/64 in.) as shown in Table 20.6.

TABLE 20.3 — Size Of Asparagus (Raw)

Size	Word Description	Actual Size
1	Very small	Less than 5/16 in.
2	Small	5/16 in. to less than 8/16 in.
3	Medium	8/16 in. to less than 11/16 in.
4	Large	11/16 in. to less than 14/16 in.
5	Very large	14/16 in. and up

TABLE 20.4 — Size Of Asparagus (Canned)

Size	Word	Approximate Diameter of Base of Spears (in Inches)	Range of Number of Spears No. 300
1	Small	3/8	39–51
2	Medium	1/2	26–38
3	Large	5/8	19–25
4	Extra large	7/8 and over	9–18

TABLE 20.5 — Size Of Bananas

Small—which shall consist of bananas of not less than 5 in. but less than 6 in. in length and not less than 4 in. in circumference

Sixes—which shall consist of bananas not less than 6 in. but less than 6½ in. in length and not less than 4 in. in circumference

Sevens—which shall consist of bananas not less than 6½ in. but less than 7½ in. in length and not less than 4 in. in circumference

Eights—which shall consist of bananas not less than 7½ in. but less than 8½ in. in length and not less than 4¼ in. in circumference

Nines—which shall consist of bananas not less than 8½ in. in length and 4¾ in. in circumference

TABLE 20.6 — Size Of Lima Beans

Midget	No. 1 sieve—Lima beans that pass through a screen with $28/64$ in. perforations
Tiny	No. 2 sieve—Lima beans that will pass through a screen $30/64$ in. perforations but not through $28/64$ in. perforations
Small	No. 3 sieve—Lima beans that will pass through a screen $34/64$ in. perforations but not through $30/64$ in. perforations
Medium	No. 4 sieve—Lima beans that will pass through a screen with $38/64$ in. perforations but not through $34/64$ in. perforations
Large	No. 5 sieve—Lima beans larger than $38/64$ in.

A 1-lb sample may be removed and placed on screens (smallest on top and largest on bottom) and then the screens shaken for 2 min. The lima beans on each screen are weighed back and the percent of each sieve size is then calculated.

Snap Beans

Snap beans are sized mechanically or by hand with a snap bean sizer according to Table 20.7. (The size of a green or wax bean is determined by measuring the shorter diameter of the bean transversely to the long axis as the thickest portion of the pod.). A 1-lb sample is removed and each bean sized into the above sizes. Each size class is weighed back and the percentage of each is determined.

TABLE 20.7 — Size Of Snap Beans

	Round Type			Flat Type (Whole, Cut or Short Cut)	
		Whole	Cut or Short Cut		
Size of Beans (Inches in Thickness)	Number Designation	Word Designation	Word Designation	Number Designation	Word Designation
Less than $14\frac{1}{2}/64$ in.	1	Tiny	Small	2	Small
$14\frac{1}{2}/64$ in. to, but not including, $18\frac{1}{2}/64$ in.	2	Small	Small	3	Medium
$18\frac{1}{2}/64$ in. to, but not including, $21/64$ in.	3	Medium	Small	4	Medium large
$21/64$ in. to, but not including, $24/64$ in.	4	Medium large	Medium	5	Large
$24/64$ in. to, but not including, $27/64$ in.	5	Large	Large	6	Extra large
$27/64$ in. or more	6	Extra large	Extra large	6	Extra large

Whole Beets
Whole beets are classified when cut through the middle as shown in Table 20.8.

Citrus
The pack of citrus in standard nailed boxes is determined by measuring the diameter in inches of the fruit. Normally a 10% tolerance is allowed by count within each crate.

Olives
Olive standards established by California list the average count per pound as shown in Table 20.11.

Peaches and Pears
Count the number of halves and interpret the size from Table 20.12 the the various can sizes.

TABLE 20.8 — Size Of Whole Beets

Size	Word Designation	Diameter	Count/#2 Can	16 oz. Glass
1	Tiny	less than 1¼ in.	44 or over	35 and over
2	Tiny	less than 1¼ in.	31–43	25–34
3	Small	over 1¼ in. up to 1½ in.	22–30	18–24
4	Small	over 1¼ in. up to 1½ in.	13–21	10–17
5	Medium	over 1¼ in. up to 1¾ in.	9–12	7–9
6	Medium	1¾ in. or more	less than 9	less than 7

TABLE 20.9 — Size Of Oranges

Pack	Minimum	Maximum
96's	3 $^{6}/_{16}$	3$^{13}/_{16}$
126's	3 $^{3}/_{16}$	3$^{10}/_{16}$
150's	3	3 $^{6}/_{16}$
176's	2$^{14}/_{16}$	3 $^{4}/_{16}$
200's	2$^{12}/_{16}$	3 $^{2}/_{16}$
216's	2$^{10}/_{16}$	3
250's	2 $^{8}/_{16}$	2$^{14}/_{16}$
288's	2 $^{6}/_{16}$	2$^{12}/_{16}$
324's	2 $^{4}/_{16}$	2$^{10}/_{16}$

TABLE 20.10 — Size Of Grapefruits

Pack	Minimum	Maximum
36's	5	$5\ 9/16$
46's	$4\ 11/16$	$5\ 4/16$
54's	$4\ 6/16$	$4\ 15/16$
64's	$4\ 3/16$	$4\ 12/16$
70's	$3\ 15/16$	$4\ 8/16$
80's	$3\ 12/16$	$4\ 5/16$
96's	$3\ 9/16$	$4\ 2/16$
112's	$3\ 7/16$	4
126's	$3\ 5/16$	$3\ 14/16$

Peas

Peas are sized mechanically with screens (9/32 to 14/32 in.) as shown in Table 20.13.

TABLE 20.11 — Size Of Olives

Classification	Size	Average count per Pound	Count Range per Pound
1	Small or standard	135	128–140
2	Medium	113	106–127
3	Large	98	91–105
4	Extra large	82	76–90
5	Mammoth	70	65–75
6	Giant		53–64
7	Jumbo		46–52
8	Colossal		33–45
9	Super colossal		Maximum of 32

TABLE 20.12 — Size Of Peach And Pear Halves

	Count		
Size	#2	#2½	#10
Small	13 or more	19 or more	67 or more
Medium	9–12	13–18	45–66
Large	6–8	8–12	27–44
Extra large	5 or less	7 or less	26 or less

SIZE, SHAPE, SYMMETRY AND STYLE

Pickles

Pickles are graded into the different sizes by the U.S. Dept. of Agriculture on the basis of count per standard gallon and/or by length in inches. The International Pickle Packer's Association gives a count for the different sizes per 45 gal. cask.

Potatoes

Potato size may be established by measuring the intermediate axis of the tuber and classifying the cultivars by the size distribution and/or by weight in ounces. The intermediate axis may be determined by passing the tubers through ring sizers designed as 4 ¼, 3 ¼, and 2 ½ inches, separating them into each size designation, that is, large, medium and small, respectively. A 10 lb. sample from each replicate should be used and the size range should be reported as the percentage of tubers which fall into each size designation. For example, assume that a give 10 lb. sample contains a total of 25 tubers. Within the sample, 3, 15, and 7 tubers were designated as very small, small, and medium respectively. Based on the following equation:

% of Tubers in Size Classification equals

$$\frac{\text{Number of Tubers in Size Classification}}{\text{Total Number of Tubers}} \times 100$$

This sample contains 12% Very Small, 60% Small, and 28% Medium tubers.

Another method of size determination for most products is to make a count of the individual units per given weight. This method is very simple, but it only provides an average size without any indication of the acutal size distribution.

SHAPE AND SYMMETRY

For fruits and vegetables of highest quality, the shape, symmetry and uniformity of same are most important attributes of quality affecting the appearance of the product. For some products, the U.S. Dept. of Agriculture's Standards for Grades have set up very specific requirements. For others, trade and industry standards based on acceptable practices are all that exist.

Good examples of the importance of shape and symmetry are found with products like apricots, pears, pickles, and French fried potatoes. With apricots, the apricot halves should be cut on the suture line. An "off-suture cut" apricot is one that has been cut more than ¼ in. at the

TABLE 20.13 — Size Of Peas

Tiny or Petite	No. 1 size peas are peas that will pass through a screen of $9/32$ in. mesh
Extra sifted or small	No. 2 size peas are peas that will pass through a screen of $10/32$ in. mesh but not through a screen of $9/32$ in. mesh
Sifted or medium small	No. 3 size peas are peas that will pass through a screen of $11/32$ in. mesh but not through a screen of $10/32$ in. mesh
Early June or sweet or medium size	No. 4 size peas are peas that will pass through a screen of $12/32$ in. mesh but not through a screen of $11/32$ in. mesh
Telephone or medium large	No. 5 size peas are peas that will pass through a screen of $13/32$ in. mesh but not through a screen of $12/32$ in. mesh
Large	No. 6 size peas are peas that will pass through a screen of $14/32$ in. mesh but not through a screen of $13/32$ in. mesh
Extra large	No. 7 size peas are peas that fail to pass through a screen of $14/32$ in. mesh

widest measurement from the suture. With pears not more than 10% variation is permitted in symmetry of the units for Grade A halves, quarters or whole; Grade B up to 15% variation and no tolerance for Grade C products. Specific tolerances have been established for frozen French fried potatoes for uniformity of size and symmetry with 15% and 30% tolerance for Grade A and B respectively. With pickles the off-shaped units are scored as defects.

TABLE 20.14 — Size Of Pickles

Word Designation	Count per Gallon	Diameter
Midget	270 or more	¾ in. or less
Small gherkin	135–269	Up to 15/16 in.
Large gherkin	65–134	Up to 1 1/16 in.
Small	40–64	Over 1 1/16 to 1 3/8 in.
Medium	26–39	Over 1 3/8 to 1 ½ in.
Large	18–25	Over 1 ½ to 1 ¾ in.
Extra Large	12–17	Over 1 ¾ to 2 1/8 in.

Blend of Sizes—A combination of any two adjacent designated sizes.
Mixed Sizes—A combination of more than two adjacent designated sizes.

In all cases models and diagrams should be available to the technologists. These can be free-hand drawings, plastic models, photographs, or "finger-print" examples. For the latter, the product can be placed as is on an inking pad and then by transferring the product to a mimeo type paper a print can be reproduced. These are excellent records for shapes of products like green beans (for curvature of pods), cavity size of stone fruits like peaches, petiole shapes of celery, and curvature and shape of pickles. Through practice, industry can standardize and develop higher quality products of greater uniformity and better appearance.

Uniformity of Size and Shape Factor in Beets

Whole Beets.—Are not more than Grade A 2¼ in., Grade C 2½ in. in diameter. The maximum diameter variation is Grade A 50%, Grade C 100%. Grade A beets may vary moderately in Shape, Grade C beets may vary considerably in Shape.

Sliced Beets.—Are not more than Grade A 5/16 in., Grade C 3/8 in. in thickness; not more than Grade A 3½ in., Grade C 3½ in. in diameter; and the maximum diameter variation does not exceed Grade A 50%, Grade C 100%.

Quartered Beets.—Are cut from beets not more than Grade A 2½ in., Grade C 3½ in. in diameter; weight variation in Grade A 50%, Grade C 100%.

Diced Beets.—Are cubes not larger than Grade A ⅜ in., Grade C ½ in. Not more than Grade A 15%, Grade C 25% by weight of the pieces are smaller than ½ cube or large and irregular in shape.

Julienne, French Style or Shoestring Beets.—Are strips with cross sections measuring not more than Grade A 3/16 in. square, Grade C not over 3/16 in. The aggregate weight of all strips less than 1½ in. long does not exceed Grade A 25%, Grade C 40% of the weight of all the strips.

Cut Beets.—For Grade A, are pieces weighing not less than ¼ oz., not more than 2 oz., and the largest piece weighs not more than 4 times the smallest; Grade C, not less than ⅛ oz nor more than 3 oz and the largest piece weighs not more than 12 times the smallest. An occasional unit not representative is excluded in determining size variation.

Definitions and Measurement of Pickles (USDA Standards for Grades of Pickles)

Curved Pickle.—A curved pickle is one that is curved at an angle of 35-60 degrees when measured as illustrated in Figure 20.1

Crooked Pickle. — A crooked pickle is one that is curved at an angle greater than 60 degrees, similar to Figure 20.2.

Misshapen Pickles.—Misshapen pickles include crooked, nubbins, and otherwise misshapen pickles. A nubbin pickle is one that is not cylindrical in form, is short and stubby, or is not well developed. Nubbins and otherwise misshapen pickles are similar to illustrations in Fig. 20.3.

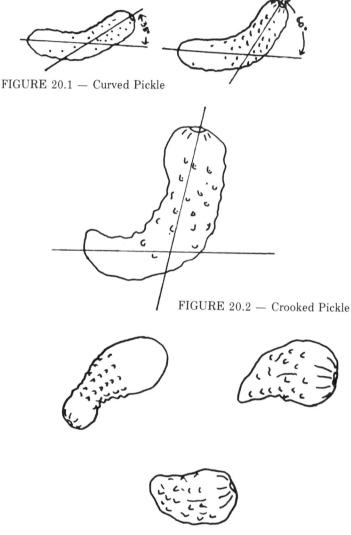

FIGURE 20.1 — Curved Pickle

FIGURE 20.2 — Crooked Pickle

FIGURE 20.3 — Mishapen Pickle

SIZE, SHAPE, SYMMETRY AND STYLE

CHANGE OF FORM OR STYLE OF PACK

Cutting, dicing, slicing, stripping and comminuting are unit operations in preparation of specific food products for the various styles of pack. Generally, this quality attribute refers to the degree of smoothness of the cut surface, uniformity or depth of cut, and the degree of freedom from loose slices or adhering surfaces. Many different thicknesses and styles of cuts can be produced on specific models of slicers. Urschel Laboratories, Inc., one of the leading manufacturers, indicates that the standard slicing head assemblies and impellers can produce slices up to 7/64 in. (2.8 mm) thickness. Slices up to ½ in. (13 mm) thick are produced on special slicing head assemblies and impellers.

Flat Slice.—The flat slice is used on potatoes and other similar produce. Slice thickness ranging from $1/32$ in. (0.8 mm) to $7/64$ in. (2.8 mm) thick are produced when a flat slicing head is used.

Crinkle Slice.—The crinkle slice is used on potatoes and other similar produce. On crinkle slices there are 3⅓ full waves to the inch and the depth of crinkle is 0.080 in. (2.0 mm). Slice thickness ranging from $1/32$ in. (0.8 mm) to $7/64$ in. (2.8 mm) thick are produced when a crinkle slicing head assembly is used.

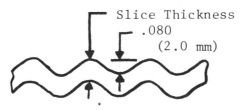

FIGURE 20.4 — Crinkle Slice

V-Cut Slice.—The V-cut slice is used on potatoes and other similar produce. V-cut slices have 8 complete "V's" to an inch and the depth of "V" is 0.050 in. (1.3 mm). Slice thickness ranging from $1/32$ in. (0.8 mm) to $7/64$ in. (2.8 mm) thick are produced when a V-cut slicing head assembly is used.

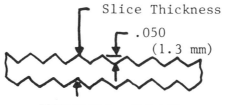

FIGURE 20.5 — V-Cut Slice

Oval Shred.—The oval shred is used on carrots, cheeses and other similar products. This type of shred is approximately ⅛ in. (3.2 mm) wide and is somewhat oval shaped in cross section.

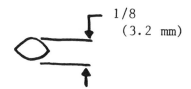

FIGURE 20.6 — Oval Shred

V-Cut Shred.—The V-cut shred is used on carrots, coconut and other similar products. This type of shred is diamond in shape and is approximately 1/16 in. (1.6 mm) on all sides.

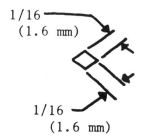

FIGURE 20.7 — V-Cut Shred

Julienne Strip.—The julienne strip cut is used on potatoes only. This type of cut is either square or rectangular in shape. The square cut will measure 3/16 in. × 3/16 in. (4.8 mm × 4.8 mm). By changing the slice thickness a 3/32 in. × 3/16 in. (2.4 mm × 4.8 mm) julienne strip can be produced, or by eliminating one of the two knives a 3/16 in. × ⅜ in. (4.8 mm × 9.5 mm) strip can be produced. The combination of reducing the slice thickness and removing a knife will produce a 3/32 in. × ⅜ in. (2.4 mm × 9.5 mm) julienne strip.

The Grade Standards and the Standard of Identity are quite specific as to the percent tolerances for the specific styles. A few examples are selected from these standards.

SIZE, SHAPE, SYMMETRY AND STYLE

Styles of Canned Mushrooms

Whole.—Caps with attached stems more than ⅛ in. long.

Buttons.—Caps with attached stems, at least 85% of which are not longer than ⅛ in., and no stems are longer than ¼ in.

Sliced Whole.—Slices of whole mushrooms. The direction of the slice, in 80% or more of the slices by weight, is approximately parallel to the longitudinal axis of the stem. Not more than 5% of units may be detached portions of stems.

Random Sliced Whole.—Sliced from whole mushrooms. Direction of slice may deviate materially from approximately parallel to longitudinal axis of stem. Not more than 15% by weight may be detached portions of stems.

Sliced Buttons.—Sliced from whole buttons. Ninety percent or more by weight of slices made in the direction approximately parallel to the longitudinal axis of stem.

Stem and Pieces.—(Or pieces and stems) do not conform to any foregoing styles and are predominately cut or broken portions of the caps and stems and may contain units of any of foregoing styles.

Styles of Canned White Potatoes

Canned white potatoes are properly peeled prior to canning; "unit" means an individual potato or a portion of a potato; and the various styles of the canned product are as follows.

Whole or *Whole Potatoes* consists of whole potatoes that are of the approximate conformation of the prepared potatoes.

Slices, Sliced, or *Sliced Potatoes* consists of units that are potato slices of practically uniform thickness.

Dice, Diced, or *Diced Potatoes* consists of units that are approximate cubes.

TABLE 20.15 — Tyler Standard Screen Scale Sieves

Tyler Standard Screen Scale 2½ Opening in Inches	For Closer Sizing Ratio 2¼ Opening in Inches	Mesh	Diameter of Wire, Decimal of an Inch	U.S. Series Equivalents (Fine Series)	
				Micron Designation	Number
3	—	—	0.207	—	—
2	—	—	0.192	—	—
1.5	—	—	0.162	—	—
1.050	1.050	—	0.148	—	—
—	0.883	—	0.135	—	—
0.742	0.742	—	0.135	—	—
—	0.624	—	0.120	—	—
0.525	0.525	—	0.105	—	—
—	0.441	—	0.105	—	—
0.371	0.371	—	0.092	—	—
—	0.312	2½	0.088	—	—
0.263	0.263	3	0.070	—	—
—	0.221	3½	0.065	5660	3½
0.185	0.185	4	0.065	4760	4
—	0.156	5	0.044	4000	5
0.131	0.131	6	0.036	3360	6
—	0.110	7	0.0328	2830	7
0.093	0.093	8	0.032	2380	8
—	0.078	9	0.033	2000	10
0.065	0.065	10	0.035	1680	12
—	0.055	12	0.028	1410	14
0.046	0.046	14	0.025	1190	16
—	0.0390	16	0.0235	1000	18
0.0328	0.0328	20	0.0172	840	20
—	0.0276	24	0.0141	710	25
0.0232	0.0232	28	0.0125	590	30
—	0.0195	32	0.0118	500	35
0.0164	0.0164	35	0.0122	420	40
—	0.0138	42	0.0100	350	45
0.0116	0.0116	48	0.0092	297	50
—	0.0097	60	0.0070	250	60
0.0082	0.0082	65	0.0072	210	70
—	0.0069	80	0.0056	177	80
0.0058	0.0058	100	0.0042	149	100
—	0.0049	115	0.0038	125	120
0.0041	0.0041	150	0.0026	105	140
—	0.0035	170	0.0024	88	170
0.0029	0.0029	200	0.0021	74	200
—	0.0024	250	0.0016	62	230
0.0021	0.0021	270	0.0016	53	270
—	0.0017	325	0.0014	44	325
0.0015	0.0015	400	0.001	37	400

[1]The Tyler Standard Screen Scale Sieves Series has been expanded to include intermediate sieves for closer sizing which gives a ratio of the fourth root of two or 1.189 between openings in successive sieves.

TABLE 20.16 — Relationship Of Tuber Size To
Various Physical Constants For Tubers
Of Different Diameters.

Physical Constants	Sphere Diameter of Potatoes in Inches				
	2"	2 1/2"	3"	3 1/2"	4"
Surface Area in Sq. Inches	12.6	20.4	28.1	39.2	50.0
Volume in Cubic Inches	4.19	9.16	14.13	23.82	33.51
Ratio of Area to Volume	3.00	2.50	2.00	1.75	1.50
Approximate # of Tubers/lb.	8.0	5.03	2.60	1.80	1.00
Approximate # of Tubers/8 lb.	64	40	21	14	8
Relative Surface Area-Sq. In/lb.	150	111.5	73	61.5	50
1/16" peel removal (% Vol. loss)	17.0	14.50	12.0	10.5	9.0

Shoestring, French Style, Julienne, Shoestring Potatoes, French Style Potatoes, or *Julienne Potatoes* are potato strips of varying lengths.

Pieces consists of units (including but not being limited to, "Orange Cuts" and "Quarters") other than those comprising any of the foregoing styles or of a mixture of units comprising any two or more styles.

Styles of Frozen French Fried Potatoes

Straight Cut.—Refers to smooth cut surfaces.

Crinkle Cut.—Refers to corrugated cut surfaces.

Strips.—Elongated pieces with practically parallel sides and of any cross-sectional shape. Further identified by the approximate dimensions of the cross section, for example: ¼ × ¼ in.; ⅜ × ⅜ in.; ½ × ¼ in.; or ⅜ × ¾ in.

Shoestring.—Strips, either straight or crinkle cut, with a cross sectional area predominantly less than that of a square measuring ⅜ × ⅜ in.

Slices.—Pieces of potato with two practically parallel sides, and which otherwise conform generally to the shape of the potato; may contain a normal amount of outside slices.

Dices.—Approximate cubes.

Rissole.—Whole or nearly whole potatoes.

Other.—Designated as to style by description of the size, shape, or other characteristic which differentiates it from the other styles.

Criteria for Length Designations

Percentages following apply to strips ½ in. in length or longer.

Extra Long.—Eighty percent or more are 2 in. or longer; and 30% or more are 3 in. or longer.

Long.—Seventy percent or more are 2 in. or longer; and 15% or more are 3 in. or longer.

Medium.—Fifty percent or more are 2 in. or longer.

Short.—Less than 50% are 2 in. in length or longer.

BIBLIOGRAPHY

COLEMAN, C. H., and DESPAUL, J. E. 1955. Measuring the size compliance of foods. Food Technol. *9,* No. 2, 94–102.

GOULD, W. A. 1955. Size and size measurement. Food Packer *36,* No. 13, 32, 36.

CHAPTER 21

Maturity/Character and Total Solids/Moisture (Specific Gravity)

MATURITY/CHARACTER

Maturity/character has reference to the degree of ripeness, the condition of the flesh of fruits, the tendency of the specific commodity to retain the original conformation without material disintegration, the tenderness, texture and the maturity of the commodity.

Commodities

Apples.—Maturity/Character refers to the texture of the slices and to the tendency to retain their conformation without material softening or disintegration: Grade A—not more than 5% consists of mushy apples; Grade C—not more than 15% consisting of slices that are markedly hard, markedly soft, or mushy.

Asparagus.—Maturity/Character refers to the degree of development of the head and bracts and to the tenderness and texture. A "well-developed" head means the appearance of the head is not materially seedy and is practically compact. "Fairly well developed" means that the head may show a seedy appearance over the surface of the head and bracts may be elongated, but not so developed or elongated as to give definite spread or branching appearance.

Grapefruit.—Character refers to the structure and condition of the cells of the grapefruit and reflects the maturity. Grade A Character is "good" meaning that the grapefruit is moderately firm and fleshy, the segments or portions possess a juicy, cellular structure free from dry cells, or "ricey" cells, or fibrous cells that materially affect the appearance or eating quality of the product; and the product is reasonably free from loose floating cells.

Clingstone Peaches.—Character refers to the degree of ripeness, texture and tenderness.

Grade A (Good Character).—This category has the following meaning with respect to quarters, slices and mixed pieces of irregular sizes and shapes. Units are pliable and possess a tender, fleshy texture typical of mature, well ripened, properly prepared and processed canned clingstone peaches; are intact and possess reasonably well-defined edges. Also, not more than 10% by count, or 1 unit, if such unit exceeds the 10% allowance, may possess a "reasonably good character," provided the appearance or eating quality, or both, is not more than slightly affected.

Grade B (Reasonably Good Character).—Units possess a texture typical of mature, properly ripened, prepared, and processed canned clingstone peaches. Reasonably good character has the following meanings with respect to halves, quarters, slices, and mixed pieces of irregular sizes and shapes. Texture is reasonably fleshy and units are reasonably tender or tenderness may be variable with the unit; units are reasonably intact with not more than slightly frayed edges and may be slightly firm or slightly soft but are not mushy. Also, not more than 10% by count, or one unit, if such unit exceeds the 10% allowance, may possess a "fairly good character," provided the appearance or eating quality, or both, is not affected materially.

Grade C (Fairly Good Character).—Units possess a texture typical of mature, properly prepared and processed canned clingstone peaches. Fairly good character has the following meaning with respect to halves, quarters, slices, and mixed pieces of irregular sizes and shapes. Units may be variable in fleshiness but the texture is fairly fleshy; the units may be soft. Also, not more than 10% by weight of the drained peaches may be mushy or be so firm as to be "not tender."

Strawberries.—Maturity/character refers to firmness, degree of wholeness, seediness, disintegration as evidenced by partial and mushy strawberries. Tolerances in percent for partial strawberries or mushy strawberries or combinations are shown below:

Grade	Whole	Sliced
A	10	20
B	20	30
C	30	50

For maturity/character of some other commodities, separation on the basis of specific gravity is most effective.

SPECIFIC GRAVITY/TOTAL SOLIDS/MOISTURE

The maturity of various commodities, notably raw, frozen and canned peas and lima beans, and raw potatoes may be determined by measuring the total solids or the moisture content. The moisture content of fruits and vegetables is an index of quality of the raw product as well as an index of processing efficiency for dehydrated products. There have been several quick methods proposed in recent years for the determination of moisture content and they all must be standardized against the A.O.A.C. vacuum oven method or for relatively low moisture content products against the toluene distillation method or the infrared method (see Test 21.1, 21.2 and 21.3).

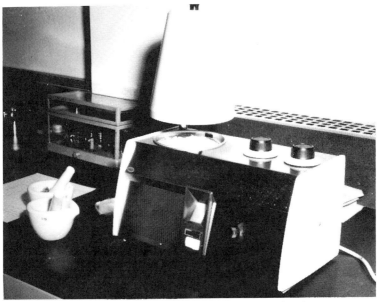

FIGURE 21.1 — Infrared Moisture Determination.

A quicker method for total solids is the floating of the specific product in salt (NaCl) solutions of various densities. The determination is based on the fundamental fact that as peas and other vegetables develop they become progressively higher in total solids or lower in moisture and have a relatively higher specific gravity or density. Another quick method to quantify the maturity for some products in the determination of alcohol insoluble solids (AIS) (see Test 21.4).

Brine Flotation

The brine flotation method has commercial applications. Brine separation equipment is used to segregate raw products into one or more maturity classifications before processing.

Certain factors may tend to make unreliable the results of the brine flotation method when applied to grading these products in the quality control laboratory. Among these, for peas and lima beans, are (1) the presence of air pockets between the skin and cotyledons and (2) the use of broken or mashed units. For potatoes, such characteristics as diseased portions, hollow heart or unclean tubers may cause erroneous separations. Also, differences in temperature between the product and test solutions should be the same for reliable results. Further, careful control of the exact concentration of the salt solution is most important. See Test 21.5 for Brine Flotation, 21.6 for Hydrometric Method for Potatoes, and 21.7 Specific Gravity for Weight Method for Whole Kernel Corn.

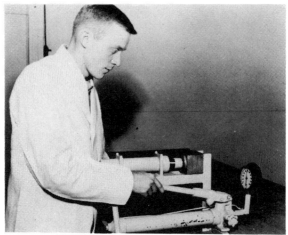

FIGURE 21.3 — Corn Succulometer

SUCCULENCE

For adequate quality control in the canning of sweet corn, it is important that the canner have available some method that will enable him to predict as well as check the final maturity of the canned product. In order to be practical, such a method must be rapid and relatively simple, and yet reasonably accurate.

Attempts to measure the maturity of corn for canning date back as early as 1917, and include tests based on sugar, starch, fiber, pericarp, alcohol insoluble solids, moisture contents, refractive index, the pressure texture and the tenderometer. For the canned product, alcohol insoluble solids, fiber and pericarp contents were suggested. Although all the above methods showed significant correlations with organoleptic ratings of maturity, only the moisture test on the raw corn, and alcohol insoluble solids test on the canned corn proved to be sufficiently accu-

FIGURE 21.4 — Alcohol Insoluble Solids

rate for prediction purposes. More recently, a rapid and accurate test on raw corn, based on the refractive index, has been described by Scott, *et al.*

The succulometer makes possible the measurement of the maturity of canned whole kernel corn as well as the measurement of the raw corn anywhere during the canning procedure after the kernels are cut. In accuracy it is approximately equal to the moisture test on the raw corn and the alcohol insoluble solids test on the canned corn, but is much more rapid and simpler to carry out than either of the other tests (see Test 21.8 Succulence — Moisture Content).

The data in Table 21.6 (see Test 21.7) show the relationship of the succulometer values to other tests of maturity. It would seem that a canner of very young (fancy) yellow whole kernel corn would strive to obtain raw material which would produce about 23 ml of extractable juice and the final product should test at least 20 ml. Similarly, raw corn for a young (extra standard) product should produce about 20 ml of extractable juice and the canned product should not fall below 16 ml. Raw corn yielding only 13 ml of extractable juice would be dangerously close to the substandard level.

These data are not applicable to cream style corn and the limits for vacuum packed corn should be lower than those for brine packed corn which are given in Table 21.1.

TABLE 21.1
Relative Succulometer, Moisture And Alcohol Insoluble Solids Values
For Raw And Canned Whole Kernel Brine Packed Golden Sweet Corn

| Canned Kernels | | Raw Kernels | | Maturity Rating |
AIS	Succulometer	Moisture	Succulometer	per AIS
15.0	23.7	77.9	25.0	A
16.0	22.9	76.0	24.5	A
17.0	22.0	74.4	23.8	A
18.0	21.1	73.0	23.1	A
19.0	20.2	71.7	22.4	A
20.0	19.3	70.4	21.7	B
21.0	18.4	69.0	21.0	B
22.0	17.4	67.6	20.2	B
23.0	16.8	66.3	19.0	B
24.0	15.6	65.1	17.5	C
25.0	14.6	64.0	15.9	C
26.0	13.7	63.0	14.1	C
27.0	12.8	62.0	12.4	C
28.0	12.0	61.0	10.7	D
29.0	11.2	60.1	9.1	D

The method for preparing the sample for the succulometer should be followed exactly, especially for the test on the raw corn. The washing and draining procedure is absolutely essential if differences caused by soil, climatic conditions, and depth of cut are to be eliminated. Subjecting the kernels to external sources of moisture at any other time will tend to result in higher values.

TEST 21.1 — Vacuum Oven Method– Moisture

Materials and Equipment

(1) Mortar and pestle
(2) Drying dish (15 × 90 mm)
(3) Vacuum oven
(4) Balance

Procedure

(1) Grind the sample. If sample cannot be ground, cut into small pieces.
(2) Spread 2–5 gm of the prepared sample as evenly as possible over the bottom of the weighing dish and weigh accurately to the second decimal place. (Run duplicate samples.)
(3) Dry at 70° for 6 hr under 26 in. of vacuum. During drying admit to the oven a slow current of air (about two bubbles per second) dried by passing through sulfuric acid.

(4) Cool in the desiccator for 30 min and reweigh.
(5) Calculate the moisture content of the sample and express as the percent of the fresh weight as follows:

$$\% \text{ Moisture} = \frac{(\text{Raw wt. of sample}) - (\text{wt. of dried sample})}{(\text{Wt. of raw sample} - \text{dish})}$$

or eliminate dish weight by using tare.

TEST 21.2 — Toluene Distillation— Moisture

Materials and Equipment

(1) Flask
(2) Stirling-Bidwell graduated collection tube
(3) Condenser
(4) Toluene

Procedure

(1) Weigh a 50-gm sample (sample should give 2–5 ml of water).
(2) Add to the flask and cover immediately with toluene.
(3) Connect the apparatus with cork stoppers (covered with tin foil) and fill the Stirling-Bidwell graduated collection tube with toluene (pouring it through top of condenser) and allow a few ml to flow over into the flask.
(4) Bring to boil and distill slowly (two drops per second) until most of the water has passed over (about 20 min). Now increase rate of distillation to four drops per second.
(5) Stop distillation when no more water collects in the collection tube. Then wash the inside of the condenser down with toluene to deliver all condensed water to the graduated arm.
(6) Cool water column to room temperature and read directly the ml of water collected. (For all practical purposes this is numerically equal to the grams of water collected.)
(7) From the weight of the sample and the weight of the water collected compute the percentage of moisture in your sample.

Interpretation

Potato Chips	3% maximum	(Potato Chip Institute)
Dried Fruits	18% maximum	(Quarter Master Corps)
Dehydrated Vegetables	5% maximum	(Quarter Master Corps)
Freeze-Dried Vegetables	3% maximum	(OSU—Gould)

TEST 21.3 — Infrared Moisture

Moisture content can be determined quickly using a top-loading balance with an attached infrared heating element. This provides continual indication of weight decrease or moisture loss throughout the drying cycle until a constant moisture loss is present.

Equipment and Materials

(1) Top loading balance
(2) Infrared heating element
(3) Drying dishes
(4) 5 gm of potato chips

Procedure

(1) Accurately weigh a representative 5 gm sample into a weighing dish directly on the balance.
(2) Turn on lamp and leave on until no further change in weight occurs.
(3) Record weight and calculate moisture content.

TEST 21.4 — Determination Of Alcohol Insoluble Solids Content Of Peas Or Corn

General

The determination of the alcohol insoluble solids in products at the present time like peas and corn is one of the most important indices of the stage of maturity for these items. The method is official and can be used on the frozen as well as the canned products. This method became effective in 1940 for the determination of borderline cases between standard and substandard merchandise. However, it has further application and that is for the calibration of quality instruments such as the tenderometer, texturemeter, consistometer and the succulometer.

Equipment, Materials and Reagents

(1) Sieve 8-mesh
(2) Pan balance
(3) Waring Blender or mortar and pestle
(4) Buchner funnel
(5) Aspirator for suction
(6) Stirring rods fitted with rubber policeman
(7) Beakers—600 ml and cover glass for same
(8) Drying dishes
(9) Air oven (100°C)

(10) Electric hot plate
(11) Filter papers
(12) Alcohol 80% by volume

Procedure

(1) Turn the electric heater on to warm up.
(2) Pour on sieve, drain the liquid from the product and rinse with a volume of water equal to twice the capacity of the container from which the product came. Drain again for 2 min and wipe the moisture from the bottom of the screen.
(3) Comminute (macerate) the product and stir to a uniform mixture and weigh two 20 gm replicate samples of the comminuted product.
(4) Add 300 ml of 80% alcohol, stir and cover beakers. (Use fresh alcohol.)
(5) Bring to a boil, simmer slowly for exactly 30 min and keep sides of beaker washed down with rubber policeman.
(6) Fit a Buchner funnel with a previously prepared filter paper of such size that its edges extend ½ in. or more up the vertical sides of a funnel. Moisten the filter paper for better fitting. Apply suction and filter the sample. Do not allow any of the product to run over the edge of the paper. Wash the material on the filter with alcohol until the washings are clear and colorless.
(7) Transfer the filter paper with the material retained thereon to the dish used in preparing the filter paper. Dry the material in a ventilated air oven for 2 hr at 100°C.
(8) Cool the dish in a desiccator and promptly weigh.
(9) From this weight subtract the weight of the dish and paper as previously found. The weight in grams thus obtained, multiplied by five gives the percent of alcohol insoluble solids.
(10) From your results determine the grade of product for maturity from Tables 21.2 and 21.3.

TABLE 21.2 — Percent AIS Determination For Peas

Type	Sieve Size	Grade A	Grade B	Grade C
Early	1 and 2	Up to 11.3	11.4–14.4	14.5–23.5
	3	Up to 15.1	15.2–18.5	18.6–23.5
	4 and 5	16.3	16.4–19.9	20.0–23.5
Sweet	2 and 3	Up to 9.8	9.9–12.2	12.3–21.0
	4	11.4	11.5–14.2	14.3–21.0
	5	12.7	12.8–15.9	16.0–21.0
	6	13.3	13.4–16.5	16.6–21.0

TABLE 21.3 — Percent AIS Determination For Corn (W.K.)
(Maximum Values)

Type	Grade A	Grade B	Grade C
Canned	20.01	24.23	27.00
Frozen	25.16	28.21	30.73

Notes

The previous preparation of the filter paper consists of drying in a flat bottomed dish for 2 hr at 100°C, covering the dish with a tight fitting cover, cooling it in a desiccator and promptly weighing.

If it is desired to determine the moisture content, the alcohol from the suction flask can be made up to volume (500 ml) with 80% alcohol and evaporated, dried to constant weight, correct for aliquot and calculate the soluble solids. The soluble solids plus the alcohol insoluble solids subtracted from 100 equals the moisture content.

TEST 21.5 — Flotation In Brine Solution

A. Peas and Lima Beans
Equipment
(1) 4–6 250-ml Beakers
(2) Brine solutions (11, 13, 13½, 15 and 16%)
(3) 2 Brine cylinders
(4) Teasing needles
(5) Spoons
(6) Thermometer (32–212°F)

Procedure.—
(1) Make up brine solutions using pure salt (NaCl). The temperature of the product and the solutions should be at the same temperature, preferably 68°F.
(2) Drain the peas and if necessary remove any adhering moisture from the product. Also, remove split or defective units.
(3) Take a representative number of units (20–25) from each sample. If frozen, remove the skins with the aid of a teasing needle.
(4) Fill a 250-ml beaker to about 2 in. with the brine solutions (start with the lowest concentration).
(5) Add the specific number of peas to the solution in the beaker.
(6) Determine the number of units that sink in the solution in ten

seconds. Only the units that sink to the bottom of the solution in this time are counted as "sinkers." The units that remain at the top of the solution or those that are in suspension are regarded as "floaters."
(7) Use increasingly higher brine solutions until the percentage of "sinkers" can be determined. The same brine solution should not be used for more than one separation and the sample should not be used more than once.
(8) Calculate the percentage of "sinkers" and interpret in accordance with Table 21.4 for canned peas and Table 21.5 for frozen peas.

TABLE 21.4 — Scoring Canned Peas For Tenderness And Maturity Using The Brine Flotation Principle For Sweet And Early Type Peas

Grade	Classification Descriptive Term	USDA Score Points	Maximum % by Count of "Sinkers"						
			Sweets			Early			
			11^1	13^1	15^1	11^1	$13\frac{1}{2}^1$	15^1	16^1
A	"Tender"	50	0	0	0	0	0	0	0
		49	4	0	0	4	0	0	0
		48	6	0	0	8	0	0	0
		47	8	0	0	12	0	0	0
		46	10	0	0	16	0	0	0
		45^2	12	2	0	20	2	0	0
B	"Reasonably Tender"	44	—	5	0	—	9	0	0
		43	—	8	0	—	16	0	0
		42	—	11	0	—	23	4	0
		41^2	—	15	4	—	30	8	0
C	"Fairly Tender"	40	—	—	6	—	—	—	6
		39	—	—	8	—	—	—	8
		38	—	—	9	—	—	—	9
		37^2	—	—	10^3	—	—	—	10^3
D	"Not Tender"	36 or less[3]	—	—	—	—	—	—	—

[1] % of salt (NaCl) in solution.
[2] These values are in accordance with the U.S. Standards for Grade of Canned Peas.
[3] Early peas that have more than 10% "sinkers" in 16% brine solutions (AIS—23.5%) or sweet peas with more than 10% "sinkers" in 15% brine solutions (AIS—21%) should be evaluated for the percent alcohol insoluble solids content.

B. Potatoes

Potatoes can be separated in a similar manner; however, the use of the National Potato Chip Institute Hydrometer is a quicker and more practical method. (See Test 21.6 and Table 21.6).

TEST 21.6 — Specific Gravity Determination

A. Hydrometric Method

Equipment.—

(1) SFA Potato hydrometer, basket and calibration bar
(2) 30–50 gal. container
(3) Scale to weigh 10 lb

Procedure.—

(1) Weigh 8 lb of potatoes which are clean, dry and representative of the lot. If not able to select potatoes to equal 8 lb, cut a potato to exactly obtain 8 lb.

TABLE 21.5 — Scoring Frozen Peas For Tenderness And Maturity Using The Brine Flotation Principle For Peas

Grade	Classification Descriptive Term	Score Points	Percentage of Salt in Solution (NaCl)	Peas With Skins Removed That Sink (By Count)
A	"Tender"	40	13%	0%
		39		Over 0% to 2%, inclusive
		38		Over 2% to 4%, inclusive
		37		Over 4% to 7%, inclusive
		36		Over 7% to 10%, inclusive
B	"Reasonably Tender"	35	15%	Over 0% to 3%, inclusive
		34		Over 3% to 6%, inclusive
		33		Over 6% to 9%, inclusive
		32		Over 9% to 12%, inclusive
C	"Fairly Tender"	31	16%	Over 0% to 4%, inclusive
		30		Over 4% to 8%, inclusive
		29		Over 8% to 12%, inclusive
		28		Over 12% to 16%, inclusive
D	"Not Tender"	27 or less	16%	Over 16%

(2) Place potatoes in hydrometer wire basket which has basket which has previously been calibrated with the calibration bar, usually to 1.070 by adjusting the scale in the hydrometer stem.

(3) Attach hydrometer to basket hook and lifting basket of potatoes in one hand while holding hydrometer in other, lower into a previously filled 30–50 gal. container of water at 68°F.

(4) As soon as hydrometer comes to rest, the specific gravity or percent total solids is read directly from the hydrometer stem at the water level.

Interpretation. Potatoes with specific gravity less than 1.070 are suitable for boiling and home fries. Potatoes with specific gravity greater than 1.070 are suitable for chip manufacture, dehydrating or baking.

B. Weight Method

During the past few years, many methods have been proposed for the evaluation of sweet corn maturity. These include AIS, succulence, and moisture tests. Some of these methods have found direct application in industry and in the official grading of fresh and processed corn. However, limitations of these methods have been observed by several workers. It would appear that a simpler and more rapid method would be in order. Further, because of the cost of some of the present equipment, a less expensive apparatus would be desirable.

With the above thoughts in mind, workers at the Ohio Agricultural Experiment Station (Crawford and Gould 1957) undertook research to develop and apply the specific gravity principle for determining corn maturity.

Several varieties of corn, each harvested at three or more maturity levels and each variety at each harvest, have been processed into canned and frozen whole kernel corn. Both the raw and processed product have been evaluated by several methods for determining corn maturity in addition to techniques developed for determining specific gravity.

TEST 21.7 — Specific Gravity Evaluation Of W.K. Corn — Weight Method

Equipment

(1) 8-Mesh 12-in. screen
(2) Scale capable of weighing, at least, 6 lb (¼ oz accuracy)
(3) 5-Mesh wire basket (12 in. in diameter by 4½ in. deep)
(4) 20 or 30 Gal. tank, at least 3 ft deep
(5) Hydrometer (0.90–1.10)

Procedure

(1) Take a representative sample of freshly cut corn.
(2) Drain for 2 min on an 8 mesh screen.
(3) Weigh a 6 lb sample.
(4) Weigh the sample in water of known specific gravity.

(5) Calculate the specific gravity of the corn as follows:

$$\text{Specific Gravity} = \frac{\text{Wt. corn in air} \times \text{Sp. Gr. of water}}{\text{Wt. corn in air} - \text{Wt. corn in water}}$$

For whole kernel, processed (canned or frozen) corn, exactly the same procedure is followed, except a 100-gm sample is used.

Equipment

(1) 1—8 mesh 8 in. screen
(2) 1—scale capable of weighing, at least 100 gm (0.01 gm accuracy)
(3) 1—14 mesh wire basket 2¼ in. in diameter by 3¼ in. deep
(4) 1—100-ml beaker
(5) 1—hydrometer (0.090–1.10)

With these specific gravity techniques, a processor should be able to evaluate raw and processed corn as to maturity more rapidly. However, it should be noted that the principle is not new since Burton as early as 1922 indicated that the method could be used to distinguish corn maturities (see Table 21.6).

TABLE 21.6 — Relationship Of Specific Gravity, Water Content, Dry Matter, and Starch in Potatoes

Specific Gravity	Percent Water	Percent Dry Matter	Percent Starch
1.040	86.4	13.6	7.80
1.045	85.4	14.6	8.75
1.050	84.5	15.5	9.60
1.055	83.6	16.4	10.46
1.060	82.6	17.4	11.41
1.065	81.7	18.3	12.26
1.070	80.8	19.2	13.11
1.075	79.8	20.2	14.06
1.080	78.8	21.2	15.00
1.085	78.0	22.0	15.76
1.090	77.0	23.0	16.71
1.095	76.1	23.9	17.56
1.100	75.1	24.9	18.51
1.105	74.2	25.8	19.37
1.110	73.3	26.7	20.22
1.120	71.4	28.6	22.01
1.125	70.5	29.5	22.87
1.130	69.6	30.4	23.72
1.135	68.6	31.4	24.67
1.140	67.8	32.2	25.43

TEST 21.8 — Succulence— Moisture Content

Preparation of a Sample of Canned Whole Kernel Corn

(1) Pour off the liquid, wash the contents of the can with twice its volume of water, and transfer to an 8-mesh screen, 8 in. in diameter.
(2) Spread evenly over the screen surface and allow to drain in a tilted position for 2 min, then weigh out a 100-gm sample for the test.

Preparation of a Sample of Raw Corn

(1) Make every effort to obtain a representative sample. Discard the extremely young, largely unusable ears.
(2) Cut with a whole-grain cutter, taking special care not to cut too deeply, not to bruise the kernels, and not to subject the material to any external source of moisture.
(3) Soak and stir the cut kernels in twice their volume of water for exactly 5 min, at the same time removing all floating material.
(4) Transfer a No. 2 canful of the thoroughly mixed soaked corn to an 8-mesh screen, 8 in. in diameter, spread evenly over the screen surface and allow to drain in a tilted position for 2 min, then weigh out a 100-gm sample for the test.

TABLE 21.7 — Specific Gravity Values for Fresh And Processed Corn By USDA

	Limits for Minimum Grades (Tenderness and Maturity Score)		
	A	B	C
Fresh cut corn for canning (6 lb values)	1.096	1.109	1.116
Fresh cut corn for freezing (6 lb values)	1.106	1.111	1.122
Canned whole kernel corn (100 gm values)	1.082	1.092	1.096
Frozen whole kernel corn (100 gm values)	1.100	1.103	1.112

Operation of the Succulometer

(1) Place 100-gm sample in sample chamber and place against backstop with plunger in place, close valve on pump tightly and pump handle up and down.

(2) After ram has made contact with plunger, pump slowly to prevent gushing of liquid.
(3) Pump until pressure of 500 lb is reached and maintain 500 lb pressure for exactly 3 min.
(4) Open valve and the ram returns automatically, then the sample unit can be removed and the plunger taken out of the sample chamber.

The volume of liquid that drains into the cylinder is the measure of maturity, young corn has more liquid, and as the maturity advances the amount of liquid that is expressed decreases.

TABLE 21.8 — Moisture Content In Fresh Fruits And Vegetables

Product	Average	Maximum	Minimum
Apples	84.1	90.9	78.7
Apricots	85.4	91.5	81.9
Avocados	65.4	68.4	60.9
Blackberries	85.3	89.4	78.4
Cherries, sweet	80.0	83.9	74.7
Figs	78.0	88.0	50.0
Grapefruit	88.0	93.1	86.0
Grapes—European	81.6	87.1	74.8
Muskmelon	92.8	96.5	87.5
Oranges	87.2	89.9	83.0
Peaches	86.9	90.0	81.9
Pears	82.7	86.1	75.9
Prunes (fresh)	76.5	89.3	61.6
Rhubarb	94.9	96.8	92.6
Watermelons	92.1	92.9	91.3
Artichokes	83.7	85.8	81.6
Asparagus	93.0	94.4	90.8
Snap Beans	88.9	94.0	78.8
Lima Beans	66.5	71.8	58.9
Beets	87.6	94.1	82.3
Cabbage	92.4	94.8	88.4
Carrots	88.2	91.1	83.1
Cauliflower	91.7	93.8	87.6
Celery, stalks	93.7	95.2	89.9
Corn, sweet	73.9	86.1	61.3
Cucumber	96.1	97.3	94.7
Lettuce	94.8	97.4	91.5
Onions	89.2	92.6	80.3
Peas, green	74.3	84.1	56.7
Potatoes	77.8	85.2	66.0
Pumpkin	90.5	94.6	84.4
Spinach	92.7	95.0	89.4
Tomatoes	94.1	96.7	90.6

Source: Taken in part from Joslyn (1950).

At the end of each test draw out the plunger, remove the corn residue, rinse with water, and dry with towel. Rinse the cylinder and hang inverted to drain. It is best to have several cylinders, to be used in rotation and an extra sample unit. A 25-ml cylinder, graduated in 0.2 ml is satisfactory. For extremely young samples it may be necessary to extend the graduations to 28 or 30 ml, or use a cylinder with a larger capacity (see Table 21.6).

BIBLIOGRAPHY

BURTON, L.V. 1938. Quality separation by differences of density. Food Ind. *10*, No. 1; 6–9, 56–58.

CALDWELL, J.S. 1939. Factors influencing quality of sweet corn. Canning Trade *61*, No. 42, 7–8.

CHASE, E.M., and CHURCH, C.G. 1927. Tests of methods for the commercial standardization of raisins. USDA Tech. Bull. *1*.

CRAWFORD, T.M., and GOULD, W.A. 1957. Application of specific gravity techniques for the evaluation of quality of sweet corn. Food Technol. *11*, No. 12, 642–647.

GOULD, W.A. 1958. Simple fast procedure tests sweet corn maturity. Food Packer *39*, No. 2, 19, 34.

GOULD, W.A., CRAWFORD, T.M., DAVIS, R.B., BROWN, W.N., and SIDWELL, A.P. 1959. A study of some of the factors affecting the grade relationship of fresh and processing vegetables. IV. Whole kernel sweet corn for canning and freezing. OAES Res. Bull. *826*.

GOULD, W.A. Quality Control Procedures. Snack Food Assn. 1982

HATTON, T.T., JR., and CAMPBELL, C.W. 1959. Evaluation of indices for Florida avocado maturity. Florida State Horticultural Society. Reprinted March 1960 by the Agricultural Marketing Service, U.S. Dept. of Agriculture.

JENKINS, R.R., and SAYRE, C.B. 1936. Chemical studies of yellow sweet corn in relation to the quantity of the canned product. Food Res. *1*. 199–216.

JOSLYN, M.A. 1950. Methods of Food Analysis Applied to Plant Products. Academic Press, New York.

LEE, F.A. 1941. Objective methods for determining the maturity of peas, with special reference to the frozen product. N.Y. State Agr. Expt. Sta. Tech. Bull. *256*.

LEE, F.A., DEFELICE, D., and JENKINS, R.R. 1942. Determining the maturity of frozen vegetables. A rapid objective method for whole kernel corn. Ind. Eng. Chem., Anal. Edition *14*, 240–241.

NICHOLS, P.F., and REED, H.M. 1932. Relation of specific gravity to the quality of dried prunes. Hilgardia *6*, 561–583.

SCOTT, G.C., BELKENGREN, R.O., and RITCHELL, E.C. 1945. Maturity of raw sweet corn determined by refractometer. Food Ind. *17*, 1030.

TWIGG, B.A., KRAMER, A., FALEN, H.N., and SOUTHERLAND, F.L. 1956. Objective evaluation of the maturity factor in processed sweet corn. Food Technol. *10*, No. 4, 171–174.

CHAPTER 22

Texture — Tenderness, Crispness, Firmness

Texture, tenderness, crispness and/or firmness are some of the more important attributes of quality in foods. Foods differ widely in their structure and physical properties. These differences are caused by (1) inherent differences within varieties or cultivars, (2) differences due to maturity-ripeness, (3) differences due to harvesting and/or handling-holding following harvest, and (4) differences caused by processing methods. For most food items, specific instruments have been developed for measuring these differences. These texture measurements are basically measured by the application of force. The force can be applied by several means: cut or shear, compression, or tensile strength. The technologist should be familiar with the available instruments and, if these are not applicable, develop instruments to measure these specific attributes. Examples of general instruments and their applications follow.

MAGNESS-TAYLOR PRESSURE TEST FOR FRUITS

Within a variety, certain factors cause changes in texture that are good indications of the maturity-softening of these fruits. An example of such a factor is whether or not the fruit is stored prior to marketing or processing. The most commonly used measuring instrument is the Magness-Taylor Pressure Tester.

The pressure tester is designed with different plungers. For apples a plunger with a $7/16$ in. diameter and penetration of $5/16$ in. is used. In making the measurement, it is customary to slice off enough of the peel so that this does not interfere with the action of the plunger. The plunger is held against the surface of the fruit and forced into the fruit with steady pressure to obtain the force required for breaking the flesh. This force is recorded on the pressure tester and used to indicate the maturity of the fruit.

Interpretation

Depending upon the variety, the data in 22.1 has been used to indicate degrees of ripeness for apples. With peaches and using the 5/16 in. plunger, Table 22.2 has been prepared by Craft (1955). With pickles and using the 5/16 in. plunger, Table 22.3 has been prepared to measure firmness.

TABLE 22.1 — Relationship Between Magness-Taylor Pressure Test, And Degree Of Ripeness For Varieties Of Apples

Degree of Ripeness	Varieties			
	Delicious	Winesap	Wealthy	Rome Beauty
Hard	17–20	19–26	16–20	18–23
Firm	14–17	15–19	13–16	15–18
Firm-ripe	11–14	12–15	10–13	12–15
Ripe	8–11	9–12	6–10	10–12

TABLE 22.2 — Range In Firmness (Magness-Taylor Pressure Test), Maturity And Ripeness Of Eastern-Grown Peaches At Harvest And Time Required For Them To Ripen At Room Temperature (70-90 degrees F)

Pressure-test Reading on				
Cheeks (Average) Pounds	Suture (Average) Pounds	Maturity at Harvest	Ripeness	Time Required to Ripen at Room Temperature (Days)
Above 16.5	Above 14.5	Immature	Hard-unripe	More than 8
16.5–16.0	14.5–14.0	Borderline	Hard-unripe	8
16.0–14.0	14.0–13.0	Mature	Hard-unripe	8–6
14.0–12.0	13.0–12.0	Shipping-mature	Firm-unripe	6–4
12.0–10.0	12.0–10.0	Advance-mature	Firm-unripe	4–2
10.0–2.0	10.0–2.0	Overmature	Firm-ripe	Less than 2
Under 2.0	Under 2.0	Overmature	Ripe	—

TEXTURE — TENDERNESS, CRISPNESS, FIRMNESS   315

FIGURE 22.1 — Magness-Taylor Pressure Tester

TABLE 22.3 —Firmness Ratings Of Whole Processed Pickles (Size 3) Using Magness-Taylor Pressure Tester (5/16" Tip)

Pounds	Rating
20–	Very firm
16–19	Firm
11–15	Inferior
5–10	Soft
0–4	Mushy

THE TEXTUREMETER FOR PEAS AND LIMA BEANS

The texturemeter, invented and manufactured by the Wm. F. Christel Co., Valders, Wisconsin, seems to meet the demands for a texture control instrument both in the field and in the laboratory. Its use has been limited primarily to peas and lima beans.

The texturemeter, like a U.S. Standard Scale, thermometer, or gauge, requires occasional checkup for accuracy. The recording of the texturemeter is gauged to conform quite closely to the U.S. Standard Scale Weight. This gives the user a basis against which to standardize. This can be done readily by using a piece of ¼ in. pipe about 6 in. long. Remove cup, place pipe on the middle punch, and, using a bathroom or other platform scale, place texturemeter with ¼ in. pipe resting on the scale (determine the weight of the texturemeter before it starts to record on the gauge) and deduct this weight from the scale recording. Press down on the texturemeter and watch the weight and see how it corresponds with the recording of the gauge. A few pounds variation are

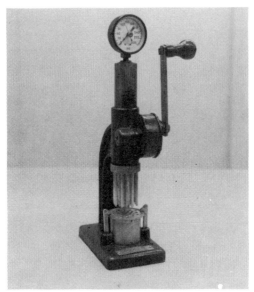

FIGURE 22.2 — Christel Texturometer

permissible. If, however, the variation is 5 lb or over, your texturemeter needs adjusting. If the recording is too low, remove gauge and fill cylinder with Lockheed Brake Fluid; if too high, let out a few drops of fluid.

Also, examine prongs and if any of them should have become sprung, which will cause friction, this will affect the recording. This can readily be remedied by using a short piece of ⅛ in. pipe; strip same over the prong that requires "truing up" and spring the prong in the direction necessary to bring it in proper alignment.

Operation

(1) Obtain a representative sample from every package or lot of product. Weigh an amount (50–100 gm) sufficient to fill the cup. This weight will vary with the size of the product being tested but should be uniform for any given product.
(2) Close the cup firmly and set indicator at zero.
(3) Turn the handle slowly and evenly until the plunger pierces the holes through the bottom of the cup. Take the reading from the gauge and readjust the indicator to zero in preparation for the next reading.

TEXTURE — TENDERNESS, CRISPNESS, FIRMNESS   317

(4) For uniformity, test three samples of each package or lot and take the average of the three readings.

Results

The following data shows the relationship between texturemeter readings of fresh Alaska peas and the grade of the processed product.

Texture Reading	Grades
Under 110	A very young—fancy
110–130	B young—extra standard
131–170	C mature standard
Over 170	D substandard

GOSUT TEXTUROMETER FOR SNAP BEANS, ASPARAGUS, CELERY AND OTHER LONG-SHAPED PRODUCTS

With snap beans, asparagus, celery, okra and other long-shaped products or products cut into longitudinal shapes, like French fried potatoes, the texture is most important to the consumer.

The texturometer was designed as a "go," "no-go" instrument with specific levels of acceptability developed for each commodity, depending on the interests of the user.

FIGURE 22.3 — GOSUT Texturometer

Operations

(1) Select sample size representative of the lot.
(2) Adjust weights on fulcrum arm for level of acceptance and raise arm to "up" position.
(3) Place sample in trough.
(4) Release fulcrum and if sample is cut, it is acceptable. If not cut, remove sample, adjust weights accordingly and use new sample and repeat evaluation. Continue until sample is cut.

TABLE 22.4 — Relationship Between GOSUT Texture Values And Sieve Size For Snap Green Beans

Sieve Size	GOSUT Texture Values
1	0.0– 5.0
2	5.1–10.0
3	10.1–15.0
4	15.1–19.0
5	19.1–23.0
6	Over 23.1

FTC TEXTURE PRESS AND RECORDER

This is different from other texture instruments in that the texture press and recorder is a multi-purpose instrument applicable to all measurements of pressure, cutting, shearing, and penetrating for a variety of foods. This instrument was developed by A. Kramer and associates. For example it measures:

(1) Maturity of raw peas, lima beans and southern beans.
(2) Firmness of raw and/or canned apple slices, beets, spaghetti, chicken, beef, and shrimp.
(3) Fibrousness of asparagus and snap beans.
(4) Succulence of sweet corn and apples.

Secondly, the texture press and recorder records complete time-force curves instead of mere indication of maximum force. For example, doughnuts with a hard crust and soft center may give a peak force reading higher than a firm doughnut with no crust.

However, when a complete curve is charted, the soft crusty doughnut will show a slowly rising curve that comes to a sharp peak toward the end of the stroke. The crustless firm doughnut will show a rapidly rising curve that levels off and gradually begins to descend toward the end of the stroke.

FIGURE 22.4 — Food Technology Corporation's Model TMS-90 Texture Management System — A Computer Controlled System For Texture measurement And Data Analysis

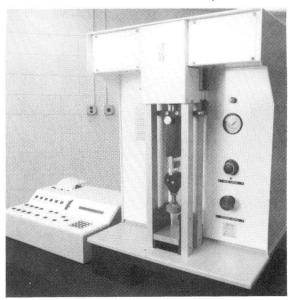

TEST 22.1 — Pericarp Content Of Sweet Corn

General

One of the attributes of the quality of canned corn is tenderness (degree of toughness). Tenderness varies according to cultivar or variety, maturity, cultural practices, harvesting, handling, and processing variables. Measurement of the pericarp content has been shown to be a reliable index of tenderness in processed corn. The procedure for pericarp is as follows.

Equipment, Materials and Reagents

(1) Sieve (8 mesh)
(2) Waring Blendor or Osterizer

(3) Monel screen (30 mesh—4 in. square by 1 in. deep)
(4) Air oven (100°C)
(5) Desiccator
(6) Balance (sensitive to three places)
(7) Water

Procedure

(1) Comminute two (duplicate) 25 gm of the product (silk, husk, cob and other material which is not corn has been removed) with 200 ml of water in a Waring Blendor for exactly 3 min.
(2) Wash the slurry from the blenders onto previously dried and weighed monel screens with water and wash until the washings are clear.
(3) Dry the screens and contents in an oven for 2 hr at 100°C.
(4) Cool in a desiccator and promptly weigh to the nearest 0.001 gm.
(5) From this weight subtract the weight of the monel screen. The weight in grams thus obtained, multiplied by four shall be considered to be the percent pericarp content. Corn that exceeds 2.00% pericarp content is definitely not Fancy corn. Pericarp content in excess of 2.30% would probably be considered too tough for even bottom standard corn. The pericarp content varies with varieties, maturities and the different growing season. See Table 22.5.

TABLE 22.5 —
Relationship Between Canned Whole Kernel Corn And % Pericarp

Grade	Pericarp Range (%)
A	1.04–1.36
B	1.37–1.78
C	1.79–2.29
D	Over 2.30

TEST 22.2 — Determination Of Fiber Content In Snap Beans And Asparagus

General

The amount of woody or fibrous material in the pods of snap beans and in the asparagus spears increases as beans and asparagus increase in maturity. Not only is there the effect of maturity, but also there are

TEXTURE — TENDERNESS, CRISPNESS, FIRMNESS

differences in fiber content at the same stages of maturity among the different cultivars or varieties. The following method is used by the F&DA in determining the fiber content of snap beans for the Minimum Standard of Quality of canned snap beans.

Equipment

(1) Double scalloped malted milk stirrer and stainless steel container fitted with handle
(2) Air oven adjusted to 100°C
(3) Balance accurate to 3 decimal places
(4) Gas burner
(5) 30-mesh monel screen—4 in. square × 1 in. deep
(6) Pyrex test tube or wooden pestle
(7) Sodium hydroxide—50% by weight
(8) A 2-octanol or capryl alcohol
(9) Phenolphthalein indicator

Procedure

(1) Weigh out 100 gm of beans. Then, remove the seeds and weigh the deseeded beans to determine seed content in percent.

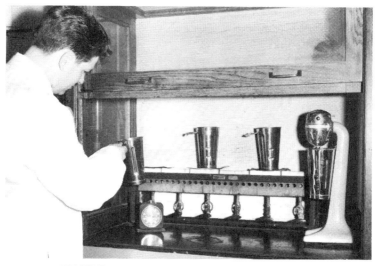

FIGURE 22.5 — Determination Of Percent Fiber

(2) Deseed more beans to give 100 gm of deseeded pods.
(3) Weigh 100 gm of deseeded beans and transfer to a metal cup of the malted milk stirrer and mash with the Pyrex test tube. For raw or frozen beans use the wooden pestle.
(4) Wash material adhering to the test tube into the malted milk cup with 200 ml of boiling water.
(5) Bring mixture nearly to a boil on a previously calibrated burner.
(6) Add 25 ml of sodium hydroxide solution and boil 5 min. (If foaming is excessive, 1 ml of capryl alcohol or 5 drops of 2-octanol may be added).
(7) Stir the contents for 5 min with a malted milk stirrer capable of a no-load speed of at least 7200 rpm.
   CAUTION: DO NOT ALLOW SLURRY TO SPLASH AROUND. DO THE STIRRING IN THE HOOD WITH THE DOOR PARTLY DOWN. THE TECHNOLOGIST MUST WEAR APRON AND GLOVES.
(8) Transfer the material from the cup to the previously weighed 30-mesh monel screen.
(9) Wash the fiber on the screen with a stream of water using a pressure not exceeding a head (vertical distance between upper level of water and outlet of glass tube) of 60 in., delivered through a glass tube 3 in. long and ⅛ in. inside diameter inserted into a rubber tube of ¼ in. inside diameter. Continue the washing until there is only the remaining fibrous material, then rinse with distilled water. Moisten with a drop of phenolphthalein. If it does not show any red color after 5 min standing, the washing is finished. Rinse to remove the indicator.
(10) Dry the screen containing the fibrous material for 2 hr at 100°C. Cool in a dessicator, weigh the screen and contents and then deduct the weight of the screen.

TABLE 22.6 —
Relationship Between Canned Snap Beans And % Fiber

Grade	% Fiber
A	Less than 0.05
B	0.10
C	0.15
D	Greater than 0.15

(11) Calculate the percent of fiber in the raw material and interpret in accordance with data in Table 22.6.

TEST 22.3 — FTC Texture Press And Recorder Using Apple Slices

Materials and Supplies
(1) Fresh, canned and frozen apple slices

Procedure
(1) Place apple slice samples in the sample box. Slide the box into the press. Apply pressure to the piston.
(2) The pressure applied to the piston is transmitted to the test cell to force it down and through the sample box. Resistance to this force is measured by subsequent compression of the proving rings.

Interpretation
See Table 22.7 for interpretation.

TABLE 22.7 —
Maturity Scores Based On Shear-Press Values For Lima Beans

Fancy Shear-Press	Score	Extra Standard Shear-Press	Score	Standard Shear-Press	Score
950	100	1400	91	1800	83
1000	99	1450	90	1850	82
1050	98	1500	89	1900	81
1100	97	1550	88	1950	80
1150	96	1600	87	2000	79
1200	95	1650	86	2050	78
1250	94	1700	85	2100	77
1350	93	1750	84	2150	76

[1]For other products, similar Shear-Press versus scores can be developed and grade or quality standards established.

BIBLIOGRAPHY

ANON. 1948. Canned vegetables: Definitions and standards of identity, quality and fill of container. Canned green beans and canned wax beans. Federal Register July 8: 13FR 3724–3728.

ANON. 1961. Lee-Kramer Comptroller. Device Tests "Feel" of Foods. Reprinted from Food Marketing International. 170 Varick St. New York.

CRAFT, C.C. 1955. Evaluation of maturity indices based on pressure test reading for eastern-grown peaches. U.S. Dept. of Agr. AMS–34.

CULPEPPER, C.W. 1936. Effect of stage of maturity of the snap bean on its composition and use as a food product. Food Res. *1*, 357–376.

DAVIS, D. and GOULD, W.A. 1961. The effect of maturity, formulation, and storage time and temperature on the consistency of canned cream style sweet corn. Ohio Agr. Expt. Sta. Res. Bull. *891*.

GOULD, W.A. 1949. Instrument to quickly reveal quality of snap and wax beans. Food Packer, *30*, No. 13, 26–27.

GOULD, W.A. 1950. What factors produce a fancy pack bean? Food Packer *32*, No. 4, 26–27, 68, 70.

GOULD, W.A. 1951. New snap beans analyzed. Food Packer *31*, No. 5, 26–27, 68, 70.

GOULD, W.A. 1951. Quality evaluation of fresh, frozen and canned snap beans. Ohio Agr. Expt. Sta. Res. Bull. *701*.

GOULD, W.A., CRAWFORD, T.M., BROWN, W.N. and SIDWELL, A.P. 1959. A study of some of the factors affecting the grade and relationships of fresh and processed vegetables. IV. Whole kernel sweet corn for canning and freezing. Ohio Agr. Expt. Sta. Res. Bull. *826*.

GOULD, W.A., JOHNSTONE, F.E., JR., BROWN, H.D., KRANTZ, F.A., JR., DAVIS, R., GUYER, R.B., and KRAMER, A. 1952. Studies of Factors Affecting the Quality of Green and Wax Beans. Maryland Agr. Expt. Sta. Bull. *A68*.

KALIA, M., and GOULD, W.A. 1974. The effect of calcium sulfite and ascorbic acid dips and storage temperature on quality of apple slices (Melrose and Golden Delicious cultivars) for processing into apple pies. Dept. Hort., Hort. Ser. *403*.

KRAMER, A. 1957. Food texture rapidly gauged with versatile shear press. Food Eng. *34*, No. 3, 44, 66, 70; No. 4, 42, 58–60; No. 5, 32, 50–51.

KRAMER, A., BURKHARDT, G.J. and ROGERS, H.P. 1951. The shear press: A device for measuring food quality. Canner *112*, No. 5, 34–36, 40.

KRAMER, A., HART, W.J., JR., DULANEY, J.H. and Son. 1954. Recommendation on procedures for determining grades of raw, canned and frozen lima beans. Food Technol. *8*, No. 1, 55–62.

PARKER, M.W., and STUART, N.W. 1935. Changes in the Chemical Composition of Green Snap Beans after Harvest. Maryland Agr. Expt. Sta. Bull. *383*.

ROWE, S.C., and BONNEY, V.R. 1936. A study of chemical methods for determining the maturity of canned snap beans. J. Assoc. Offic. Anal. Chemists *19*, 620–628.

SHERMAN, P. 1972. Structure and textural properties of foods. Food Technol. *26* No. 3, 69–79.

STARK, F.C., JR., and MAHONEY, C.H. 1942. A study of the development of the fibrous sheath in the side wall of edible snap bean pods with respect to qaulity. Proc. Am. Soc. Hort. Sci. 41:351–359.

WEGENER, J.B., and BAER, B.H. 1950. Quality comparison favors top crop bean for freezing. Food Packer *31*, No. 11, 54–56.

CHAPTER 23

Rheology — Viscosity — Consistency

DEFINITIONS

Viscosity is a measure of the resistance offered by a fluid to relative motion of its parts. More precisely, it is the ratio of resistance to shear to rate of shear.

A *Newtonian Liquid* is a liquid for which resistance to shear is directly proportional to rate of shear.

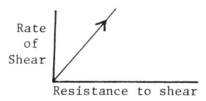

A *Non-Newtonian Liquid* is a liquid for which resistance to shear is not linearly related to rate of shear. Apparent viscosity is a measure of the "viscosity" and must be related to a particular rate of shear for Non-Newtonian Liquids.

Types of Non-Newtonian Products:
 (a) *Plastic* materials show a decrease in apparent viscosity as the rate of shear increases. (Examples are chocolate liquor, hot catsup, cream style corn.)

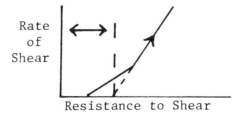

Ignore curve below yield value (i.e., ignore apparent viscosity) at low rates of shear because this is caused by "plug flow" and is not a measure of product viscosity at higher rates of shear. Plug flow occurs when the material moves through a pipe as though it were a solid, i.e., without relative motion of different parts of the fluid.
(b) *Pseudoplastic* materials show that the apparent viscosity decreases as the shear at which the material is tested increases.

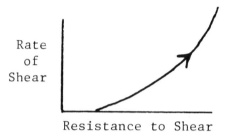

(c) *Dilatant* materials show an increase of apparent viscosity (thickening) as the rate of shear increases.

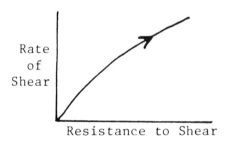

(d) Non-Newtonians which are plastic may also be *Thixotropic*. This curve demonstrates a hysteresis effect, in that the apparent viscosity at any particular rate of shear will depend on the amount of previous shearing to which it has been subjected (Brookfield Engineering Laboratory 1961).

CLASSIFICATION OF INSTRUMENTS

Flow Rate Instruments

All rate measurement devices have an inherent defect; their accuracy is proportional to the time consumed in making the measurements.

Gravity Flow.—*Capillary Constriction.*—These types were designed for very precise measurement of low viscosity systems. Determination of whether the liquid under observation is Newtonian or not can be made by measuring the flow under different initial heads or different diameter capillaries (i.e., vary shear rate). If the flow is not Newtonian, the viscosity will differ as the conditions are varied. For determining the characteristics of non-Newtonian flows, these instruments are ineffective for the following reasons.

(1) They can be used only with low viscosity materials.
(2) They have a wide variation of shear gradient across any cross section of flow.
(3) The driving force or head, and thus the instantaneous flow rate, is continuously changing.

These instruments have the advantage that the hydrodynamics of their flow have been subjected to extensive and rigorous mathematical analysis, and the interpretation of flow data obtained with capillaries has been well described and documented.

The original instrument was that designed by Ostwald and may be obtained either as an *Ostwald* viscometer or the *Fenske* modification.

Another modification of the original Ostwald viscometer is that by *Cannon & Fenske* which utilizes a reverse flow principle. This is necessary for materials that cling to the viscometer walls or otherwise have characteristics that make end point and volume measurements inaccurate.

The *Ubbelohde* variation of the capillary viscometer employs an extra tube, connected just below the capillary, which provides for free drainage from the capillary, thus avoiding skin and head effects causing complications with the other arrangements.

The *Fisher-Irany* modification of the capillary viscometer employs a spiral capillary which gives a maximum of capillary path with a minimum of head, thereby increasing the sensitivity for very fluid liquids.

Orifice Constriction.—These are the simplest possible types and are usually employed for comparative purposes without an attempt at strict physical interpretation. They may be enclosed within a bath for close temperature control or they may be simple suspended devices exposed to the ambient temperature of the room. Volume and proportions of the cup, as well as the size and configuration of the orifice, are subject to infinite variation. Measurements may be made of the total efflux whereby the volume is determined by calibration of the receiving container.

Since there is a very sharp change in shear rate as the material being measured enters the orifice, and since the time interval during which a portion of the material is being subjected to this high shear rate is very small, the measurements with orifice type instruments approximate conditions close to zero time with regard to the effect of major shearing forces on the material being examined. This type of instrument can be used only with fairly fluid products and suffers from the drawback that the material being measured is subjected to a continually varying head and, therefore, varying rate of shear. These instruments do, however, have the advantage of simplicity and can be useful for comparative measurements.

A number of these instruments have been devised as industry standards.

(1) The *Saybolt* is the standard for the petroleum industry. The cup is built into a temperature-controlled bath. Efflux of a definite volume is measured.

(2) The *Engler* and *Scott* are both minor variants of the Saybolt.

(3) The *Ford* cup is basically a Saybolt without a bath.

(4) The *Zahn* cup is a convenient dip cup with an orifice from which the time for total flow is measured. This is the simplest and most convenient of the orifice type viscometers but since it has no built-in means for temperature control, it cannot be employed for very precise work.

(5) The *Dudley* pipette is only a normal transfer pipette with a volume calibration line on the stem below the bulb. Since precise duplication of bulb size and orifice are difficult to obtain with glass apparatus, comparisons should be made with the same pipette.

(6) The modified Efflux-tube viscometer (GOSUC consistometer) was developed at The Ohio State University Processing and Technology Laboratory. It consists of a blown-glass reservoir sloped into a 2-mm orifice on the efflux and a ¾ in. orifice at the top, which enables the pouring of the sample into the reservoir through a funnel or the filling of the reservoir by vacuum. A piece of rubber

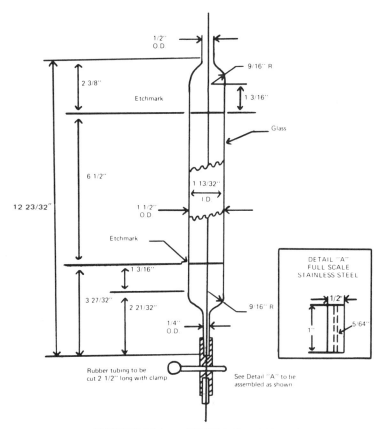

FIGURE 23.1 — GOSUC Consistiometer

tubing attaches a metal plug with a 5/64 in. (2 mm) precision-bore orifice. This allows the flow to be stopped by a pinch-clamp and the instrument to be accurately standardized by adjusting the length of rubber tubing. Consistency is measured in seconds required for the efflux of 150 ml of juice. The instrument is standardized with water by varying the length of the rubber tubing so that a flow of 150 ml is completed in 32 sec. The water and product should be at 70°F for all readings. No standards for Grades have been promulgated; however, our studies show that thin juice is in the range of 38-44 sec and thick juice is over 70 sec (See Fig. 23.1).

Flow Under Pressure.—Through the use of compressed gases or hydraulic rams, constant pressure substantially greater than gravity can be obtained. The greater induced pressure also allows for the examination of more highly viscous materials. For the same reasons that apply to the gravity instruments, the pressure types will measure the rheological properties of the material at close to zero time.

Falling Body Instruments

Falling Ball Types.—The *Hoeppler* falling ball viscometer will provide a measure of the initial flow rate under constant stress. There is an inherent upper limit in the stresses that can be applied due to the density of materials that can be used for balls. The low limit is zero since the density of the ball can be balanced against the density of the liquid to be measured. In this type of instrument, as with all ball types, the density of the material, along with the viscosity, is a factor in determining the rate of fall or rise of the ball. This type of device is limited to essentially clear materials unless a nonvisual sensing means is used for following the position of the ball.

Falling Plate Devices.—The *Fisher-Gardner* and the *SIL Koehler Mobilometers* are instruments in which the orifice moves instead of the sample. They will determine the initial shear rate under a constant stress. Unlike the falling ball type, there is no limit to the loading that can be applied and opaque materials that can be measured, since the movement of the plate is followed through a guiding rod extending above the sample. This guiding rod produces one of the major problems in the use of this type of equipment since it involves an uncertain source of friction.

The *Gardner Mobilometer* was developed in 1927 by Gardner and Van Heucheroth. Gray and Southwick (1930) reported on the development and the use of the Mobilometer. This instrument measures the consistency of products by determining the time in seconds for the plunger to fall in the food product through the distance between two reference points (usually 10 cm) on the plunger rod.

As previously mentioned for other consistency measuring instruments, the consistency of any product must be taken on any product at controlled temperatures. The Mobilometer is constructed with a water-cooling jacket which surrounds the sample and is specifically designed to maintain uniform temperatures when determining the consistency of any sample.

The Mobilometer has been used to evaluate the consistency of vegetable and other oils, sirups, heavy cream products, mayonnaise and tomato products. The results obtained with this instrument may be expressed as the product of the time (seconds) and load (grams) divided by the distance traveled (centimeters). With the distance and the load remaining constant, consistency or viscosity as measured with the Gardner Mobilometer then becomes a function of time.

This instrument consists of five basic parts: (1) base plate with bubble leveling device; (2) sample cylinder; (3) plunger (several types of discs are available) and weight pan; (4) a bracket to support the weight pan; and (5) a water-cooling jacket. The general procedure for determining the consistency of food products is as follows:

(1) Place the instrument on a horizontal shelf and level it by adjusting the leveling screws.
(2) Remove the cylinder and fill to a uniform height and then replace the cylinder into its position inside the water jacket.
(3) Attach the desired disc to the plunger.
(4) Adjust the plunger in the cylinder until the disc is just immersed (approximately ½ in.) in the sample and add the weights to the weight pan.
(5) Release the plunger and record the time with a stopwatch for the plunger to pass through the desired distance (usually 10 cm).

The number of seconds required for the plunger with uniform driving weight to travel through 10 cm distance is used as a measure of the consistency of the food sample. For each product in which consistency is a factor of quality, the quality control technologist should establish the range in seconds or the desired consistency level for each quality.

Limit-Of-Flow Devices

These devices are applicable only to products of fairly heavy body. They provide a quantitative measure of a subjective attribute of the product. The techniques employed are generally simple and rapid so that they lend themselves to control operations.

Cone Penetrometer.—The most common penetrometer employed is the Universal. When a cone attachment is used it penetrates the mass under test until the effective cross-section area times the yield value of the material balances the weight applied. In this way, the distance the cone penetrates bears a relation to the initial yield value.

Spreading Mass.—These are simple devices where a pile of material allowed to flow under its own weight spreads out until there is insufficient head to overcome the resistance to flow. Since the rates of flow involved are relatively low, and the limiting flow used is very close to zero, the measurements obtained are close to the initial yield value of the product.

The Bostwick Consistometer.—Developed by E.P. Bostwick of the U.S. Department of Agriculture in 1938, the Bostwick consistometer is officially used by the USDA in establishing the score points for tomato catsup. The Bostwick consistometer is an instrument used to determine the consistency of a viscous material by determining how far the material will flow under its own weight along a level surface in a given period of time. The consistometer consists of a metal trough closed off near one end by a gate which can be opened almost instantaneously.

The end closed off by the gate, the reservoir, has walls which are carefully measured and leveled along the top. The longer end is graduated in ½ cm steps starting 1 cm from the gate. The graduations are numbered at each centimeter. It is equipped with a two-way level and two leveling screws on the reservoir end. The gate slides vertically between two grooved posts. It is pushed up by two springs in the posts, and is held down in a closed position by one end of an L-shaped trigger which hooks over the top of the gate. When the free end of the trigger is

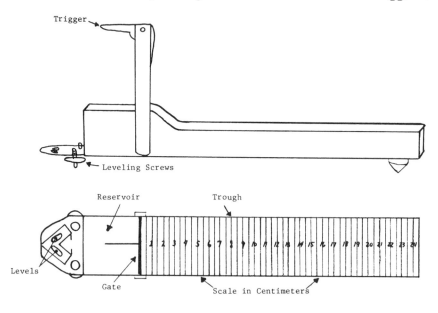

Drawing Scale: ½" equals 1"

FIGURE 23.2 — Bostwick Consistometer

pushed down, it rotates around its swivel and the gate opens up. Necessary accessories for making consistency measurements with the Bostwick consistometer are a stopwatch or other timing device, straightedge such as a spatula, and a thermometer. All measurements should be made at given temperatures. Also, before making consistency measurements the product must be at the temperature specified for the test. The product should be kept at that temperature (usually 20°C or 68°F) for several hours before the test so that the entire lot has been adjusted to that specific temperature.

The Adams Consistometer.—This device was designed and constructed by Adams and Birdsall of the Kroger Food Foundation. This instrument has been described in the Food and Drug Administration's minimum standards of quality for canned cream style corn for determining the consistency of cream style corn. Adams and Birdsall state that the Bostwick Consistometer "was not found to be applicable for cream style corn consistency measurement." Graders could not duplicate their results satisfactorily for several reasons. Its capacity was too small for an adequate sample. The small mass when released did not possess sufficient weight to extend the flow over a range adequate for good differentiation. Thus, the device lacked sensitivity. The measuring scale etched in the bottom of the channel is composed of straight, parallel lines spaced 1 cm apart, yet the advancing line of flowing corn forms an arc. To determine what gradation under this arc to record as the measure of consistency led to inaccuracies and encouraged divergent results by different operators. Another inherent limitation is "that the sample of corn is allowed to flow in one direction only, and is restrained in all directions." The above statements pertain to the reason for the development of the Adams consistometer.

The essential parts of the Adams consistometer are: leveling screws; a hollow, truncated cone with an inside bottom diameter of 3 in., inside top diameter of 2 in., and a height of $4^{27}/_{32}$ in.; and a polished circular measuring plate with a minimum diameter of 12 in. This plate is marked from the center, starting at 3 in., out to 12. Some models are marked in ¼ in. intervals while others are marked in ½ in. intervals. In the design of one model, the measuring plate is made of thick plexiglass with the numbers etched on the bottom sides. This has the added advantage of ease of cleaning and the prevention of friction in the flow of the product.

Ticher at the Iowa Experiment Station reported on the use of inexpensive papers instead of the polished metal plate or the plexiglass plate. He states that "the paper consistometer may be used in any reasonable

FIGURE 23.3 — Adams Consistometer

location to determine the consistency of cream style corn and similar products without the necessity for washing facilities. In addition to the elimination of the time and expense of washing, the paper consistometer may be printed in any quantity at the discretion of the canner and should be especially useful to the small canner who may not wish to make the investment necessary to purchase more expensive equipment."

Rotational Viscometers.—Any of the commonly available rotating devices can be equipped with pins, vanes, discs, or cylinders for sensing the viscosity of the product under observation. The differences between the various devices available involve cost, ease of operation, sensitivity of the measuring portion and reproducibility of the sensing situation. Some are made primarily for use with one or another of the possible sensing devices. Others are adaptable to many types. For fundamental work the sensing device is important since this determines the geometry of the flow being observed.

Constant Force Type.—The constant force type of rotational viscometer is not adaptable to thixotropic materials since measurements are made over a time interval during which the consistency is changing, and the measurement is confined to a finite time.

The *Stormer Viscosimeter* is used to measure objectively the viscosity or consistency of certain food products and to give an index of the resistance of the sample to flow. It is a precision instrument in design and is rather simple in operation.

When using the Stormer Viscosimeter, the operator should periodically check the instrument against known standards to make certain the instrument is always in calibration. Also, when making viscosity measurements all the samples should be at the same temperature, as the viscosity decreases with elevated temperatures.

With most food products, the consistency or viscosity of the sample, as taken with the Stormer Viscosimeter, is reported as the relative viscosity. The relative viscosity is "obtained by dividing the time required for the rotor to make 100 revolutions in distilled water, using the identical procedure, temperature, and driving weight. Distilled water may be replaced by another standard such as castor oil or glycerol if the liquid under test is beyond the viscosity range of water."

If desired to express the viscosity time in seconds for 100 revolutions of the rotor, in terms of centipoises (absolute viscosity), this can be done by referring to calibration curves of liquids of known viscosities. The viscosity of water at 68°F is very nearly 1 cp. However, one must remember that some samples of food products may not be truly viscous, that is, the revolutions per second are not always directly proportional to the driving weight. This is referred to as that "apparent viscosity" of the sample. In making viscosity determinations, the quality control technologist should determine a consistency or viscosity curve; that is, plot increased driving weight versus flow (revolutions/second) for each type of food product.

The Stormer Viscosimeter consists of six basic parts: (1) test cup, for many food products a No. 2 can is most satisfactory; (2) rotor, a cylindrical rotor is satisfactory for many food products, although the paddle or propellor types may have certain specific uses; (3) driving weight, depending on the type of product, the weight may be increased from 0–850 gm or more; (4) winding drum; (5) brake mechanism; and (6) revolution counter. The gears and pinion mechanisms are mounted inside a dustproof cap mounted on the top of the upright supports.

Constant Speed Devices.—The *MacMichael Viscometer,* one of the early models developed in this country, is adaptable to a large number of measuring situations since it is equipped with means for infinite speed variation over a limited low speed range and a variety of sensing and force-measuring accessories, so that a wide range of products can be measured. Inherent mechanical defects cause difficulty in maintaining calibration. Temperatures below room temperature are difficult to control. The mechanical setup limits the variety of sensing devices that can be used. Because the sample is rotated, centrifugal effects limit the rotational speeds that can be employed.

The *Fisher Electroviscometer* was recently developed by the Fisher Scientific Company. This instrument is of the rotational type and measures the viscosity of the sample by means of a patented torque-magnetic-electrical system. The instrument also has a great advantage because the viscosity is measured directly in the absolute units, that is, centipoises.

The sample to be tested is rotated about a stationary bobbin. The sample cup rotates at a constant speed. The rotating sample exerts a torque on the bobbin, and the viscosity is determined. Since the bobbin is attached directly to a patented coil, its torque tends to turn the coil. The coil, however, is in a magnetic field and resists the tendency to turn when current is passed through it. The restoring torque from a regulated voltage power supply tends to swing the coil back to its original position. The force necessary to keep the coil from turning registers on a meter which is calibrated directly in centipoises. The operation is as follows: (1) fill the sample cup with the sample and put into the holder; (2) start the motor; (3) adjust the control; and (4) take the reading directly in centipoises.

The *Brabender Recording Viscometer* (Brabender Corp., Rochelle Park, N.J.) is another device which rotates the sample. It has the same defects as the MacMichael and Fisher instruments with regard to speed and sensing devices, lacks their sensitivity, but does have the recording feature desirable for observing change inconsistency as work is applied.

The *Drage Viscometer* (Drage Products, Union City, N.J.) is essentially a force measuring device on a controlled speed turntable. The principle seems to be the same as that of the *Brookfield Viscometer*.

The *Brookfield Synchro-lectric Viscometer* is an instrument that has received wide acceptance in many segments of the food field. The instrument was developed by D.V. Brookfield in 1933. Today the Brookfield Engineering Laboratories specialize in the manufacture of several models of viscometers and viscosity control instruments. Their instruments are used to measure the viscosity and consistency of food products, such as custards, pie fillings, starches, mustards, catsups, mayonnaise, salad dressings, cream style corn and dairy products. The instrument has several outstanding advantages. One is that the viscosity is measured in centipoises; thus, no conversion tables, etc., are needed. The instrument is also versatile, rugged in design, and easy to clean.

When using the Brookfield instrument the operator need only select the specific spindle for the food product, place the spindle in the food product (the product may be left in a No. 2 can or any other container having a capacity of approximately 1 lb), and turn on the motor. The clutch lever is then released and the spindle is allowed to make several revolutions. The spindle is then stopped, motor turned off and the dial reading taken. This is recorded as the viscosity of the sample. The different models are equipped with different spindles and a wide range of speeds. For various products, specific spindles and speed are recommended.

The FMC Corporation has recently developed an *FMC Consistometer* as a control instrument for use in the application of the FMC continuous pressure cooker for cream style corn. The instrument has appliation with other food products, although suitable changes in the paddle design, size and shape may be necessary.

The FMC consistometer consists of two basic parts: (1) the turntable which rotates at a constant speed; and (2) a measuring head calibrated in arbitrary units from 1 to 100. The general operating procedures are as follows: (1) the food sample is placed in a 303 × 406 can which is placed on a turntable; (2) the measuring head is lowered into the sample; (3) the turntable is started; (4) a reading is made at the end of 30 sec and is referred to as the dynamic reading; (5) the turntable is stopped and another reading is made at the end of 60 sec. This reading is referred to as the static reading. By checking against known samples, it is possible to control viscosity with this instrument. The instrument is used to measure the consistency of corn before and after processing and the readings have given good correlation with those obtained from the Adams consistometer.

Other high shear rate viscometers are the *Hagan* (Hagan Corp., Pittsburgh, Pa.) and the *Hercules* (Martinson Machine Co., Kalamazoo, Mich.) Both seem to be versatile, well-built mechanical systems, but lack the temperature control required in high speed devices.

Another viscometer for making measurements with closely defined shear rates at high speed is the *Feranti-Shirley* (Feranti Electric, New York). This viscometer makes use of the cone and plate principle for maintaining a uniform rate of shear across the sample being measured. It also provides a jacketed system for temperature control. Rotational speeds indicated are up to 1,000 rpm, although comments from people who have experimented with cone and plate devices indicate that at high rotational speeds, centrifugal forces can cause difficulties.

One drawback of both cone and plate and narrow angular sensing elements is that introduction of the sample subjects it to considerable manipulation so that if measurements starting with zero time conditions are desired, the instrument or some essential part of it is tied up waiting for thixotropic recovery.

There are many other instruments that work on principles somewhat similar to the above instruments. Many of these may find direct application to specific food products either for control purposes or for quality evaluation. Some of these instruments that have been used in the past for specific purposes are: (1) The MacMichael viscosimeter and (2) Searle's viscosimeter.

TEST 23.1 — Consistency Of Food Products With The Bostwick Consistometer

Equipment

(1) Bostwick consistometer
(2) Spatula
(3) Stopwatch

Procedure

(1) The consistometer is first carefully leveled by adjusting the leg screws in order to obtain a uniform flow over the scale.
(2) The gate is then closed and held in this position by engaging the trigger release mechanism; this forms a holding compartment at one end of the consistometer.
(3) The product is poured into the holding compartment and leveled off even with the sides of the compartment with the aid of a spatula. Care should be exercised to avoid air pockets in heavy products such as fruit jams, applesauce, canned cream style corn and similar products having a heavy consistency.
(4) The gate is then released, allowing the product to flow over the centimeter scale for 30 sec. The instrument should be held steady to prevent jumping when the gate is released. As the flow at the center of the scale exceeds slightly the flow at the sides of the consistometer, the maximum distance of flow is taken as the average of the two extremes. Further, the Bostwick consistometer is a very satisfactory instrument for factory control and quality evaluation of catsup.

Also, the Bostwick is a small, very portable, and easily operated instrument, and does not require large samples.

Interpretation

See Table 23.1.

TABLE 23.1 — Tomato Catsup Consistency Score Points With Bostwick

Grade	Point Score	Flow in Centimeters
A	25	7 or less
	24	8
	23	9
	22	10
C	21	Over 10 to 12
	20	14
	19	16
	18	18
D		Over 18

TEST 23.2 — Consistency Of Food Products With The Adams Consistometer

Equipment

(1) Adams consistometer
(2) Spatula
(3) Stopwatch

Procedure

(1) First, level the consistometer by adjusting the leg screws in order to obtain a uniform flow over the disk.
(2) The hollow truncated cone is held down on the center of the disk.
(3) The sample (for cream style corn, the sample should stand at least 24 hr at a temperature of 68°F to 85°F) is filled into the cone and leveled off with a spatula. Care should be taken to avoid air pockets in heavy consistency products.

(4) The cone is then raised, allowing the product to flow over the measuring disk for 30 sec.
(5) The consistency of the product is determined by recording the extent of flow of the product at four equidistant points on the disk. The average of these readings is then taken and this is the consistency reading for the particular product.

Interpretation

According to the F&DA minimum standards of quality for canned cream style corn, for samples which contain less than 20% alcohol insoluble solids the spread does not exceed 12 in. and for corn over 20% alcohol insoluble solids, the spread does not exceed 10 in. (See Table 23.2).

TABLE 23.2 —
Cream-Style Consistency Score Points

Grade	Point Score	Flow in Centimeters (30 sec flow)
A	25	3
	24	4
	23	5
	22	6
B	21	over 6 or 7
	20	8
	19	9
C	18	over 9 or 10
	17	11
	16	12
D		over 12

TEST 23.3 — Viscosity With The Stormer Viscosimeter

Equipment

(1) Stormer viscosimeter
(2) Stopwatch

Procedure

(1) Place the instrument on a horizontal shelf or table so that the driving weight can drop through a distance of approximately 40 in.
(2) Attach the rotor to the chuck and secure it with the screw on the chuck.

(3) Adjust the driving weight to the desired weight for the particular food product (cream style corn, 750 gm; tomato juice, 50 gm; and tomato pulp, 400 gm).
(4) Place the can of food product (well stirred and at the desired temperature) on the platform. Raise the platform to immerse the rotor centrally into the sample until the top of the rotor is covered to a depth of ½ in.
(5) With the brake "on," raise the driving weight by turning the handle of the rewinding drum counterclockwise until the weight nearly touches the pulley. Set the dial by releasing the brake by a quarter turn of the brake control knob, allowing the rotor to revolve until the pointer on the dial is located between 80 and 90; then set the brake.

The purpose of locating the pointer 10 to 20 graduations to the right of zero is to permit the rotor to turn a sufficient number of revolutions to get a running start before checking with the stopwatch. If in setting the dial it has caused an appreciable lowering of the weight, again rewind the drum to raise the weight box to the pulley before making the test.

(6) With stopwatch in hand, release brake and measure the time in seconds required for 100 revolutions of the rotor as indicated by the revolution counter. Set the brake and record the number of seconds.
(7) Make at least two readings on the same sample and average the results. The seconds reading can be made on the same sample by rewinding the cord on the drum, which acts independently of the gear wheel when the brake is set. Before each successive reading, the rotor should again be permitted to turn 10 to 20 revolutions before starting the stopwatch.

After each use all the parts of the instrument should be thoroughly cleaned and dried.

Interpretation

The number of seconds required for the rotor to make 100 revolutions has been used to measure the consistency of some food samples. For cream style corn, the instrument has been used on the hot brine-corn mix (using a 250-gm driving weight) to predict the finished product consistency. It has also been used to grade the finished products of canned cream style corn and tomato products. For corn (using the 750-gm driving weight), light consistency products will require approximately 15 sec, while heavy consistency products will require over 175 sec. With tomato pulp (12% solids), reading from 10 sec to over 300 sec have been recorded when using a 400-gm driving weight. Tomato pulp of 12% solids manufactured with the hot break process will give a different reading at the same Bostwick or Adams value than tomato pulp manufactured by the cold break process. Thus, the Stormer Viscosimeter measures other factors than just the ability of the product to flow under a given set of conditions.

TEST 23.4 — GOSUC Procedure For Determining Consistency

Materials and Supplies
(1) GOSUC reservoir
(2) Ring stand and clamp
(3) Tomato juice samples—200 ml

Procedure
(1) Secure the GOSUC in a vertically upright position by means of a ring stand and clamp.
(2) Slip pinch-clamp on rubber tubing at bottom of the consistometer so that it will restrict the flow of product.
(3) Carefully fill the reservoir from the top.
(4) With stop watch in hand, remove the pinch clamp and initiate flow.
(5) Start timing the flow when the juice is level with the top etch mark.
(6) Measure the time in seconds for juice to flow between etch marks and record as such.

Interpretation
Thin juice = < 44 seconds
Good consistency juice = 44 to 70 seconds
Thick juice = > 70 seconds

BIBLIOGRAPHY

ADAMS, M.C., and BIRDSALL, E.T. 1946. New consistometer measures corn consistency. Food Ind. *18*, No. 6, 844–846.

ARTHUR H. THOMAS CO., 1948. Directions for Use of Stormer Viscosimeter. Arthur H. Thomas Co., Philadelphia, Pa.

HENRY A. GARDNER LAB. 1949. The Original Gardner Mobilometer. Henry A. Gardner Laboratory, Bethesda, Md.

BROOKFIELD ENGINEERING LAB. 1961B. Solutions to Sticky Problems. Brookfield Engineering Laboratory, Stoughton, Mass.

GEDDES, J.A., and DAWSON, D.H. 1942. Calculation of viscosity from stormer viscosimeter data. Ind. Eng. Chem. *34*, No. 2, 163.

GOULD, W.A. 1953. Measuring Consistency in Foods. Food Packer *34*, No. 3, 44, 66, 70; No. 4, 42, 58–60; No. 5, 32, 50–51.

GRAY, D.M. and SOUTHWICK, C.A., JR. 1930. Scientific research affords means of measuring stability of mayonnaise. Successful quality control in food processing. Food Industries.

KINNEY, P.W. 1941. Measurement of flow properties with the gardner mobilometer. Ind. Eng. Chem. Anal. Edition *13*, 178–185.

LANE, E.W., and BOSTWICK, E.P. 1939. Consistency test made by tea garden on preserves. Western Canner and Packer *31*, 43.

PERRY, J.H. 1950. Chemical Engineers' Handbook, 3rd Edition. McGraw Hill Book Co., New York.

TISCHER, R.G. 1952. Measuring the consistency of cream style corn. The Canner *114*, No. 15, 12–13.

CHAPTER 24

Defects — Imperfections or Appearance

Defects in food products are generally due to (1) raw materials or (2) inefficient unit operations in processing. A classification of the various types of raw product defects includes the following.
- (A) Genetic or physiological—abnormalities due to heredity, or unfavorable environmental conditions caused by extremes in temperature, water or nutrients. Generally, these resultant products are misshapen or off-colored.
- (B) Entomological—direct damage to the tissues by the feeding, oviposition, and/or sting by the insect, and/or indirect damage resulting from a disease organism being introduced into the product. Generally holes, scars, lesions or curled leaves result.
- (C) Pathological—the crop is disfigured, deformed, or lesions on the tissues due to bacteria, fungi, yeast or virus growth within the tissues.
- (D) Mechanical—damage to the tissues from harvesting or handling of the product, with resultant bruises which are discolored from enzymatic action or actual cracking and breaking of the tissues.
- (E) Extraneous or foreign material present in the product—leaves, pods, stems, thistle buds, seeds, silk, bark, stones, dirt, rodents, insects, animals, etc. In other words, material that is not considered part of the edible portion of the product.

Inefficient unit operations may impart specks from burn-on or incomplete removal of peel, core, stems and/or incomplete washing to remove soil, spray residues, insect or pathologically damaged tissues. Further, metal pickup or extraneous material may enter the product during the various unit operations if extreme care is not taken in processing. Defects of any kind are not desired, yet it may be excessively costly to completely remove all types of defects from the products. The FDA Minimum Standards of Quality and the U.S. Department of Agriculture Standards for Grades set forth maximum tolerances for products that are standardized. For nonstandardized products, industry may have specific tolerances for their own products. Further, specific firms and organizations may have tolerances for their products not in accordance with the established tolerances.

Examples of tolerances as established by the USDA and FDA for some products are given in Tables 24.1-24.5. The technologist should refer to the U.S. Grade Standards or FDA Minimum Standard of Quality for definitions of the types of defects. Generally, they may be removed by dilution, flotation, elution, etc., and measured by counting the number or area of defective units or the actual percentage of area or units involved.

TABLE 24.1 —
Apricots — Tolerances For Defects

	A	B	C	D
Halves and Whole				
		1 ea.	1 ea	1 ea
Harmless and extraneous materials	None	100 oz	30 oz	30 oz
Short stems (ea 30 oz)	1	2	3	3
Sq in. of peel per lb (peeled)	$\frac{1}{8}$	$\frac{1}{4}$	$\frac{1}{2}$	$\frac{1}{2}$
Loose pits (whole) ea 30 oz	2	3	4	4
Crushed or broken percent[1] by count	5	5	5	Unlimited
Minor blemishes and blemishes				
percent[1] by count	10	20	40	40
including maximum blemishes	5	10	20	20
Slices or Mixed Pieces of Similar Size and Shape				
		1 ea	1 ea	1 ea
Harmless extraneous materials	None	100 oz	30 oz	30 oz
Short stems (ea 30 oz)	1	2	3	3
Sq in. of peel per lb (peeled)	$\frac{1}{8}$	$\frac{1}{4}$	$\frac{1}{2}$	$\frac{1}{2}$
Minor blemishes and blemishes;				
percent by weight[2]	5	10	20	20

[1]Or 1 unit, whichever is larger. Percentage holds in averaging all containers comprising sample.
[2]Provided appearance or edibility not (A) more than slightly, (B) materially, (C) and (D) seriously affected.

TEST 24.1 — Procedure For Evaluation Of Quality Of Canned Cream Style Corn For Defects

Equipment

(1) Grading tray
(2) 8-Mesh 8-in. sieve

DEFECTS — IMPERFECTIONS OR APPEARANCE

(3) Ruler
(4) 100-ml Graduated cylinder

Procedure

(1) Determine the gross weight of the can.
(2) Open and transfer the contents into a pan, and mix thoroughly in such a manner as not to incorporate air bubbles. (If the net contents

TABLE 24.2 — Green Or Wax Beans
Maximum Allowances For Defects Or Defective Units

	A Classification	B Classification	C Classification
All Styles			
Total—all defects other than loose seeds and seed pieces	10% of drained wt	15% of drained wt	20% of drained wt
Blemished and seriously blemished	2% total, but no more than 1%, by count, seriously	4% total, but no more than 2% by count, seriously	8% total, but no more than 4% by count, seriously
Unstemmed units and detached stems	4 total, but no more than 2 long stems, per 12 oz drained wt	5 total, but no more than 3 long stems, per 12 oz drained wt	Unstemmed units: 6 total, but no more than 4 long stems, per 12 oz drained wt
Leaves and other extraneous vegetable	1 piece per 12 oz drained wt	2 pieces per 12 oz drained wt	Including detached stems, 0.6 oz per 60 oz drained wt
All Styles Except as Indicated Otherwise			
Loose seeds and pieces (except in sliced lengthwise style)	3% of drained wt	4% of drained wt	5% of drained wt
Small pieces of pod (in cuts; and mixed-cut and short cut styles only)			
(a) 240 count or less per 12 oz drained wt	40 pieces per 12 oz drained weight	60 pieces per 12 oz drained wt	60 pieces per 12 oz drained wt
	or	or	or
(b) More than 240 count per 12 oz drained wt	15% by count of all units	25% by count of all units	25% by count of all units
Ragged—cut units and/or damage by mechanical injury (except in sliced lengthwise style)	5 total per 12 oz drained wt	10 total per 12 oz drained wt	No limits (but included in total allowance of 20% of drained wt)

of a single container is less than 18 oz, determine the gross weight, open, and mix the contents of the least number of containers necessary to obtain 18 oz.
(3) Dry and weigh each empty container and subtract the weight so found from the gross weight to obtain the net weight.
(4) Transfer the material from the pan onto the 8-mesh sieve.
(5) Set the seive in a pan. Add enough water to bring the level within ⅜ in. to ¼ in. of the top of the sieve. Gently wash the material on the sieve by combined up-and-down and circular motion for 30 sec. Repeat washing with a second portion of water. Remove sieve from pan, incline to facilitate drainage and drain for 2 min.
(6) From the material remaining on the 8-mesh sieve count, but do not remove, the brown or black discolored kernels or pieces of kernel and calculate the number per 2 oz of net weight.

TABLE 24.3 — Pears —
Maximum Allowance For Defects Or Defective Units

	GRADE A	GRADE B	GRADE C
All Styles Except as Otherwise Stated			
Peel	¼ sq in. per 16 oz (average)	½ sq. in per 16 oz (average)	1 sq in. per 16 oz (average)
External stems[1]	1 per 100 oz	1 per 100 oz	1 per 100 oz
Interior stems	1 per 60 oz	1 per 30 oz	1 per 15 oz
Units of core material[2]	1 per 60 oz	1 per 30 oz	1 per 15 oz
Loose seeds	1 per 30 oz (average)	1 per 15 oz (average)	1 per 10 oz (average)
Broken or crushed[3]	5% by count or 1 unit	10% by count or 1 unit	10% by count or 1 unit
Seriously trimmed[4]	None	None	None
Moderately trimmed[4]			
Minor blemished; and blemished (combined)	Total: 10% by count	Total: 20% by count	No limit for moderately trimmed
		With further limitations of	
Minor blemishes; and blemished	10% by count but no more than 5% by count blemished or 1 unit (average)	20% by count but no more than 10% by count blemished or 1 unit (average)	Total: 30% by count but no more than 20% by count blemished or 1 unit (average)

[1]Does not apply to whole uncored with stems.
[2]Does not apply to whole uncored.
[3]Does not apply to diced or mixed pieces.
[4]Does not apply to sliced, diced or mixed pieces.

(7) Remove pieces of silk more than ½ in. long, as well as husk, cob, and other material which is not corn (i.e., peppers).
(8) Measure aggregate length of such silk and calculate the length per ounce of net weight.
(9) Spread the husk flat and measure its aggregate area and calculate the area per 20 oz of net weight.
(10) Place all pieces of cob under a measured amount of water in a cylinder which is so graduated that the volume may be measured to 0.1 cc. Take the increase in volume as the aggregate volume of the cob and calculate the volume of cob per 20 oz of net weight.

The corn remaining on the tray is then used to evaluate the other factors of quality according to the USDA scoring system (use QC Data Form 24.1).

TABLE 24.4 — Pickles —
Maximum Allowances For Defects Or Defective Units

Defects or Defective Units (in All Styles and Types Unless Stated Otherwise)	Grade A	Grade B
Curved pickles (in whole style or whole units in other styles)	10%, by count, of whole units	20%, by count, of whole units
Misshapen pickles (in whole style or whole units in other styles)	5%, by count, of whole units	15%, by count, of whole units
Units with attached stems (longer than ⅜ in.)	10%, by count, of all cucumber units but no more than 1%, by count, of all cucumber units with "long stems"	20%, by count, of all cucumber units but no more than 4%, by count, of all cucumber units with "long stems"
End cuts (in sliced crosswise style or units sliced crosswise)	5%, by wt, of all cucumber units	15%, by wt, of all cucumber units
Damaged by mechanical injury	10%, by count, of all pickle units including vegetable ingredients other than cucumber	15%, by count, of all pickle units including vegetable ingredients other than cucumber
In cured type:		
Minor blemish	Reasonably free	
Major and serious blemish	10%, by count, but no more than 1%, by count, may be serious	20%, by count, but no more than 3%, by count, may be serious
In fresh-pack type:		
Minor blemish	Fairly free	
Major and serious blemish	20%, by count, but no more than 5%, by count, may be serious	30%, by count, but no more than 10%, by count, may be serious

TEST 24.2 — Determining Kernel Count And Percent Defects Of Raw Popcorn

Equipment:
1. Tare weigh sample cup.
2. Select a sample of corn from the lot to be checked.
3. Weigh out ten grams of corn into sample cup.
4. Count the number of kernels in the ten gram sample and record as kernels/ten grams.
5. Count the number of defective kernels in the ten gram sample and record as defective kernels.
6. If desired the defective kernels can be weighed and defects reported as a percentage of total weight.

Note: To determine the extent of defective kernels mix the kernels with isopropyl alcohol (80%) iodine (20%) solution for approximately one minute. Pour off the iodine solution and wash the kernels on an eight or smaller mesh screen thoroughly with water. Examine for discolored kernels indicating damaged or defective ones.

TEST 24.3 — Bruise Detection In Raw Potatoes

General Statement:
Bruises developed during harvesting, handling and storage are directly related to inferior potatoes for chip manufacture. The bruised or defective area shows up as darkened area on the chip unless cut out during preparation for chipping. Chemical methods for detection of surface bruises are available and should be used to pinpoint damage. Black spot and discolored areas may be observed after peeling of the tuber.

Chemical Equipment:
1. Catechol solution (1 oz catechol/1 gallon of water and 1 teaspoon of detergent).
2. Knife.
3. Batch Peeler.

Procedure:
1. Select a representative sample (10 tubers as minimum).
2. Wash the tuber.
3. Immerse the tubers in the catechol solution for 1 minute.
4. Drain the tubers and allow to dry.

DEFECTS — IMPERFECTIONS OR APPEARANCE

5. Determine the number and extent of the damaged tuber by noting the dark red or purplish areas on the surface of the tubers.
6. Evaluate severity of the bruise as follows:
    a. **Skinning**—one stroke of peeler removes all visible damage.
    b. **Slight bruise**—two strokes of peeler removes all visible damage.
    c. **Serious bruise**—more than two strokes are required to remove all the damage.

Bruise levels can be expressed as either:

Bruise free	80%
Allowable bruise	5%
Non-allowable bruise	15%
Total	100%

or as an index where the index is calculated as 7 x% serious bruise by weight plus 3 x% slight bruise by weight. A damage index of 0-50 is considered Excellent; 50-100, Good: 100-150, Fair; 150-200, Poor; 200, Reject.

Since the catechol bruise detection method only identifies bruises that break the skin, such a method does not reflect total damage or defects. Internal bruising (blackspot) can only be detected after peeling. Therefore, representative samples of peeled tubers should be selected and evaluated as to actual percentage of bruised tubers or bruised areas of tubers.

1. Select 10 tubers after conventionally peeled and weighed.
2. Examine each tuber for damage and count number of tubers damaged. Record as percent damaged tubers.
3. Cut out damaged area and weigh.
4. Calculate percentages as follows:

$$\frac{\text{Weight in 3}}{\text{Weight in 1} \times 100} = \% \text{ defective areas}$$

Remember if waste is 1 inch in diameter and the slice thickness is 1/16 of an inch (0.064) you have 16 defective slices.

TEST 24.4 — Testing For Freedom From Defects By Visual Examination

General:

Defects in potato chips refer to pieces of peel, green discoloration, internal discoloration of the slice, or other harmless extraneous material.

Equipment:
1. Two-quart sample container.
2. Major and minor defect sample display (select your own).

Procedure:
1. Fill sample container with chips to be tested.
2. Examine chips visually and separate any chips that show defects.
3. Separate defects into the following classes and calculate the percentage of chips for each class:

	Defect Classification	
	Minor	Major
Group I Defects Discolored appearance which adversely affects the chip to a noticeable degree, that is, ¼ square inch or less.	X	
Discolored appearance which adversely affects the chip to a degree that is objectionable; that is, more than ¼ square inch in area.		X
Group II Defects Blemished area including peel, internal discoloration or harmless extraneous material which adversely affects the chip; that is, ¼ square inch or less.	X	
Blemished area including peel, internal discoloration or harmless material which seriously affects the chip; that is, more than ¼ square inch in area.		X

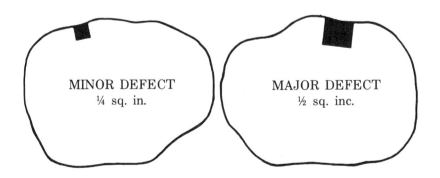

MINOR DEFECT
¼ sq. in.

MAJOR DEFECT
½ sq. inc.

DEFECTS — IMPERFECTIONS OR APPEARANCE

4. Do not count more than two major defects on any one chip; for example, a chip with two major defects is counted as two, a chip with three or four major defects is counted as two.
5. If a chip has a wet center and a major defect, it is counted once for each test. The wet center is not counted as a major defect.
6. Score the chips according to the following:

TABLE 24.5 — Potato Chips — Defect Evaluation

Class	Score	Maximum Number of Defects by:	
		Minor	Major
1	18-20	0-5	0-3
2	16-17	6-10	4-5
3	11-15	11-15	6-8
4	6-10	16-20	9-12
5	0-5	over 20	over 12

TABLE 24.6 —
Maximum Defects For Tomatoes

Grades	Defect (Aggregate Area)	In Cans of Less Than 2 lb Total Contents	In Cans of 2 or More Pounds Total Contents	In Cans of Any Size
		In any container	Equivalent Amount Per Pound of Contents of Any Container	Per Pound of Total Contents of All Containers (Average)
A and A Whole	Peel	2 sq in.	1 sq in.	½ sq in.
	Blemished areas	⅛ sq in.	$1/16$ sq in.	$1/16$ sq in.
	Discolored portions	½ sq in.	¼ sq in.	¼ sq in.
	Objectionable core material: Practically none			
	Harmless plant material: Not more than a trace			
B	Peel	3 sq in.	2 sq in.	1 sq in.
	Blemished areas	¼ sq in.	⅛ sq in.	⅛ sq in.
	Discolored portions	1 sq in.	½ sq in.	½ sq in.
	Objectionable core material: Slight amount			
	Harmless plant material: Slight amount			
C	Peel	no limit	no limit	1 sq in.
	Blemished areas	½ sq in.	¼ sq in.	¼ sq in.
	Discolored portions	1½ sq in.	¾ sq in.	¾ sq in.
	Objectionable core material: Moderate amount			
	Harmless plant material: Moderate amount			

DEFECTS — IMPERFECTIONS OR APPEARANCE

QC DATA FORM 24.1 — Work Sheet For Evaluating Quality Of Canned Whole Kernel Sweet Corn

Date and hour of cutting							
Code							
Vacuum							
Gross weight in ozs.							
Drained wt in ozs.							
% liquid in net wt.							
# of brown or black kernels/2 oz.							
Length of silk/can							
Calculate length of silk/1 oz.							
Vol. of cob/can							
Calculate vol. of cob/14 oz.							
Area of husk/can							
Calculate area of husk/14 oz.							
% AIS							
% Pericarp							
Succulence (cc/100gm)							
USDA Score Points							

Color (10)   A 9-10
               B 8
               C 6-7[1]
Substandard 0-5[2]							

Cut (10)   A 9-10
               B 8
               C 6-7[3]
Substandard 0-52							

Absence Defects (20) A 18-20
                       B 16-17[2]
                       C 14-15[2]
Substandard 0-13[2]							

Maturity (40) A 36-40
                B 32-35[2]
                C 30-31[2]
Substandard 0-20[2]							

Flavor (20)   A 18-20
                B 16-17
                C 14-15[2]
Substandard 9-13[2]							

Total Score (100) A 90-100
                   B 80-90
                   C 70-80							

Grade							
Remarks							

[1] Partial limiting rule. Score of 6 limits grade to std; 7 to extra standard.
[2] Limiting rule. In this class can't receive higher grade.
[3] Limiting rule. Can't receive grade higher than extra standard.

Quality Control Technologist _____

CHAPTER 25

Drosophila and Insect Control

Drosophila is the generic name of a group of small flies that feed on and breed in plant material (Wilson 1952). Although the fly is a constant menace to all fruit canners, most emphasis is placed herein on their detrimental effects on tomatoes and tomato byproducts.

The *Drosophila* fly may be referred to as a vinegar gnat, fruit fly, sour fly or pomace fly. Common to the species are populations of species causing the most problem, *Drosophila melanogaster*.

The fly, as such, is not the subject of as much concern as the eggs it deposits and subsequent stages of development of the fly in the life cycle. For many years it was erroneously assumed that the *Drosophila* fly laid eggs only on fermenting and rotting plant products where the adults were feeding (Wilson 1952). Field observations have revealed that eggs are deposited on fresh growth cracks in sound tomatoes on the vine and in cracks resulting from the picking operation. The eggs are deposited by means of an adhesive substance common to the fly, and they are not easily or readily removed. Because of this, the egg stubbornly resists removal by all known washing methods. As such, Drosophila represents a potential threat of ultimate seizure of the finished product in conjunction with Section 402 (a) (3) of the Federal Food, Drug and Cosmetic Act, which defines a food as adulterated if it consists in whole or in part of any filthy, putrid or decomposed substance (Judge 1972).

LIFE CYCLE, HABITS, AND OTHER FACTORS

The life cycle of *D. melanogaster* varies from 5–8 days, with an average summer cycle of 7 days. Other cycles for species at 85°F are: *D. immigrans,* 9 days; *D. busckii,* 11 days; and at 70°F, *D. pseudoobscura,* 14 days. Starting with the adult, the life cycle progresses as follows.

FIGURE 25.1. Life Cycle of *DROSOPHILA MELANOGASTER*

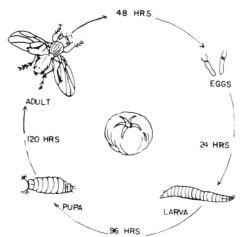

Adult

The adult of the *D. melanogaster* species is a transparent-winged fly, about ⅛ in. long. The body is yellowish, and the abdomen is crossed by dark bands (USDA 1962). Females live longer than males; higher temperatures markedly decrease the life span of each. The female fly may begin laying eggs during the second day of adult life and continue at the rate of about 26 per day. A single fly may lay as many as 2000 eggs (Wilson 1952).

Eggs

The eggs are pearly white, elongated, and about $1/50$ in. long. Eggs of *D. melanogaster* have two appendages or filaments attached near the head end. Eggs of other species may be slightly longer and have 2–4 appendages.

Individual eggs are too small to be seen easily by the unaided eye. Eggs usually are laid with their appendages or filaments above the surface of the medium in which they are placed (USDA 1962). These serve as respiratory organs (Pepper *et al.* 1953). The eggs may then hatch into larvae in about 24 hr.

This time may be drastically reduced, however, because if flies do not find suitable media on which to lay, the eggs will be retained in the body. The fly is capable of holding eggs and depositing them shortly before hatching or actually depositing live larvae which hatch within the fly itself (Wilson 1952).

Larvae

The larvae, commonly referred to as maggots, may be quite transparent, or they may appear to be cream-colored or some other shade caused by food in the gut (USDA 1962). When first hatched, larvae measure about 1/30 in. (Wilson 1952). They shed their skin three times and may mature into the pupa stage in about four days, at which time they are about ¼ in. long. Larvae may confine their feeding to the cracks in fruit and free juice. Upon reaching maturity, they seek a dry place in which to pupate.

Pupae

The pupae are about ⅛ in. long. At first they are yellowish-white, then turn amber, and finally turn brown within a few hours. The anterior end of a pupa is broader and flatter than the posterior end, and it has two stalk-like structures that bear the respiratory organs (USDA 1962). The pupal period may require five or more days during transformation into an adult fly (Pepper *et al.* 1953).

As described in the following paragraphs, five factors affect the activity of the *Drosophila* fly: (1) temperature, (2) moisture, (3) food, (4) light intensity, (5) air movement (Michelbacker 1958).

A temperature of approximately 55°F is close to the lower threshold of activity, an optimum temperature is 75–80°F, and a temperature over 90°F eliminates most activity. It has been found that fruit exposed to the sun on a hot day may reach 125°F or higher; this is sufficient to kill the larvae. The flies usually are more active in the morning and late afternoon. During the heat of the day, they seek shelter in the shade of rank tomato vines and in grass and weeds (USDA 1962).

Moisture favors adult feeding and maintains moist surfaces for egg deposition. Further, it prevents the rapid drying out of breeding sources (Michelbacker 1958). Rainy periods at any time during the season are favorable to activity by directly lowering the temperature and light intensity.

Drosophila, like most other insects, are attracted from considerable distances by the odors of plant materials, expecially after fermenting begins. Adults feed on fermenting plant materials; apparently, yeast constitutes an important part of their diet (Wilson 1952). There is a marked difference in the food habits of the adults and larvae. Ideal breeding media for the larvae are sound, ripe fruits with fresh moist cracks or breaks in the skin (Michelbacker 1958).

Adult flies are most active when the light intensity approaches that of about sundown. As such, most activity occurs between the hours of 6 to 8 A.M. and 4 to 8 P.M. when the light and temperature are lower. The rate of egg deposition during these periods has been found to be 25–35 times the rate of egg deposition during the rest of the day (Natl. Canners Assoc. 1960).

Most flight occurs when it is calm, and the direction of flight is into the drift. Strong winds tend to pin down the flies (Michelbacker 1958). Wind velocities above 15 mph curtail flight, but flies may remain active, feeding, mating, and egg-laying in the mass of tomato vines, provided other conditions are favorable (Wilson 1952).

METHODS OF DROSOPHILA DETECTION

The purpose of performing a fly egg and larvae count is to accurately determine their presence and, as such, to ascertain exactly how many are present. The ideal method for making this determination should be simple, fast, and accurate; i.e., the method should show a definite number of eggs or larvae, if indeed, such exist.

There are a number of methods available for the detection of *Drosophila* eggs and larvae. The tolerance for eggs and larvae as set forth by the Food and Drug Administration are listed on Table 25.1.

TABLE 25.1 —
FDA Defect Action Level For Drosophila Eggs And Larvae

Product	Sample Size (gm)	No. of Eggs	No. of Larvae
Tomatoes	500	10 or 5 and	1 or 2
Tomato juice	100	10 or 5 and	1 or 2
Tomato puree	100	20 or 10 and	1 or 2
Tomato paste, pizza, and other sauces	100	30 or 15 and	1 or 2

TABLE 25.2 —
FDA Defect Action Level Of Aphids And/Or Thrips

Product	Sample Size (gm)	Tolerance (number)
Broccoli	100	80
Brussels sprouts (frozen)	100	30
Spinach (canned)	100	50

Different methods for the detection of *Drosophila* eggs and larvae are described as follows (see Tests 25.1-25.4).

TABLE 25.3 —
FDA Defect Action Level For Insect Fragments In Vegetables

Product	Sample Size	Tolerance
Corn (sweet, canned)	24 lb	Two 3 mm or longer larvae, cast skins, larval or cast skin fragments of corn ear worm or corn borer, and aggregate length of such larvae, cast skins, larval or cast skin fragments exceeds 12 mm
Peas, black-eyed, canned (cowpeas, field peas)	No. 2 Can	5 Cowpea curculio larvae
Asparagus (canned or frozen)	15% by count	Spears are infected with 6 asparagus beetle eggs or egg sacs

TEST 25.1 — GOSUL Method

This is a simple and fast method of detection that can be applied either to the processed or unprocessed product. The method consists of the application of a 100-watt long-wave ultraviolet light source to a series of filters that have had the sample placed upon them. Moisture is removed from the sample by means of a Buchner funnel with the aid of applied vacuum. As mentioned, advantages include simplicity and speed; the entire test can be conducted in 20–30 min, along with a very high percentage of accuracy. Expenditure for equipment is nominal (Geisman and Gould 1963).

Equipment

(1) 100-watt Long-wave (3660°A) ultraviolet light source
(2) #5 Buchner funnel
(3) Several 18.5 or 20-cm "sharkskin" filter papers
(4) 2000-ml Filtering flask
(5) Two 500-ml beakers
(6) 9 in. × 9 in. Glass or plastic plate
(7) Aspirator pump or other vacuum source
(8) Binocular wide-field microscope (10X) or 10X hand lens
(9) 100-ml Graduate cylinder
(10) Teasing needle
(11) One pair tweezers

Reagents

(1) The reagents required are vinegar (40 or 50 grain) or 4% or 5% acetic acid

Procedure

(1) Thoroughly mix sample. This can be accomplished by vigorously shaking the sample in a container in an up-and-down motion 200 times.

(2) Open container and pour a 100-ml aliquot of sample into a graduated cylinder.

Tomato Juice.—(a) Seat #5 Buchner funnel securely into a 2000-ml filtering flask and attach flask to aspirator pump or other vacuum source. Moisten an 18.5 or 20-cm "sharkskin" filter paper with water and spread evenly in funnel.

(b) Pour approximately 50-ml aliquot onto the seated filter paper. This can be accomplished by carefully pouring sample into center of paper and rotating flask and filter to spread juice to sides. Apply vacuum to filtering flask to remove moisture. Paper should be nearly dry before it is removed from funnel. Repeat this operation with the rest of the samples and apply rinse water from graduated cylinder to last paper.

(c) Remove dry filter paper from flask with aid of tweezers and place on a previously-lined glass or plastic plate. Plate may be lined into 1-in. square to aid in examination.

(d) Using a 100-watt long-wave ultraviolet light source and binocular wide-field microscope, examine the paper for eggs and larvae. For best results, the light source should be positioned so that beams strike the paper at approximately a 45° angle. Eggs and larvae appear blue-white and may be easily detected.

(e) Record number of eggs and larvae separately for each sample and report as number per 100 gm.

Tomato Pulp and Paste.—A 100-ml aliquot of tomato pulp should be diluted with 200 ml warm (100°F) water and 100 ml tomato paste diluted with 200 ml of warm water. This aids in filtering the sample.

Catsup.—(a) Dilute catsup with 250 ml vinegar (40 or 50 grain) or 4 or 5% acetic acid solution (100°F water may be used if acid is not available). Stir mixture thoroughly. The acid aids in the removal of the sugar present and thus prevents a haze which would otherwise form. (b) Follow steps (a) to (e) as for juice. (Requires 6–8 filter papers.)

Unprocessed Tomato Juice.—(a) Pour only 34 ml of raw juice onto filter paper at one time. Spread juice evenly, as before. (Requires 3 filter papers.) (b) Follow steps (a) to (e) as for juice, except use 34-ml portions of aliquot instead of 50-ml portions.

Recommendations

(1) Sample should be thoroughly mixed before analysis.
(2) Aliquot must be spread evenly and thinly over surface of the filter paper in order to prevent some eggs and larvae from being covered with tomato fibers.
(3) Best counting results are obtained in a darkened room.
(4) After an analyst has become familiar with the counting technique it is possible to use a 10X hand lens instead of the microscope without loss of accuracy.

TEST 25.2 — AOAC Method

This method relies on the principle that fly eggs and larvae will settle out of a two-medium solution, whereas most insects or hair float. Being relatively dense, the eggs and larvae will settle in a water-gasoline mixture. A separatory funnel is used instead of trap flasks that are utilized to recover insects or insect parts. The gasoline-water mixture is vigorously stirred and allowed to settle, permitting most of the plant tissue to rise and thereby releasing the eggs and larvae. At regular intervals, a 15- to 20-ml portion is drained off. After an adequate amount of draining, the water phase plus any eggs and/or larvae is filtered through a 10XX bolting cloth. The cloth is used instead of paper because the openings are fine enough to catch even small larvae but not large enough to trap tomato tissue and fiber that would obscure the microscopic examination (Wilson 1952).

The equipment and procedure (Gould 1971) are as follows.

Equipment

(1) 2000-ml Separatory funnel
(2) Ringstand
(3) No. 10 sieve
(4) 600-ml Beaker
(5) 10XX Bolting cloth
(6) Hirsch funnel
(7) 10X Binocular wide field microscope
(8) Teasing needle

Reagents

(1) Gasoline

Procedure

Comminuted Products.—
(1) Mix sample thoroughly and transfer 100 gm to a 2-liter separatory funnel.
(2) Add 20–30 ml gasoline and shake thoroughly, releasing pressure as necessary.
(3) Fill separator with water in such a manner as to produce maximum agitation.
(4) Place separatory funnel in a ringstand and let settle. At 15-min intervals during 1 hr, drain 15–20 ml from funnel, and gently shake funnel with a rotary motion to facilitate settling out of the fly eggs and larvae.
(5) If drained liquid contains seeds, pass it through a No. 10 sieve, and rinse seeds and sieve thoroughly, recovering both liquid portion and rinse water in beaker.
(6) Filter through 10XX bolting cloth in Hirsch Funnel.
(7) Examine for eggs and larvae at about 10X. If fly eggs or larvae are found in this examination, continue separation, and draining as above for additional hour.

Canned Tomatoes.—
(1) Pulp entire contents of can in such a way that a minimum number of eggs and larvae are crushed or broken. This may be done by passing the material through a No. 6 or No. 8 sieve and adding seeds and residue remaining on the sieve to the pulp.
(2) Place 500 gm of the well-mixed pulped tomatoes in a 6-liter separatory funnel.
(3) Add 125–150 ml gasoline and about 1 liter of water.
(4) Shake vigorously, releasing pressure as necessary.
(5) Fill funnel with water.
(6) Place funnel in ringstand and let layers separate.
(7) At 15-min intervals during 1 hr drain 25–30 ml from bottom of funnel and gently shake funnel with rotary motion to facilitate settling of fly eggs and larvae. Each portion may be examined at once or combined with subsequent portions.
(8) Pass drained portions through a No. 10 sieve and rinse seeds and sieve thoroughly, recovering both liquid portion and rinse water in a beaker.
(9) Filter through 10XX bolting cloth in Hirsch Funnel.
(10) Examine cloth for eggs and larvae at about 10X.
(11) If eggs or larvae found in this examination, continue separation and draining as above for an additional hour.

TEST 25.3 — Staining Method

This is another relatively simple method that is based on the staining of tomato solids with crystal violet dye in preference to fruit fly eggs and maggots. They appear pearly white in comparison with the blue-colored juice.

Equipment

(1) Beaker, 600-ml or 500-ml
(2) Glass funnel with rubber tubing and pinch clamp
(3) Glass plate with under surface painted flat black
(4) Flood lamp or strong light source
(5) 10X Magnifying lens
(6) Thin spatula

Reagents

(1) 1% crystal violet dye; dissolve 1 gm of dye in 99 gm alcohol

Procedure

Tomato Juice.—
(1) Weigh a 100-gm sample of juice in a beaker and dilute with an equal volume of water.
(2) Add 4 ml of 1% crystal violet dye to the sample. Mix and allow to stand for 1 min.
(3) Pour the dyed juice into the glass funnel with rubber tubing and pinch clamp.
(4) Allow the sample to slowly flow downward over the glass pane with light source on. Position the light to eliminate reflected light and glare.
(5) Accurately count eggs and/or larvae present. The unstained eggs and larvae appear pearly white and are obvious in contrast with the dark blue tomato material and black surface upon which they are viewed.
(6) Remove any eggs or larvae with a thin spatula in question and observe under a 10X magnifying lens.
(7) Wash the beaker and funnel thoroughly after being emptied and inspect in same manner.
(8) If there is a question relating to the sample upon completion of the assay, it is repeated by pouring the sample back over the viewing surface.

TEST 25.4 — Insect Fragment Determination

The Wildman trap flask method utilizes the principle that oil will float on water as insects rise with the oil. The amount of lift exerted on an insect fragment depends partly on the specific gravity of the adhering oil compared to that of the aqueous medium (Harris and Reynolds 1960). The oil phase can then be examined under a microscope after filtering with the aid of vacuum. The equipment and procedure are as follows (Gould 1971).

Equipment

Comminuted Tomato Products.—
(1) 2-liter Wildman trap flask
(2) Buchner or Hirsch funnel
(3) 1 Box of rapid suction filter paper with hard finish and 1–2 cm wider than the interval diameter of the funnel
(4) Petri dishes
(5) Wash bottle
(6) Castor oil
(7) 95% Alcohol
(8) 20–30X Wide field microscope
(9) 500-ml Beaker

Procedure

Comminuted Tomato Products.—
(1) Place 200 gm of any tomato product except paste (where 100 gm is used) in Wildman trap flask.
(2) Add 20 ml of castor oil and mix well.
(3) Add enough warm tap water (approximately 100°F) to fill flask. At first the bubbles may tend to bring up tomato tissue, but gentle rotary stirring with the metal rod will cause material to settle.
(4) Let stand, with occasional stirring, for 30 min; then trap off oil layer into beaker.
(5) Wash out the neck of the flask with alcohol to remove all castor oil.
(6) Add additional warm water to fill Wildman flask, stir and let stand 10 min; then repeat trapping operation.
(7) Pour oil layer from beaker on filter paper which has been fitted into filter funnel and apply vacuum to filter flask until paper is dry. Be sure to wash beaker and filter paper with alcohol to remove oil.
(8) Examine filter paper under 20–30X wide field microscope and record any insect fragments.

Examination of Sweet Corn for Insect Fragment.—

(1) Make macroscopic examination of whole can, or equivalent of No. 2 can where larger sized can is being examined. Determine number of worm-eaten kernels and of rotten kernels.

(2) Place 200 gm of well-mixed sample in a 2-liter Wildman Trap Flask.

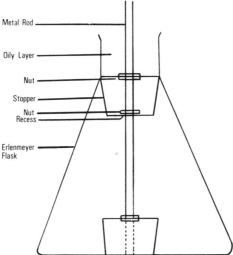

FIGURE 25.2 — Wildman Trap Flask

(3) Add 20 ml of castor oil and mix well.
(4) Add sufficient hot water (50°C) to fill flask.
(5) Let stand for 30 min with occasional gentle stirring.
(6) Trap off into beaker the oil and water layer and any corn debris that may have risen to the neck of flask.
(7) Wash out the adhering oil in the neck of the flask with hot alcohol.
(8) Add a little more hot water, stir, let stand 10 min and trap off again.
(9) Set a No. 10 sieve into beaker and pour oil and water mixture from trappings onto sieve.
(10) Wash corn debris thoroughly with hot alcohol.
(11) The material is then filtered through a rapid filter paper in a Buchner funnel with suction applied. The sieve, beaker, sides of funnel and paper should be thoroughly washed with hot alcohol to dissolve the oil and speed filtration.
(12) Examine the paper with a Greenough microscope with a 10X magnification.

BIBLIOGRAPHY

ANON. 1962B. Post harvest *Drosophila* control programs for tomato processors. Canning Trade *85*, No. 2, 12, 13, 17.

BICKLEY, W.E., and DITMAN, L.P. 1953. *Drosophila* problem in the canning of tomatoes. Publ. *190*, Contribution *2513*, Maryland Agr. Expt. Sta., College Park, Md.

GEISMAN, J.R., and GOULD, W.A. 1963. Detection of *Drosophila* eggs and larvae in tomato products. The Dart, *2-763*. Ultra-Violet Products, San Gabriel, Calif.

GOULD, W.A. 1971. Suggestions for *Drosophila* Control in Canning Plants. Tomato Processor Quality Control Handbook. Dept. of Horticulture, Ohio State Univ., Columbus.

GOULD, W.A. 1974. Tomato Production, Processing and Quality Evaluation. Avi Publishing Co., Westport, Conn.

HARRIS, K.L., and REYNOLDS, H.L. 1960. Microscopic-Analytical Methods in Food and Drug Control. Food and Drug Control. Food and Drug Tech. Bull. *1*. U.S. Dept. of Health, Education and Welfare. FDA, Washington, D.C.

JUDGE, E.E., and SONS. 1987. The Almanac of the Canning, Freezing, and Preserving Industries, E.E. Judge and Sons, Westminster, MD.

MICHELBACKER, A.E. 1958. Biology and Ecology of *Drosophila*. *Drosophila* Conf., Dec. 8–9, Western Regional Res. Lab. Albany, Calif.

NATL. CANNERS ASSOC. 1960. *Drosophila* Control. National Canners Assoc., Washington, D.C.

NATL. CANNERS ASSOC. 1961. Research Information. Bull. *49*, National Canners Assoc., Berkeley, Calif.

PEPPER, B.B., REED, J.P., and STARNES, O. 1953. *Drosophila* as a Pest of Processing Tomatoes. Extension Bull. *266*, Extension Serv. Coll. of Agr. Rutgers Univ., New Brunswick, N.J.

SIEGEL, M. 1958. The tomato menace. Canning Trade *81*, No. 20.

USDA. 1962A. Controlling *Drosophila* flies on tomatoes. Farmer's Bull. *2189*.

WILSON, J. J. 1952. Chairman of Special Committee on Tomato Products Sanitation. Suggestions for Improving Sanitary Conditions in the Harvesting, Transportation and Processing of Tomatoes. Natl. Canners Assoc., Washington, D.C., July 15.

CHAPTER 26

Enzyme Activity

Enzymes are specific thermolabile protein catalysts which occur in all living plant and animal tissues and in microbial cells. Being protein in nature, they have colloidal properties. They may act in the presence of living cells or in the absence of cells. Their action may be desirable or undesirable.

Although only present in small quantities, they are capable of causing important chemical changes in all foods in the raw state. Changes of flavor and texture associated with ripening of fruits are caused by enzymatic actions.

In frozen products the inherent enzymes in fruits and vegetables remain active even in low temperatures and cause deterioration. In order to keep frozen or dehydrated products, enzymes should be inactivated before freezing by means of blanching, use of chemicals or by the exclusion of air.

Enzymes may be classified in different ways but the most used method is based on the functions they perform.

(1) Proteinases or protein-decomposing enzymes (tenderizing meat, modification of dough, cheesemaking)
(2) Diastases or starch-digesting enzymes (sirups, brewing, cereal modification)
(3) Lipases or fat-splitting enzymes (cheese, chocolate and mayonnaise production)
(4) Pectic enzymes which hydrolyze pectins (clarifying fruit juice or wine, jellies)
(5) Oxidases or oxidizing enzymes (discoloration of foods or removal of oxygen in foods)
(6) Invertase (soft cream candy centers)

Many authorities have shown that oxidative and proteolytic enzymes are common in most cells, and these enzymes are probably the chief groups of interest to food processors on account of the undesirable changes they catalyze in different products. Examples of these changes are decomposition of ripe fruits in storage, discoloration of cut surfaces of many products due to oxidation of some tannins by enzymes, loss of Vitamin C in cut vegetables by oxidation, and disintegration of some fruits in maturing, for instance, tomatoes from the action of pectic enzymes. Determinations of enzyme activities in fresh, canned and frozen products are desirable to determine the time and temperature of blanching operation as well as an index of keeping quality.

Catalase is not an oxidizing enzyme but is closely related to it. It is most abundant in leafy tissues. Peroxidase is an oxidizing enzyme which is found in abundance, especially in root crops. Catechol is an autooxidizable substance which is found mostly in fruits.

In Tables 26.1-26.4 from Joslyn and Heid (1963) are shown the characteristics and uses of some enzymes.

The control of enzyme activity is of importance to the food processor. The most common method is by blanching, since a temperature of 80°C will denature most proteins. In addition, a change of the substrate by adjusting the pH, lowering the temperature of the product, or by oxidizing the enzyme are examples.

The food technologist's main concern on processed foods is a knowledge of the extent of inactivation of the enzymes. Many methods have been published and are in use. The following are suggested for quality evaluation of processed fruits and vegetables.

TEST 26.1 — Enzyme Determination

Materials and Equipment
(1) 0.08% H_2O_2 (2.8 cc of 30% H_2O_2/L.) solution
(2) 3% H_2O_2 solution
(3) 0.5% Guaiacol in 50% ethyl alcohol
(4) 1% Catechol solution
(5) Mortar and Pestle
(6) Fermentation tube
(7) Ignited sand
(8) Test tube, ¾ in. and 7/7 in.
(9) 3-in. Funnels
(10) Graduated cylinder
(11) Pipettes

Catalase Procedure

(1) Take a 10-gm representative sample of vegetable, 1 gm of calcium carbonate, and 1 gm of sand and place in a mortar.
(2) Add 15–20 cc of water to the mortar and grind all to a slurry.
(3) Filter through a cotton milk filter.
(4) Remove a 10 cc aliquot from the filtrate and place in a flask.
(5) Place 22 cc of 3% hydrogen peroxide in a small vial. Using a pair of long tweezers, place the vial of hydrogen peroxide upright into the flask.
(6) Attach the flask to the manometer and suspend it in the constant-temperature water-bath (20°C).
(7) Allow the contents of the flask to reach bath temperature, then close the manometer stopcock and adjust the water level in the manometer to zero reading.
(8) Shake the sample with the hydrogen peroxide for 2 min.
(9) At the end of 3½ min read the water level on the manometer.
(10) The difference between initial and final readings on the manometer indicates the amount of gas produced from the sample. If no more than 0.2 cc of gas is liberated from the sample, the test is considered negative. Any volume of gas in excess of 0.2 cc is considered a positive test for catalase.

Peroxidase Procedure

(1) Remove ½–1 gm portion of the residue from the catalase filtration or prepare the sample as prescribed in steps (1) to (2) for the catalase test.
(2) Place the sample on a clean spot plate.
(3) Add ½ cc of 0.5% guaiacol solution to sample. Guaiacol solution is prepared in 50% ethyl alcohol.
(4) Add ½ cc of 0.08% hydrogen peroxide to sample.
(5) Mix sample, guaiacol, and hydrogen peroxide and then allow to stand for 3½ min.
(6) If no color develops in 3½ min, the peroxidase test is considered negative. The development of a reddish-brown color is considered a positive peroxidase test.

TABLE 26.1 — Characteristics Of Enzymes

Name	Substrates	End Products	Optimum, pH	Nature of Change Produced
Amylase				
(a) Liquefying	Starch	Dextrins and maltose	5.0–7.0	Conversion of relatively insoluble into soluble starch
(b) Saccharifying	Starch	Maltose and limit dextrins	5.0–7.0	Conversion of starch to fermentable sugar
Anthocyanase	Anthocyanin	Anthocyanidin and glucose	3.0–4.5	Decoloration of berry and grape juices by conversion of natural pigment into less soluble form
Ascorbic acid oxidase	Ascorbic acid and molecular oxygen	Dehydroascorbic acid, H_2O	—	Oxidative destruction of vitamin C
Catalase	Hydrogen peroxide	Water, oxygen	6.8	Decomposition of hydrogen peroxide
Chlorophyllase	Chlorophyll	Chlorophyllide, phytol alcohol	6.0	Stabilization of green color of spinach leaves
Dextrinase	Dextrin	Maltose	6.0	Conversion of soluble starch into fermentable sugar
Glucose oxidase (and catalase)	Glucose and molecular oxygen	Gluconic acid, water	5.6	Removal of sugar from egg whites or of dissolved oxygen from juices
Glucosidases	Glucoside	Sugar, nonsugar residue	4.4–6.0	Hydrolysis of bitter glycosides which may lead to loss of bitterness
Invertase	Sucrose	Glucose, fructose	4.6–5.0	Conversion of sucrose into invert sugar
Lipases	Triglycerides	Glycerol, fatty acids	5.0–8.6	Conversion of neutral oils and fats into fatty acids
Lipoxidase (Lipoxyperrase)	Unsaturated oils + molecular oxygen	Peroxides of linolenic acid	6.5	Oxidative rancidity of oils and induced destruction of carotene
Maltose	Maltose	Glucose	4.5–7.2	Hydrolysis of disaccharide into reducing sugars
Pectin esterase	Pectinates	Methyl alcohol, and pectates	4.0–7.0	Conversion of soluble pectinic acids into calcium precipitable pectic acid
Pectin poly-galacturonase	Pectates	Oligogalacturonides and monogalacturonic acid	3.5–8.0	Degradation of pectic acid into lower molecular weight constituents

TABLE 26.1 — Continued

Name	Substrates	End Products	Optimum, pH	Nature of Change Produced
Polymethyl-galacturonase	Esterified pectic substances	Oligogalacturonides and β-unsaturated galacturonides	6.0	Degradation of polymethyl galacturonides into lower molecular weight constituents
Peptidases	Peptides	Amino acids	6.0–7.4	Complete hydrolysis of proteins
Peroxidases	Hydrogen peroxide and phenolic compound	Quinone and other oxidation products	5.0–6.0	Oxidation of chromogens in absence of oxygen
Phosphatases	Esters of phosphoric acid	Phosphoric acid esters	3.0–10.0	Hydrolysis of phosphate
Polyphenoloxidase	Molecular oxygen and phenolic constituents	Quinone and other oxidation products	5.0–7.0	Oxidation of chromogens leading to browning and induced oxidation of ascorbic acid
Proteases (Bromelin, papain, pepsin, trypsin, rennin)	Proteins	Polypeptides and dipeptides	1.5–10.0	Partial protein hydrolysis
Thiaminase	Thiamin	Pyrimidine + thiazole	5.0–8.0 (varies with purity and source of enzymes)	Destruction of vitamin B-1 activity
Urease	Urea	Carbon dioxide, ammonia	7.0	Hydrolysis of urea into ammonia
Cellulase	Cellulose	Cellulodextrins and glucose		Hydrolysis of cellulose to produce additional fermentable sugar, improve texture and appearance

TABLE 26.2 —
Promotion Of Desirable Enzyme Reaction

(1) By providing favorable conditions.
   (a) Controlling ripening of pears in storage rooms.
   (b) Tenderizing beef by natural enzymes under controlled storage temperatures.
   (c) Releasing flavor compounds from precursors during rehydrating dried onions and garlics.
   (d) Coloring citrus fruits with ethylene to stimulate respiration.
(2) By adding enzyme preparations or microorganisms which secrete enzymes.
   (a) Using proteinases (pepsin, pancreatin, trypsin, ficin, papain, bromelin, rennet, etc.) to tenderize meat; or form a curd in producing cheese, or modify the texture of dough in bakeries, or clarify beer against chillhaze; or to produce protein hydrolysates as soy sauces, seasonings, amino acids, etc.
   (b) Using carbohydrases (such as amylase, maltase, invertase, lactase, dextrinase, etc.) to convert starch to dextrose, maltose, and dextrins, maltose to dextrose, sucrose to dextrose and fructose, or lactose to dextrose and galactose, etc. (Commercial applications include malting of grain in fermentation industries, production of corn sweeteners and gums, production of breakfast cereals, and in baking, candy making and ice cream production.)
   (c) Using pectic enzymes to aid in extracting juices (as from grapes) and in clarifying juices (as grape and apple); to prevent jellying of concentrated juices, to remove the coating from coffee beans, etc.
   (d) Using enzymes to release flavor constituents from precursors in reconstituting dried vegetables (such as cabbage) in which enzymes have been inactivated by blanching.
   (e) Using glucose oxidase to avoid darkening in dried egg white during storage by converting glucose to gluconic acid.
   (f) Using enzymes to remove oxygen from carbonated beverages, juices, cheese, and other products after packaging.
   (g) Using cellulase to produce additional fermentable sugars from grain mash in brewing and distilling, to remove cellulose fibers from vegetables in canning and freezing.
   (h) Using lipase to improve whipping quality of dried egg albumin and to improve flavor of cheese and milk chocolate.
   (i) Using selected microorganisms which secrete enzymes to promote fermentation in producing beer, wine, vinegar, pickles, kraut, etc., converting fermentable carbohydrates into alcohol, acetic and lactic acids, etc.

TABLE 26.3 —
Prevention Of Undesirable Enzyme Changes

(1) By providing storage conditions favorable for minimizing enzymic changes.
  (a) Avoiding excessively low (near freezing) temperatures in the storage of potatoes, papayas, bananas, grapefruit in which normal ripening processes are irreversibly disturbed by such conditions.
  (b) Using cold or frozen storage to slow or substantially halt undesirable enzyme changes, as in hydrocooling of shelled peas and lima beans to retard deterioration prior to processing.
  (c) Dehydrating, reducing moisture to below levels favorable for enzyme action.
  (d) Handling perishable foods, such as sweet corn, rapidly.
  (e) Avoiding mechanical, insect or freeze injury to raw products. Damaged tissue may be badly discolored or off in flavor. Freeze injury in oil seeds and fruit (peanuts, olives, etc.) may result in excessive amounts of free fatty acids. Enzymes activated by such damage may cause undesirable changes in sound material.
  (f) Excluding fruit or vegetables subjected to microbial attacks. Enzymes secreted by bacteria and fungi contaminate sound material; for example, damage of pickles by pectic present in infected blossoms attached to cucumbers.
(2) By using heat to inactivate enzymes.
  (a) Using hot break in extraction of tomato juice to prevent degradation of pectic compounds.
  (b) Blanching (steam or hot water scalding) of vegetables to be frozen, canned or dehydrated.
  (c) Blanching cut, peeled fruits such as peaches to minimize surface discoloration during canning.
  (d) Flash heating of citrus juices to stabilize cloud.
(3) Adding permissible inhibitors—sulfur dioxide, ascorbic acid, lemon juice, rhubarb juice, etc.—to inhibit enzyme action and thereby prevent darkening of cut fruits such as apples, peaches, bananas, and oxidation discoloration in fruit juices.
(4) Excluding constituents essential to undesirable changes by excluding oxygen from the surface of cut fruit by covering with water, brine or syrup. By vacuum closing packaged products, or by use of inert gas in head space, or by use of oxygen acceptors.
(5) Selecting varieties of fruits and vegetables, and harvesting at stages of maturity, in which undesirable enzyme action is at a minimum. Enzyme and substrate content both vary with variety and maturity.
(6) Taking proper care to minimize contamination of fruit and vegetable tissues with plant constituents containing high concentrations of enzymes such as wheat germ in flour, sugar cane tassels in cane milling and peel extractives in citrus juice extraction.
(7) Avoiding excessive mechanical bruising during peeling, pitting, halving or slicing.

TABLE 26.4 —
Available Commercial Enzymes And Their Applications

Type	Typical Use
Carbohydrases	Production of invert sugar in confectionery industry; production of corn syrups from starch; conversion of cereal starches into fermentable sugars in malting, brewing, distillery, baking industry; clarification of beverages and sirups containing fruit starches
Proteases	Chill-proofing of beers and related products; tenderizing meat; production of animal and plant protein hydrolysates
Pectinases	Clarification of fruit juices; removal of excess pectins from juices such as apple juice before concentration; increase of yield of juice from grapes and other products; clarification of wines; de-watering of fruit and vegetable wastes before drying
Glucose oxidase (Catalase)	Removal of glucose from egg white before drying; removal of molecular oxygen dissolved or present at the surface of products wrapped or sealed in hermetic containers
Glucosidases	Liberation of essential oils from precursors such as those present in bitter almonds, etc.; destruction of naturally occurring bitter principles such as those occurring in olives and the bitter principle glycosides in cucurbitaceae (cucumber and related family)
Flavor enzymes (flavoreases)	Restoration and enrichment of flavor by the addition of enzymes capable of converting organic sulfur compounds into the particular volatile sulfur compounds responsible for flavor in garlic and onions, e.g.; conversion of alliin of garlic into garlic oil by alliianase; conversion of sulfur containing flavor precursors of cabbage and related spices (watercress, mustard, radish) by enzyme preparations from related rich natural sources of enzymes; addition of enzyme preparations from mustard seeds to rehydrated blanched dehydrated cabbage to restore flavor; production of natural banana flavor in sterilized banana puree and dehydrated bananas by naturally occurring banana flavor enzyme; improvement in flavor of foods by an enzyme preparation from fresh corn
Lipases	Improvement in whipping quality of dried egg white and flavor production in cheese and chocolate
Cellulase	Mashing of grain and brewing, clarification and extraction of fruit juices, tenderization of vegetables

ENZYME ACTIVITY

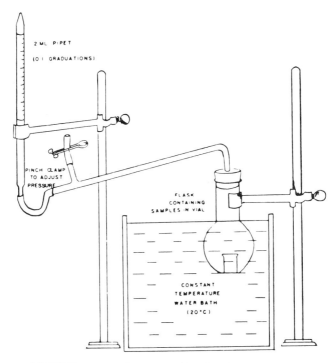

FIGURE 26.1 — Apparatus For Catalase Determination

Catechol Procedure

(1) To the cut surface of a fruit material add sufficient amount of 1% catechol solution to cover the entire surface.
(2) Allow to stand 5–10 min. A black color indicates a positive test. (The test is negative if no color appears in 20 min.)

BIBLIOGRAPHY

COBEY, H. S., and MANNING, G. R. 1953. Catalase vs. peroxidase as indicator for adequacy of blanch of frozen vegetables. Quick Frozen Foods *15*, No. 10, 54, 160–161.
GOULD, W. A. 1955. Blanching of vegetables before freezing—necessary for high quality. Food Packer *36*, No. 8, 22–23.
JOSLYN, M. A., and HEID, J. L. 1963. Food Processing Operations, Vol. 2. Avi Publishing Co. Westport, Conn.
MASURE, M. P., and CAMPBELL, H. 1944. Rapid estimation of peroxidase in vegetable extracts—An index of blanching adequacy for frozen vegetables. Fruit Prod. J. Am. Food Mfr. *23*, 369–375, 383.
THOMPSON, R. R. 1942. Catalase determination. A simplified apparatus. Ind. Eng. Chem., Anal. Edition *14*, 585.

CHAPTER 27

Alcohol in Foods and Beverages

Alcohol is present in some flavors at less than ½ of 1% as a carrier. It is a natural ingredient of many fermentation processes with Beer, Wine, and Spirits as the end results.

Beers are classified into light and regular beers with the light beers having less than 3% alcohol and the regular beers up to 6% by alcohol content.

Wines may be natural wines, that is, the alcohol content less than 14% alcohol content. Wines that are fortified (addition of alcohol before or after fermentation) may have an alcohol content up to 24%. Wines can be broadly classified into the following groups:

> Appetizer—Vermouth—Alcohol content is excess of 15%.
> White Table Wines—Alcohol content from 9 to 14%.
> Red Table Wines—Alcohol content from 9 to 20%.
> Sparkling Wines—Alcohol content from 10 to 14%.
> Dessert Wines—Alcohol content from 17 to 21%.
> Spirits have alcohol content ranging up to 50 or 100% proof.

TEST 27.1 — Determining The Alcoholic Content Of Beverages By Use Of Ebulliometer

General

It is very important to keep the alcoholic content of natural beverages lower than the legal limit. A knowledge of a method of determining the alcoholic content of beverages is, therefore, likely to prove of value.

Equipment, Materials and Reagents

Ebulliometer and Alcohol Tables.

Procedure

Rinse the instrument twice with ordinary water. In doing so pour out an approximate glass tube full of water through one of the openings into the Ebulliometer. Take the instrument in your left hand and close the two openings on top of the instrument with the index and middle finger, shake slightly and empty. Now measure 20 cc in the glass tube as much water as is indicated by the mark and pour it into the ebulliometer. Fasten the empty condenser to the Ebulliometer. Place the thermometer into the opening in front of the condenser. The lamp, which is to be filled with ordinary denatured alcohol (procurable at any drug store) is now lit and put under the instrument. In order to protect the thermometer from heat, the little shield should always be fastened to the top of the heating tubes. Within about 8 to 9 minutes the mercury in the thermometer commences to rise. When the water boils and steam comes out of the top of the condenser watch the thermometer until the mercury remains stationary. The temperature, at which the water boils, is then noted.

Determination of the Boiling Point of the Beverage—After the boiling point of the water has been determined, remove the thermometer, empty the ebulliometer, drain it through the draining cock and blow out any liquid which may have remained in the condenser and in the instrument. Now rinse the ebulliometer twice with an approximate glass tube full of the beverage which is to be tested for alcohol.

Drain the instrument through the draining cock. Now measure in the glass tube as much of the beverage (which latter must be well decarbonated) as is indicated by the mark and pour it into the ebulliometer

Example

The boiling point of the water was found to be at 99.9 and boiling point of the beverage 97.5.

Subtract the one figure from the other and look up the percentage of the alcohol in the tables.

For Instance: $99.9\,°C$ (Boiling Point of Water)
$\underline{-97.5\,°C}$ (Boiling Point of Beverage)
$2.4\,°C$ (Difference)

$2.4 = 2.57\%$ Alcohol by Volume
2.046% Alcohol by Weight
4.502% Alcohol by British Proof Spirits

through the thermometer opening. Insert the thermometer. Fill the condenser with ordinary faucet water. Light the alcohol lamp. When the mercury remains stationary, which will be after 10 to 12 minutes from the time the alcohol lamp is lit, the temperature is read off again and noted. Usually the time for reading the temperature of the thermometer has arrived when the lower cylindrical part of the condenser has become warm.

Important

(1) Keep the instrument in perfectly clean condition. Always rinse the ebulliometer with water after using. Always determine the boiling point of the water first.

(2) Since the atmospheric presure changes from one day to the other and sometimes during the same day, the boiling point of the water and that of the beverage is dependent therefrom.

It is, therefore, necessary to determine the boiling point of the water every day, when it is desired to make an alcohol determination.

If the ebulliometer is used continually during the day, the boiling point of the water should be determined from time to time.

Much time may be saved and still greater accuracy may be had with the employment of two ebulliometers, one with which to determine the boiling point of the water and the other with which to determine the boiling point of the beverage. It is important that the beverage of which the alcohol is to be determined be thoroughly free of carbonic acid gas.

References

GALLANDER, J.F. and A.C. Peng. Wine Making for the Amateur, Cooperative Extension Service. The Ohio State University, Bulletin 549, August 1972.

AMERINE, M.A., Berg, H.W. and Creuss, W.V. Technology of Wine Making, 3rd Edition. The AVI Publishing Co., Westport, Conn. 1972.

AMERINE, M.A. and Ough, C.S. Wine and Must Analysis. John Wiley & Sons, 1974.

TABLE 27.1 — Simplified Alcohol Tables

Difference Between the Boiling Points	% Alcohol by Volume	% Alcohol by Weight	% Alcohol by British Proof Spirits	Difference Between the Boiling Points	% Alcohol by Volume	% Alcohol by Weight	% Alcohol by British Proof Spirits	Difference Between the Boiling Points	% Alcohol by Volume	% Alcohol by Weight	% Alcohol by British Proof Spirits	Difference Between the Boiling Points	% Alcohol by Volume	% Alcohol by Weight	% Alcohol by British Proof Spirits
0.05	0.05	0.040	0.087	0.36	0.36	0.288	0.631	0.67	0.68	0.544	1.191	0.98	1.00	0.790	1.752
0.06	0.06	0.048	0.105	0.37	0.37	0.296	0.648	0.68	0.69	0.552	1.209	0.99	1.01	0.800	1.769
0.07	0.07	0.056	0.123	0.38	0.38	0.304	0.665	0.69	0.70	0.560	1.226	1.00	1.02	0.806	1.787
0.08	0.08	0.064	0.140	0.39	0.39	0.312	0.683	0.70	0.71	0.568	1.243	1.10	1.12	0.890	1.962
0.09	0.09	0.072	0.157	0.40	0.40	0.320	0.701	0.71	0.72	0.576	1.261	1.15	1.17	0.950	2.049
0.10	0.10	0.080	0.175	0.41	0.41	0.328	0.720	0.72	0.73	0.584	1.278	1.20	1.23	0.990	2.155
0.11	0.11	0.088	0.192	0.42	0.42	0.336	0.740	0.73	0.74	0.592	1.296	1.25	1.28	1.030	2.242
0.12	0.12	0.096	0.210	0.43	0.43	0.344	0.750	0.74	0.75	0.600	1.314	1.30	1.34	1.062	2.347
0.13	0.13	0.104	0.227	0.44	0.44	0.352	0.760	0.75	0.76	0.608	1.332	1.35	1.39	1.102	2.435
0.14	0.14	0.112	0.245	0.45	0.45	0.360	0.780	0.76	0.77	0.615	1.352	1.40	1.45	1.150	2.540
0.15	0.15	0.120	0.263	0.46	0.46	0.369	0.806	0.77	0.78	0.623	1.372	1.45	1.50	1.190	2.628
0.16	0.16	0.128	0.280	0.47	0.47	0.377	0.822	0.78	0.79	0.631	1.393	1.50	1.56	1.238	2.733
0.17	0.17	0.136	0.297	0.48	0.48	0.386	0.840	0.79	0.80	0.638	1.414	1.55	1.61	1.278	2.820
0.18	0.18	0.144	0.315	0.49	0.49	0.395	0.857	0.80	0.81	0.645	1.436	1.60	1.67	1.326	2.925
0.19	0.19	0.152	0.332	0.50	0.51	0.404	0.893	0.81	0.82	0.654	1.454	1.65	1.72	1.366	3.013
0.20	0.20	0.160	0.350	0.51	0.52	0.413	0.911	0.82	0.83	0.662	1.472	1.70	1.78	1.414	3.118
0.21	0.21	0.168	0.368	0.52	0.53	0.422	0.929	0.83	0.84	0.670	1.489	1.75	1.83	1.454	3.200
0.22	0.22	0.176	0.385	0.53	0.54	0.431	0.947	0.84	0.85	0.678	1.507	1.80	1.89	1.502	3.311
0.23	0.23	0.184	0.403	0.54	0.55	0.440	0.963	0.85	0.86	0.686	1.524	1.85	1.94	1.542	3.393
0.24	0.24	0.192	0.420	0.55	0.56	0.448	0.981	0.86	0.87	0.694	1.542	1.90	1.99	1.582	3.486
0.25	0.25	0.200	0.438	0.56	0.57	0.456	0.999	0.87	0.88	0.702	1.559	1.95	2.04	1.622	3.574
0.26	0.26	0.208	0.455	0.57	0.58	0.464	1.016	0.88	0.89	0.710	1.577	2.00	2.10	1.670	3.679
0.27	0.27	0.216	0.473	0.58	0.59	0.472	1.034	0.89	0.90	0.718	1.590	2.05	2.15	1.710	3.760
0.28	0.28	0.224	0.490	0.59	0.60	0.480	1.052	0.90	0.91	0.726	1.612	2.10	2.21	1.758	3.871
0.29	0.29	0.232	0.508	0.60	0.61	0.488	1.068	0.91	0.92	0.734	1.630	2.15	2.27	1.806	3.977
0.30	0.30	0.240	0.525	0.61	0.62	0.496	1.086	0.92	0.93	0.742	1.646	2.20	2.33	1.854	4.082
0.31	0.31	0.248	0.544	0.62	0.63	0.504	1.104	0.93	0.94	0.750	1.664	2.25	2.39	1.902	4.187
0.32	0.32	0.256	0.560	0.63	0.64	0.512	1.121	0.94	0.95	0.758	1.682	2.30	2.45	1.950	4.292
0.33	0.33	0.264	0.578	0.64	0.65	0.520	1.139	0.95	0.96	0.766	1.699	2.35	2.51	1.998	4.390
0.34	0.34	0.272	0.595	0.65	0.66	0.528	1.156	0.96	0.97	0.774	1.717	2.40	2.57	2.046	4.502
0.35	0.35	0.280	0.613	0.66	0.67	0.536	1.171	0.97	0.98	0.782	1.735	2.45	2.63	2.092	4.607

ALCOHOL IN FOODS AND BEVERAGES 383

TABLE 27.1 — Simplified Alcohol Tables (Continued)

Difference between Boiling Points	% Alcohol by Volume	% Alcohol by Weight	% Proof Spirits by British
2.50	2.68	2.13	4.69
2.55	2.74	2.18	4.80
2.60	2.79	2.23	4.88
2.65	2.85	2.27	4.99
2.70	2.90	2.31	5.08
2.75	2.96	2.36	5.18
2.80	3.02	2.41	5.29
2.85	3.09	2.46	5.41
2.90	3.15	2.51	5.51
2.95	3.21	2.56	5.62
3.00	3.27	2.61	5.72
3.05	3.34	2.66	5.85
3.10	3.40	2.72	5.97
3.15	3.45	2.76	6.04
3.20	3.50	2.80	6.13
3.25	3.56	2.85	6.23
3.30	3.64	2.91	6.37
3.35	3.70	2.96	6.48
3.40	3.76	3.01	6.58
3.45	3.82	3.06	6.69
3.50	3.88	3.10	6.79
3.55	3.94	3.15	6.90
3.60	4.00	3.20	7.01
3.65	4.07	3.25	7.13
3.70	4.14	3.31	7.25
3.75	4.21	3.36	7.37
3.80	4.27	3.42	7.48
3.85	4.34	3.47	7.60
3.90	4.41	3.52	7.77

Difference between Boiling Points	% Alcohol by Volume	% Alcohol by Weight	% Proof Spirits by British
4.05	4.60	3.68	8.06
4.10	4.67	3.73	8.18
4.15	4.74	3.78	8.30
4.20	4.80	3.84	8.41
4.25	4.87	3.89	8.53
4.30	4.93	3.95	8.64
4.35	5.00	4.00	8.76
4.40	5.08	4.06	8.90
4.45	5.15	4.12	9.02
4.50	5.22	4.17	9.14
4.55	5.29	4.23	9.27
4.60	5.36	4.29	9.39
4.65	5.43	4.34	9.51
4.70	5.50	4.40	9.63
4.75	5.57	4.45	9.76
4.80	5.64	4.51	9.88
4.85	5.71	4.57	10.00
4.90	5.78	4.62	10.13
4.95	5.85	4.68	10.30
5.00	5.92	4.74	10.37
5.05	5.99	4.79	10.59
5.10	6.06	4.85	10.62
5.15	6.13	4.90	10.74
5.20	6.20	4.96	10.86
5.25	6.27	5.02	10.98
5.30	6.34	5.07	11.11
5.35	6.41	5.13	11.23
5.40	6.48	5.19	11.35
5.45	6.55	5.25	11.47

Difference between Boiling Points	% Alcohol by Volume	% Alcohol by Weight	% Proof Spirits by British
5.60	6.76	5.42	11.84
5.65	6.83	5.48	11.97
5.70	6.90	5.53	12.09
5.75	6.98	5.59	12.23
5.80	7.05	5.65	12.35
5.85	7.12	5.71	12.47
5.90	7.19	5.77	12.59
5.95	7.26	5.83	12.73
6.00	7.34	5.89	12.86
6.05	7.42	5.95	12.99
6.10	7.49	6.01	13.13
6.15	7.57	6.07	13.26
6.20	7.65	6.14	13.40
6.25	7.73	6.21	13.54
6.30	7.81	6.27	13.68
6.35	7.89	6.34	13.83
6.40	7.97	6.40	13.97
6.45	8.06	6.47	14.12
6.50	8.14	6.54	14.26
6.55	8.22	6.60	14.41
6.60	8.30	6.67	14.55
6.65	8.38	6.73	14.70

Difference between Boiling Points	% Alcohol by Volume	% Alcohol by Weight	% Proof Spirits by British
6.70	8.47	6.80	14.84
6.75	8.55	6.87	14.93
6.80	8.63	6.93	15.13
6.85	8.72	7.00	15.27
6.90	8.80	7.07	15.42
6.95	8.89	7.14	15.56
7.00	8.97	7.20	15.71
7.05	9.05	7.27	15.85
7.10	9.13	7.34	16.00
7.15	9.22	7.41	16.15
7.20	9.30	7.48	16.29
7.25	9.39	7.55	16.44
7.30	9.47	7.62	16.58
7.35	9.55	7.68	16.73
7.40	9.63	7.75	16.87
7.45	9.72	7.82	17.03
7.50	9.80	7.88	17.17
7.55	9.89	7.95	17.33
7.60	9.98	8.02	17.48
7.65	10.07	8.10	17.64
7.70	10.16	8.17	17.80
7.75	10.25	8.24	17.96

CHAPTER 28

Fats and Oils

Fats and oils are recognized as important food substances. They provide the most concentrated source of calories of any food product. They contribute greatly to the feeling of satiety after eating. They are carriers of the fat soluble vitamins and they serve to make many foods more palatable. In the fatty tissues, they are the principle means for the storage of energy in the body and they serve to cushion certain internal organs against physical shock. Further, they provide thermal insulation for the body and contribute to the body contours.

Fats and oils are predominantly the tri-fatty acid eaters of glycerine, commonly called triglycerides. They are insoluble in water, but soluble in most organic solvents, such as, ethyl ether, petroleum ether, and carbon tetrachloride. They have lower densities than water and at room temperatures they range in consistencies from liquid to solids appearing substances. When solid appearing they are referred to as fats and when liquid appearing they are called oils.

Fats and oils are produced from the seeds of annual plants (corn, soybean, cotton, peanut, etc.) and from perennial plants (coconut, palm, etc.). They are also rendered from animal tissues after slaughter. Fats and oils are used extensively in the formulation of foods and in the cooking of foods.

The oil content of foods varies quite widely. For example potato chips may have an oil content from 30 to 45% depending on the type of oil used in their manufacture, the frying temperature, the thickness of the slice, the solids content of the potato slice, and the final moisture content of the chip. Other deep fried foods will vary in oil content depending on frying temperatures and times, the amount of breading that is used, and the unit size of the product. Generally, the liquid oils will be adsorbed less than the solid fats.

Iodine Value—An expression of the degree of unsaturation of a fat. Measured by the amount of iodine which will react with a natural or processed fat under prescribed conditions. The iodine reacts with the chemically unsaturated "hydrogen short" groups and hence is indicative of the degree of hydrogenation. Unsaturation is reduced and iodine value is lowered by hydrogenation.

Peroxide Number—A measure of the extent to which a fat or oil has already reacted with oxygen. To interpret properly the peroxide number, the history of the fat sample must be known because, during oxidation, the peroxide content increases rapidly, reaches a peak and then decreases sharply during the course of fat breakdown. A high peroxide number indicates advanced oxidation. A low perixode number in the absence of any other information does not necessarily mean oxidation has not taken place.

Refractive Index—The refractive index of a substance is a numerical expression of the ratio of the speed of light in a vacuum to the speed of light in the substance. For practical measurements the scales of instruments indicate refractive indices respective to air rather than vacuum. The refractive index is characteristic within limits for each kind of oil, but it is related to the degree of saturation and is affected by other factors such as free fatty acid, oxidation and heat treatment. The refractive index increases during frying and oxidation.

Saponification Number—Saponification is a process of breaking down or degrading a neutral fat into glycerine and fatty acids by treatment of the fat with hot caustic or alkali. The fatty acids thus produced exist in the form of the alkali salts or soaps, hence the term sponification. The sponification number is the amount of alkali necessary to react completely with a definite quantity of fat sample. The saponification number or value is an indication of the type of fatty acids in the fat. Long chain fatty acids have low saponification numbers and short chain fatty acids have high saponification numbers.

Schaal Test—An accelerated rancidity test for speeding up the determination of stability of a fat or fat containing food product. The test involves the heating of a given amount of fat or food product in a covered glass container at 145 degree F. Results are reported as the time ellapsing until rancidity is detected by odor. Rancidity development may be correlated with determination of peroxide value on the fat sample.

Smoke Point—The temperature of which a fat sample heated under a prescribed set of conditions first gives off a thin continuous stream of bluish smoke. The smoke point of a good frying fat should be as high as possible. A high smoke point on a fresh fat indicates proper processing but the length of time used for frying before excessive smoking is a better indication of stability than the smoke point. An oil has as high a smoke point as a hydrogenated shortening but the oil will breakdown much more rapdily during frying than the shortening.

TEST 28.1 — Testing For Percent Oil In Foods

The oil or shortening used in the manufacture of potato chips or other fried foods may be one of the most costly ingredients. Excessive oiliness in fried foods is generally an indicator of poor quality products. There are several methods to determine the oil content in foods. Some require more expensive equipment and laboratory layouts than others. The following are several options:

A. BAILEY-WALKER METHOD

Equipment and Materials:
1. Balance.
2. Mortar and pestle.
3. Cylinder.
4. Petroleum ether.
5. Electric air oven.
6. Condenser vials.
7. Paper thimbles.
8. Electric hot plate with water circulating system.

Procedure:
1. Weigh oven-dried thimble.
2. Using a mortar, grind the sample up.
3. Weigh out a ten gram sample.
4. Place the ten gram sample in the preweighed thimble.
5. Add approximately 40 ml of petroleum ether to condenser vial and then place the thimble with ten gram sample in it into the vial. **Use enough ether to fill up the vial just below the top of the thimble** (approximately ¼").
6. Turn hood on and start the water circulating through the condenser. **Important:** Check to make sure the hood is operating and the water is circulating through the condenser.
7. Turn the hotplate on high until the samples begin to boil; then turn the hotplate to a low temperature and reflux for one hour.
8. At the end of one hour, remove the thimbles and dry them for thirty-minutes in the air circulating oven at 100 degrees C.
9. Weigh the thimbles and samples and calculate the percent of oil absorption.

10. Divide weight of oil by weight of sample used and report in percentage of oil absorbed.

$$\frac{\text{Grams of Oil}}{\text{Grams of Chips}} \times 100 = \text{Percent of Oil Absorption}$$

Interpretation:

The QMC has a maximum tolerance of 46%. Good quality chips have an oil content of between 30-40%.

B. TESTING FOR PERCENT FAT BY USING THE PRESS METHOD
Carver Press Method (Modified)
Equipment:
1. Carver Press.
2. Paper towels.
3. Triple beam balance.

Preliminary Preparation:
1. Grind twenty to thirty grams of the sample.
2. Thoroughly mix this sample to make sure the sample is consistent throughout.
3. Put three folded paper towels (standard size) in the bottom of the chamber to absorb the expressed oil.
4. Weigh out ten grams into the sample chamber.

Test Procedure:
1. Place the piston portion of the assembly on the warm sample and chamber.
2. Place the entire assembly in the press and pump it up at a rate of about one stroke every two seconds until 15,000 pounds pressure is reached.
3. Allow twenty seconds for the pressure to drop off in the press and then pump it back up to 15,000 pounds pressure.
4. Set the timer for exactly three minutes (180 seconds).
5. **Do not** pump the press up again even though the pressure may drop off.
6. At the end of three minutes release the pressure and remove the sample cake from the bottom side of the chamber using care to include all cake portions and to **not** include any oil portions in the weighing process.
7. Weigh the sample cake.
8. Read the oil content on the chart provided **for the type oil used.**
9. Record your results and indicate the time and date the test was performed.

FATS AND OILS

Sample Used	Sample Cake Weight	% Oil from Std. Curve	Comments

Date of the Test: _____ Time of the Test: _____

Sample Size = grams, Pressure used = pounds, Time = minutes.

C. QUICK DETERMINATION OF FAT CONTENT BY REFRACTOMETER

Equipment:
   1. Bausch and Lomb "Abbe 56" refractometer or equivalent. A constant temperature water bath maintained at 30 degrees C. If it is not a circulating bath, a pump is needed. Water at 30 degrees C should be circulated through the refractometer for at least thirty minutes before readings are taken.
   2. Torsion balance sensitive to 0.01 grams or equivalent.
   3. Waring blender or equivalent.
   4. Glass funnels, 2.5 inch in diameter.
   5. Whatman #1 filter paper, 12.5 cm.
   6. 50 ml Erlenmeyer flasks or equivalent as filtration receivers.
   7. Glass eye dropper.
   8. 50 ml pipette, burette, or equivalent to measure and deliver n-Heptane.

Reagents:
   n-Heptane—Phillips Petroleum Company "Pure Grade," 209.1 degrees F boiling point, refractive index 1.3840 at 30 degrees C. Flammable—use with care. **Refractive index of each batch of n-Heptane should be checked to be sure that it is normal** and calibrated by following the procedure below under standard curve.

Procedure:
   1. Weigh fifty grams of representative sample of chips: transfer to Waring Blender jar.
   2. Add 50 ml n-Heptane to jar.
   3. Mix slowly for one minute or until chips are chopped up and mixed into solvent. If trouble is encountered with too low volume at this stage, sixty grams of chips and 60 ml of n-Heptane may be used or any other quantities preferred as long as the ratios are kept one to one. Do not spill any solvent on blender motor. Be sure blender jar bearings do not leak.

4. Mix at high speed two minutes with loose lid on blender.
5. Decant some of the extract into filter paper in funnel. Have funnel in Erlenmeyer and place a watch glass on the funnel during filtration to minimize evaporation.
6. If first part of filtrate is cloudy, discard— collect a few ml of clear filtrate.
7. Place two or three drops of clear extract on refractometer prisms to cover.
8. Close prisms, allow extract approximately twenty seconds to come to temperature and read refractive index..

Calculations:

From the standard curve or chart prepared with the frying fat in use, convert refractive index directly to percent fat of chips.

Standard Curve (See Figure 28.1)

Calibration curve can be done directly by adding to 50 ml n-Heptane an equivalent amount of fat found in a fifty gram sample of chips. Example: 40% fat would require twenty grams of fat added to 50 ml n-Heptane. A number of solutions of different fat content may be prepared in this way. The refractive index of each solution is plotted against the percentage of fat based on fifty grams of chips in each solution as the source of fat. The relationship of percent fat to reflective index is as follows:

% Fat	Refractive Index, 30 degrees C
25%	1.3993 degrees
30	1.4010
35	1.4042
40	1.4062
45	1.4079
50	1.4094

*Temperature Corrections: Add four units in the fourth place for each degree less than 30°C. Subtract four units in the fourth place for each degree over 30°C.

A chart may be prepared by listing a series of percentages of fat with the corresponding refractive indices.

FATS AND OILS

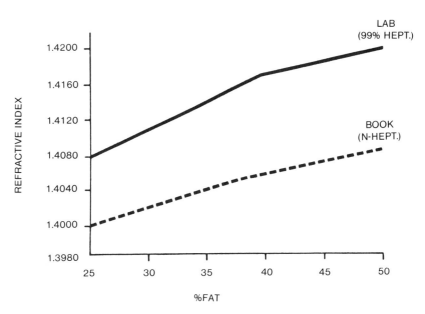

FIGURE 28.1 Standard Curve

Note: If temperature control is not used, a correction factor of 0.0004 should be subtracted from the refractive index reading of an unknown extract or each degree centigrade in temperature of the extract below that of the temperature used in establishing the standard chart and curve. If the temperature of the extract is higher than that used in establishing the curve, 0.0004 should be added to the refractive index of the extract for each degree centigrade difference.

A separate standard curve or chart is needed for each frying fat, as fats may differ in refractive index.

TEST 28.2 — Testing For Free Fatty Acid Content

The percentage of uncombined acids in the frying media indicates the degree of breakdown in the oil or shortening. Since the shelf-life of the product, such as potato chips, is influenced by the degree of rancidity of the frying oil, it would be desirable for the manufacturer to know when a breakdown of the oil is imminent. Fresh oil usually has a low F.F.A. content (.03-.06). Upon cooking this content rises and a 10-fold increase indicates that a breakdown is imminent.

FIGURE 28.2 — Free Fatty Acid
Testing Apparatus

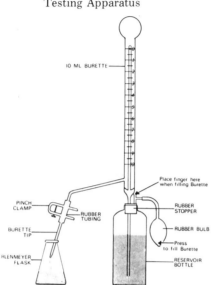

Objectives:
1. Evaluate quality of oil as purchased.
2. Evaluate quality of oil as used.
3. Evaluate oil quality of chips for shelf life.

Equipment and Chemicals
1. Automatic burette assembly, complete with 10 ml capacity burette, glass reservoir bottle (for sodium hydroxide solution), two-hole rubber stopper, rubber bulb, glass burette tip, pinch clamp and rubber tubing for connecting tip to burette. (See sketch for proper set up.)
2. Erlenmeyer flask, wide mouth, 250 ml capacity.
3. Graduate, 50 ml capacity.
4. Dropping bottle for indicator solution.
5. Sodium hydroxide solution (0.1N).
6. Isopropyl alcohol.
7. Phenolphthalein (one gram of phenolpthalein in 50 cc of alcohol and add 50 cc H_2O).

Procedure:
1. Fill the graduate to the 50 ml mark with isopropyl alcohol and empty the alcohol into the Erlenmeyer flask.

2. Add three to five drops or more of phenolpthalein from the dropping bottle to the isopropyl alcohol in the Erlenmeyer flask.

3. While shaking the flask add 0.1 N sodium hydroxide solution from the burette, one drop at a time, until the color of the isopropyl alcohol first changes from white to pink.

Caution: Add only enough to cause color to change from white to pink. Only a few drops will be required and the color change will take place with one drop.

4. Fill the graduate to the 32.5 ml mark with oil (200 degrees F) to be tested for F.F.A. content. Empty the oil completely into the pink colored isopropyl alcohol in the Erlenmeyer flask. Shake the flask vigorously to mix the oil and alcohol. Color will change to color of oil.

5. Fill the burette to the top mark with 0.1 N sodium hydroxide solution. Be sure that this is done before each test is made.

6. Titrate the oil sample in the Erlenmeyer flask with sodium hydroxide solution from the burette by opening the pinch clamp and shaking the Erlenmeyer flask vigorously until the color changes to pink. Continue to shake the flask and add the sodium hydroxide solution until the color does not disappear. When using hot oil samples no extra heating is required. However, if analysis is made on cold oil samples the alcohol and oil mixture should be warmed to 150 degrees F before titration for best results. (A small electric hot plate is convenient for this.)

7. Read the burette to the nearest 0.1 ml to determine the amount of sodium hydroxide used in the titration. Each whole number (1.0 ml) is equal to 0.1% F.F.A.; each small division (0.1 ml) is 0.01% F.F.A. For example, 5.2 ml equals 0.52% F.F.A.

Cleaning the Apparatus:

1. After each test, wash Erlenmeyer flask and graduate thoroughly with a hot detergent solution. Rinse three times with clear water and once with a small quantity of alcohol. Store them upside down to drain. Drying is not necessary.

2. When in regular use the burette and reservoir are kept full of sodium hydroxide solution. If it is not to be used for several days, empty the burette and reservoir and rinse thoroughly three times with clear water. Before refilling, rinse with a small quantity of sodium hydroxide solution.

TABLE 28.1 —
Relationship Of Free Fatty Acid With Smoke
Point Of Frying Oils

Free Fatty Acid %	Smoke Point Temp. (°F)
0.04%	425°
0.06	410
0.08	400
0.10	390
0.20	375
0.40	350
0.60	340
0.80	330
1.00	320

TEST 28.3 — Testing For Peroxide Value Of Fats And Oils

General:
Chemical tests for rancidity correlate only roughly with organoleptic tests. The most frequently used test is for peroxide value. Peroxide value of the drying oil as received or in storage does give some information on oil quality. If the value is over 2.5, it can indicate excessive oxidation and potential lack of stability. If the value is over 7.5, it can indicate sufficient breakdown of the oil to yield fatty aldehydes which will result in a definitely rancid flavor. These peroxide values are approximate since correlation between peroxide values and actual rancidity varies with the type of oil and the specific lot of oil involved. Some individual oils with peroxide values of two to five units may have no apparent rancidity. Heat and the moisture evolved during frying will destroy peroxides so that the oil in the kettle under continuous frying conditions will have essentially zero peroxide value. Potato chips themselves cannot be analyzed for peroxides; however, the oil extracted from the press cake method may be used for peroxide value analysis. The oil may oxidize during extraction and give abnormally high values. The peroxides may break down further during extraction and give misleadingly low values— these peroxide values from the potato chips may be misleading unless full standardization of the methods and techniques are followed.

A. OFFICIAL AOCS METHOD
Equipment:
1. Several 250 ml Erlenmeyer flasks.
2. Balance.
3. Burette with stand.
4. Acetic acid-chloroform solution (3:2 by volume).
5. Saturated solution of potassium iodide.
6. Distilled water.
7. 0.1 sodium thiosulfate solution.
8. Starch indicator.

Procedure:
1. Weigh give grams of fat into a 250 ml Erlenmeyer flask—oil from the press cake method may be used if care is fully exercised to prevent oxidation.
2. Add 30 ml acetic acid-chloroform solution.
3. Swirl to dissolve the fat.
4. Add 0.5 ml of saturated potassium iodide.
5. Shake solution occasionally for exactly one minute.
6. Add 30 ml distilled water.
7. Titrate this solution with 0.1 N sodium thiosulfate from the burette, using starch indicator until the blue color just disappears with stirring. Peroxide value, expressed as milliequivalents per kilogram of fat, is calculated from the formula:

$$\text{ml of titration} \times \text{N thiosulfate} \times \frac{1000}{\text{weight of sample}}$$

B. LEA'S RAPID METHOD
Equipment:
1. Analytical balance.
2. 25 mm test tubes.
3. 25 ml graduated cylinder.
4. 125 ml Erlenmeyer flasks.
5. Starch (three grams soluble potato starch dissolved in 100 ml distilled water).
6. 2:1 Glacial acetic acid—chloroform solution (by volume).
7. Potassium iodide (crystalline).
8. Potassium iodide solution (five percent).
9. Sodium thiosulphate solution (.002 N).
10. Boiling water bath.
11. Burette assembly.

Procedure:
1. Accurately weigh approximately one gram of oil or fat into a boiling tube (25 mm test tube).
2. Add one gram of crystalline potassium iodide and 20 ml of acetic acid-chloroform mixture to tube.
3. Place tube in boiling water bath for sixty seconds.
4. Pour the hot contents of tube into a flask containing 20 ml of five percent potassium iodide solution. Wash out tube with one 15 ml and one 10 ml portion of water and add washings to flask.
5. Add three to five drops starch indicator to flask.
6. Titrate with 0.002 N sodium thiosulphate to a blue end point.

Peroxide value in mili-equivalents peroxide oxygen per kilogram of oil $= \dfrac{2T}{W}$

Where T = Number of milliliters sodium thiosulphate.
Where W = Weight of oil sample.

Note: Fresh oils normally show no peroxide value. However, a low peroxide value does not necessarily indicate oil freshness. Indeed, a relatively low value of 6 mg/kg is likely to immediately precede a marked increase, indicating oxidative rancidity. Values on the order of twenty to forty usually indicate rancidity.

References

GOULD, Wilbur A. Quality Control Procedures For The Manufacture Of Potato Chips And Snack Foods. Snack Food Association, Alexandria, Va. 1986
WEISS, Theodore J. Foods Oils and Their Uses. Avi Publishing Company, Inc., Westport, Conn. 1983.

CHAPTER 29

Total Acidity and pH

Practically all foods contain an acid or a mxiture of acids. The organic acids present in fruits and vegetables influence flavor, brightness of color, stability and keeping quality. These acids may occur naturally, may be produced by action of microorganisms, or may be added in such products as catsup or chili sauce during their manufacture. In all cases, the acids present are largely responsible for the tart or sour flavor. Total-acidity determinations are useful as a measure of this tartness. In fresh plant tissues, the acidity may vary from 6% in lemons down to 0.1% in other products. Total acid is usually determined by titrating an aliquot of sample with a base of known strength using a suitable indicator to determine the end point. In the case of highly colored foods, such as tomatoes, accurate determination of the end point is very difficult when using an indicator; thus it is easier and more accurate to use electrometric methods. The titration is usually calculated and reported in terms of the predominant acid: in the case of tomatoes, citric acid.

Total acid may be expressed on three different bases which are outlined below:

(1) As ml of $0.1N$ NaOH per 100 ml of sample:
  (a) Equation: $A = \dfrac{B \times C}{D}$

  where: A = ml of $0.1N$ NaOH per 100 ml of sample
  B = 100 (used because the calculation is based on a 100-ml sample)
  C = volume of $0.1N$ NaOH used
  D = volume of sample used

(b) Example: If 12.26 ml of 0.1N NaOH are required to titrate a 10-ml sample, it would take 12.26 × 10 or 122.6 ml of 0.1N NaOH per 100 ml of sample.

$$A = \frac{100 \times 12.26 \text{ ml}}{10 \text{ ml}} = 122.6 \text{ ml}$$

(2) As grams of acid in a sample aliquot:
  (a) In general, the predominant acid in the juice or product is used in expressing grams of acid present. It is customary to express acids in the tomato products as citric acid.
  (b) Equation: $W = V \times N \times$ Meq. Wt.
    where: W = gm of acid in aliquot
    V = volume in ml of NaOH titrated
    N = normality of NaOH (0.1N)
    Meq. Wt. = milliequivalents of acid, 0.064 for citric acid
  (c) Example: If 12.26 ml of 0.1N NaOH are required to titrate a 10 ml sample of a tomato product, the acid content of the sample aliquot would be 0.0785 gm.
    $W = 12.26 \times 0.1 \times 0.064 = 0.0785$ gm

(3) As percent of acid in a sample aliquot:
  (a) Equation: $Z = \dfrac{V \times N \times \text{Meq. Wt.}}{Y} \times 100$
    where: Z = % of acid in sample
    V = volume in ml of NaOH titrated
    N = normality of NaOH (0.1N)
    Meq. Wt. = milliequivalents of acid
    Y = volume in ml or weight in gm of sample
  (b) Example: If 12.26 ml of 0.1N NaOH are required to titrate a 10 ml sample of tomato product, the % acid in the sample would be 0.785. (0.064 for citric acid).
    $$Z = \frac{12.26 \times 0.10 \times 0.064}{10} \times 100 = 0.785\%$$

TABLE 29.1 —
Common Acids Found And Used In Foods And Their Weights
And Factors

Acid	Formula Wt	Equivalent Wt	Acid Milliequivalent Factor
Acetic	60.05	60.05	0.060
Butyric	88.10	88.10	0.088
Citric	192.12	64.04	0.064
Lactic	90.08	90.08	0.090
Malic	134.09	67.05	0.067
Oleic	282.46	282.46	0.282
Oxalic	90.04	45.02	0.045
Succinic	118.09	59.05	0.059
Stearic	284.47	284.47	0.284
Tartaric	150.08	75.04	0.075

pH DETERMINATION

The term pH is the symbol for hydrogen-ion concentration. The hydrogen-ion concentration of a food is a controlling factor in regulating many chemical and microbiological reactions. Because the hydrogen-ion concentration expressed in moles is cumbersome, the pH scale as developed by Sorenson is commonly used.

The pH scale ranges from 0 to 14. A neutral solution has a pH of 7.0. A lower scale reading indicates an acid solution and a value above 7.0 indicates an alkaline solution. The pH scale is logarithmic rather than linear in character. Therefore a pH of 5.0 is 10 times as acid as a pH of 6.0 See Table 29.2.

MEASUREMENT OF pH

Colorimetric Method

There are two principal methods used to measure pH. One is the colorimetric method. This depends on the use of an indicator solution that produces a characteristic color at a given pH. Because there is no one indicator that produces characteristic colors for the entire pH range, it is necessary to use several indicators to cover the entire range from pH 0 to pH 14.

TABLE 29.2 —
Relationship Of pH Value To Concentration
Of Acid (H+) Or Alkalinity (OH−)

pH Value	Concentration	
0	10,000,000	
1	1,000,000	
2	100,000	
3	10,000	Acidity
4	1,000	
5	100	
6	10	
7	0	Neutral
8	10	
9	100	
10	1,000	
11	10,000	Alkalinity
12	100,000	
13	1,000,000	
14	10,000,000	

Most of these indicators are weak organic acids that are capable of existing in two tantomeric forms in equilibrium with each other. When placed in a solution on the acid side of their neutral point, the equilibrium will shift to the acid form, and the acid color of the indicator will appear. The reverse is true when the indicator is placed in a solution on the alkaline side of the netural point. Table 29.3 lists some of the indicators used for pH measurement.

To measure the pH of a solution, the appropriate indicator is added to an aliquot of the solution. The color of this solution is then compared with the color of the indicator at known pH's. The pH of the solution will be the same as that of the indicator standard that has the same color. A serious objection to colorimetric pH determinations is that they are not reliable when used with highly colored samples, as the natural color will obscure the color of the indicator. See Table 29.3.

Glass-Calomel System

The other method used to measure pH is to measure the potential developed between two electrodes when immersed in a solution. Several types of electrodes have been developed for this purpose. The most useful type used at this time is the glass-calomel system and an associated potential-measuring device. This device measures the voltage development between the glass electrode and the calomel electrode. The voltage is a measure of the hydrogen-ion activity and each change of 1 pH unit corresponds to a change of 0.059 volt. This developed voltage is measured by balancing the potential of the electrode system against a known potential provided by the interval battery or line voltage and a calibrated slide-wire potentiometer. The scale of the electric meter is calibrated in terms of pH rather than millivolts.

While the glass-calomel system is the one most widely acceptable, it should be remembered that other electrode systems are useful in certain operations. The choice of electrode system will depend on the use intended and the character of the material being tested.

Taking the Measurement

In order to measure the pH of an unknown it is necessary to reduce the sample to liquid form. Extraction of the liquid by means of a hand-operated juicer will provide sufficient liquid to make the pH determination. This can be made directly on the product where an electric pH meter is used. When using a colorimetric determination, it is necessary to first filter the puree to obtain a solution free of solid particles.

TABLE 29.3 —
Indicators Showing Both pH Ranges And
Color Changes

Indicator	pH Range	Color
Acid cresol red	0.2- 1.8	Red-yellow
Acid meta cresol purple	1.2- 2.8	Red-yellow
Benzo yellow	2.4- 4.0	Red-yellow
Brom phenol blue	3.0- 4.6	Yellow-blue
Brom cresol green	3.8- 5.4	Yellow-blue
Methyl red	4.4- 6.0	Red-yellow
Chlor phenol red	5.2- 6.8	Yellow-red
Brom thymol blue	6.0- 7.6	Yellow-blue
Phenol red	6.8- 8.4	Yellow-red
Cresol red	7.2- 8.8	Yellow-red
Meta cresol purple	7.6- 9.2	Yellow-purple
Thymol blue	8.0- 9.6	Yellow-blue
Phthalein red	8.6-10.2	Yellow-red
Tolyl red	10.0-11.6	Red-yellow
Acyl red	10.0-11.6	Red-yellow
Parazo orange	11.0-12.6	Yellow-orange
Acyl blue	12.0-13.6	Red-blue

The actual mechanics of making a pH measurement are relatively simple. Using an electrode system, a sample of the unknown liquid is placed in a small beaker, the electrode inserted, and the pH measurement made. The adjustments necessary and the operating instructions will vary with the apparatus used. Usually concise operating instructions are furnished with the instrument.

When using a colorimetric procedure, place a definite volume of liquid in a calibrated tube and then add a definite amount of indicator. To another tube, the unknown is added. This second tube is aligned in back of the tubes containing the standard indicator solution, in order to compensate for any turbidity or slight amount of color in the sample which would cause a change in the observed color. There are several types of colorimetric pH apparatus available at moderate cost. Any chemical supply house catalog will list several of them.

The pH of a solution is a measurement of the free hydrogen (H) ion or hydroxyl (OH) ion concentration. In pure solutions of an acid or base, it is proportional to the normal concentration of the acid or base. However, in solutions of fruits or vegetables, this is not the case. Any such solution will contain colloids and buffer salts which will influence the pH reading. For this reason, it is possible to have two solutions with the same pH but with widely different total acid concentrations. Therefore, because of the presence of buffers in food products, the total acid content does not have as much significance as pH measurements in determining the types of reaction that will take place.

pH AS A QUALITY-CONTROL CHECK

In the canning of foods, one of the most important factors affecting the sterilization times and temperatures is the actual pH value of acidity in the food, the lower the degree of heat required for sterilization. Examples of typical pH values for some canned foods classified with respect to their pH values is shown in Table 29.4.

It is usually considered that a pH of 4.6 is the dividing line between acid and nonacid foods. This usually means, with a product having a pH of 4.6 or less, that the growth of bacterial spores from organisms such as *Clostridium botulinum* will be inhibited after proper sterilization.

Also, the pH of a food is important, for it affects many other functional properties such as color, flavor, and texture of food. Thus, the importance of pH to the food processor is obvious, and consequently its accurate measurement.

TABLE 29.4 —
Fruits And Vegetables Classed According To Acidity

Group No.	Group Description	pH	Examples of Food Products
I	Nonacid	7.0–5.3	Corn, lima beans, and peas
II	Low or Medium acid	5.3–4.6	Beets and pumpkin
III	Acid I	4.6–3.7	Apricots, pears, and tomatoes
IV	Acid II	3.7 and below	Applesauce, grapefruit, and pickles

Even within any given product the pH may vary considerably. Some of the most important factors believed to have an effect on the actual pH values of a product are listed.
(1) variety or cultivar
(2) maturity
(3) seasonal variations due to growing conditions, etc.
(4) geographical areas
(5) handling and holding practices prior to processing
(6) processing variables

Therefore, two important points to remember are:
(1) A processor should have knowledge of the variety of product he is processing. It may be necessary to adjust the process times and temperatures to provide adequate sterilization. Using careful control procedures, he could blend varieties and perhaps rest assured that his sterilization conditions are safe, thus preventing spoilage. Or he could use FDA-permitted organic acid additions to products to adjust the pH of the product to a level below 4.6.
(2) Quality of the raw material may seriously affect the sterilization conditions unless known acid is carefully controlled.

In conclusion, pH measurement is one of the quality-control checks many food processors have not fully relied upon, even though it is a quality-control check that will assist in the prevention of spoilage. It is a simple measurement requiring little time to accomplish. Further, little cost is required to provide the adequate equipment. It is one of the more important factors accounting for flavor changes in many products.

TEST 29.1 — pH Measurement Of Fruits And Vegetables

The Beckman Model G instrument measures the voltage developed between a glass electrode and a calomel electrode. This voltage is a measure of hydrogen ion activity and each change of one pH unit corresponds to a change in potential of 0.059 volts. This developed voltage is measured by balancing the potential of the electrode system against the known potential provided by the internal battery and a calibrated slide-wire potentiometer. Balance is determined with a vacuum tube amplifier and a rugged electrical meter rather than by use of the conventional galvanometer.

Operation Procedure for the Beckman Model G pH meter

(1) Allow meter to warm up for at least 15 min to ensure accurate readings.
(2) Accurately measure the temperature of the solution to be tested and set the manual temperature control at this temperature. Make certain that the MANUAL push-button is depressed at all times when using the manual temperature adjustment control.
(3) The instrument is standardized with a buffer solution which is as close to the sample as possible and has temperature within 10°C of the unknown. The electrodes are immersed in the buffer solution and depress the READ push-button. Adjust asymmetry control (ASYMM) until meter needle indicates correct value of buffer solution. For most accurate work, check standardization of the instrument before and after each series of measurements.
(4) Depress standby push-button and remove the buffer solution from the electrodes. Wash any remaining buffer solution from the electrodes with distilled water from a wash bottle and dry the electrodes with lens paper. Collect the rinse water in a beaker specially reserved for this purpose.
   CAUTION: Never remove the electrodes from a solution when the STANDBY push-button is not depressed.
   Extreme care must be taken to ensure that the buffer solution does not become contaminated. If the solution is believed to be contaminated do not reuse. Obtain new buffer from the stock bottle.
(5) To obtain the pH of the sample immerse the electrodes in the unknown and depress the READ push-button. Read the pH value of the sample to the nearest hundredth from the meter. Repeat preceding step (4). The pH meter is now ready for the next sample. It is

TOTAL ACIDITY AND pH

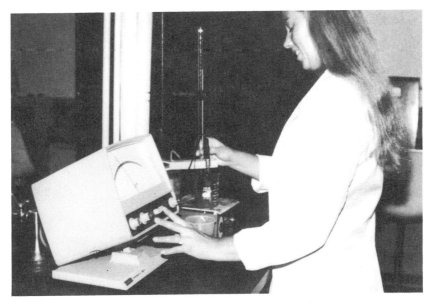

FIGURE 29.1 — pH And Total Acid Evaluation

recommended that a magnetic stirrer be used during all pH determinations to help ensure accurate results.

(6) Leave instrument connected to power line at all times. This will ensure stable, trouble-free performance and will protect the instrument from adverse effect of excessive humidity. Make sure STANDBY push-button is depressed when instrument is not in use. Also, always keep the electrodes immersed in water to prevent drying out and consequent harm to the electrodes.

Operation Procedure for the Beckman Zeromatic SS-3 pH meter

Standardization.—

(1) Depress the STANDBY push-button.
(2) Accurately measure the temperature of the buffer solution and set the manual control to this temperature.
(3) Set the function control to pH.
(4) Standardize the instrument with a buffer solution that has a pH as close to the pH of the sample as possible. (Buffer solution with pH of 4.00 for tomato or orange juice should be used.)
(5) Pour buffer solution into a beaker and immerse the tip of the electrodes into the solution.
(6) Depress the READ push-button. Allow meter reading to stabilize.
(7) Adjust the standardize control until the meter reads the pH of the buffer. Be sure to relock the control in place.

TABLE 29.5 —
pH Values Of Some Commercially Canned Foods

Canned Product	pH Values		
	Avg.	Min.	Max.
Apples	3.4	3.2	3.7
Apple cider	3.3	3.3	3.5
Applesauce	3.6	3.2	4.2
Apricots	3.7	3.6	3.9
Apricots, strained	4.1	3.8	4.3
Asparagus, green	5.5	5.4	5.6
Asparagus, white	5.5	5.4	5.7
Asparagus, pureed	5.2	5.0	5.3
Beans, baked	5.9	5.6	5.9
Beans, green	5.4	5.2	5.7
Beans, green, pureed	5.1	5.0	5.2
Beans, lima	6.2	6.0	6.3
Beans, lima, pureed	5.8	—	—
Beans, red kidney	5.9	5.7	6.1
Beans, and Pork	5.6	5.0	6.0
Beans, wax	5.3	5.2	5.5
Beans, wax, pureed	5.0	4.9	5.1
Beets	5.4	5.0	5.8
Beets, pureed	5.3	5.0	5.5
Blackberries	3.6	3.2	4.1
Blueberries	3.4	3.3	3.5
Carrots	5.2	5.0	5.4
Carrots, pureed	5.1	4.9	5.2
Cherries, black	4.0	3.9	4.1
Cherries, red sour	3.3	3.3	3.5
Cherries, Royal Ann	3.9	3.8	3.9
Cherry juice	3.4	3.4	3.4
Corn, W.K., brine packed	6.3	6.1	6.8
Corn, cream style	6.1	5.9	6.3
Corn, on cob	6.1	6.1	6.1
Cranberry juice	2.6	2.6	2.7
Cranberry sauce	2.6	2.4	2.8
Figs	5.0	5.0	5.0
Gooseberries	2.9	2.8	3.2
Grapes, purple	3.1	3.1	3.1
Grape juice	3.2	2.9	3.7
Grapefruit	3.2	3.0	3.4
Grapefruit juice	3.3	3.0	3.4
Lemon juice	2.4	2.3	2.6
Loganberries	2.9	2.7	3.3
Mushrooms	5.8	5.8	5.9
Olives, green	3.4	—	—
Olives, ripe	6.9	5.9	7.3
Orange juice	3.7	3.5	4.0
Peaches	3.8	3.6	4.1
Pears, Bartlett	4.1	3.6	4.7
Peas, Alaska, (Wisc.)	6.2	6.0	6.3
Peas, sweet wrinkled	6.2	5.9	6.5
Peas, pureed	5.9	5.8	6.0
Pickles, dill	3.1	2.6	3.8
Pickles, fresh cucumber	4.4	4.4	4.4

TABLE 29.5 — (Continued)

Canned Product	pH Values		
	Avg.	Min.	Max.
Pickles, sour	3.1	3.1	3.1
Pickles, sweet	2.7	2.5	3.0
Pineapple, crushed	3.4	3.2	3.5
Pineapple, sliced	3.5	3.5	3.6
Pineapple tidbits	3.5	3.4	3.7
Pineapple juice	3.5	3.4	3.5
Plums, green gage	3.8	3.6	4.0
Plums, Victoria	3.0	2.8	3.1
Potatoes, sweet	5.2	5.1	5.4
Potatoes, white	5.5	5.4	5.6
Prunes, fresh prune plums	3.7	2.5	4.2
Pumpkin	5.1	4.8	5.2
Raspberries, black	3.7	3.2	4.1
Raspberries, red	3.1	2.8	3.5
Sauerkraut	3.5	3.4	3.7
Spaghetti in tomato sauce	5.1	4.7	5.5
Spinach	5.4	5.1	5.9
Spinach, pureed	5.4	5.2	5.5
Strawberries	3.4	3.0	3.9
Tomatoes	4.4	4.0	4.6
Tomato juice	4.3	4.0	4.5
Tomato, pureed	4.2	4.0	4.3

(8) Depress the STANDBY push-button and remove the buffer solution. Wash any remaining solution from the electrodes with distilled water from a wash bottle and dry them with soft paper.
CAUTION: Never remove the electrodes from a solution when the STANDBY push-button is not depressed.

pH Determination

(1) Depress STANDBY push-button.
(2) If the sample temperature differs from that of the buffer solution, measure the sample temperature and adjust the temperature control accordingly.
(3) Immerse the tips of the electrodes in the prepared sample. Turn on the magnetic stirrer (be sure the rod does not hit the electrodes), depress the READ push-button, and allow the meter reading to stabilize.
(4) Read the pH value of the sample.

(5) Depress the STANDBY push-button.
(6) Rinse the electrodes with distilled water and immerse in a beaker of tap water or pH 4.0 buffer.

Leave the instrument connected to the power line. Keep the electrodes immersed in distilled water.

Samples

(1) Tap water
(2) Orange juice
(3) One or more tomato juices

Procedure

(1) Pipette 10 ml of orange or tomato juices into a 250-ml beaker with 90 ml of distilled water. Tap water does not need any preparation.
(2) Place a clean magnetic stirrer into the beaker and allow the solution to mix for a few seconds.
(3) Follow directions of the operating procedure of the pH meter to obtain the pH of the sample.

NOTE: For the measurement of solution above 25°C or for measurement of oxidation-reduction potential or for solutions above pH 11 or for electrometric titrations see the operating instructions with the instrument and the manufacturer's technical bulletins.

TEST 29.2 — Determination Of Acidity—Titration

Reagents

Sodium Hydroxide, 0.1N.—To prepare carbonate-free sodium hydroxide, first make a stock solution of concentrated base: dissolve 100 gm of sodium hydroxide in 100 ml of distilled water in a Pyrex Erlenmeyer flask. Dilute to one liter about 4.0 ml of concentrated, carbonate-free base with freshly boiled CO_2-free water, cooled to room temperature. This solution should be stored in Pyrex bottles and should be protected from carbon dioxide by using a soda-lime tube.

The diluted sodium hydroxide solution is standardized with potassium acid phthalate. Weigh out about 0.5000 gm of potassium acid phthalate. This primary standard must be of known purity and weighed accurately to 0.1 mg. Transfer the potassium acid phthalate into a 250-ml Erlenmeyer flask and dissolve with 75 ml of recently boiled distilled water. Add two drops of phenolphthalein and titrate the primary standard with the sodium hydroxide solution. This means filling the 50-ml burette with the sodium hydroxide solution and adding the base to the flask until the appearance of a faint pink color which persists for 15 sec. The flask should be placed on a white background (sheet of white paper) and the flask should be constantly rotated. Record the number of milliliters of sodium hydroxide needed to reach the end point, pink color.

Calculate the normality of the hydroxide solution by the following general equation.

$$N = \frac{W}{\text{Meq. Wt.} \times V}$$

where: N = normality of standard solution
W = weight of primary standard
Meq. Wt. = milliequivalent weight of primary standard
V = milliliters of standard solution

The equivalent weight of potassium acid phthalate ($KHC_8H_4O_4$) is 204.22 and therefore its milliequivalent weight is 0.20422. Thus, if 25.00 ml of sodium hydroxide solution was required to titrate 0.500 gm of potassium acid phthalate, primary standard, then the normality of the NaOH is as follows:

$$\text{Normality of NaOH} = \frac{0.5000}{0.20422 \times 25.00} = 0.098N$$

Phenolphthalein Indicator.—Dissolve 10 gm of indicator in 600 ml of 95% ethyl alcohol and dilute to 1 liter with distilled water.

Preparation of Sample

(1) Juices: All fruit juices should be thoroughly mixed by shaking to ensure uniformity of sampling. Filter the juices through cheesecloth or filter paper to remove coarse particles which would result in an inaccurately measured sample.

(2) Fresh Fruits: In order to secure a sample of the water soluble acids, the fresh fruits must be crushed. This can be accomplished by placing the fruit in a small hand-operated press type juice extractor or a layer of cheesecloth and squeezing it tightly until the pulp is fairly dry. If the juice sample contains fibers and particles, the liquid should be centrifuged or filtered until the juice is clear.

If the specific gravity of the sample is approximately 1.0 (same as water) it is not necessary to weigh out the sample for titration, and a pipette will give good accuracy. However, for sample of specific gravity greater or less than 1.0, the aliquot should be weighed. (Example: tomato juice and puree).

TEST 29.3 — Electrometric Method For Highly And Slightly Colored Solutions

Equipment

(1) pH Meter
(2) Magnetic stirrer
(3) 50-ml Burette with stand and clamp
(4) 10-ml Pipette
(5) 250-ml Beaker
(6) 100-ml Graduated cylinder

Procedure

(1) Pipette 10 ml of a sample into a 250-ml beaker and add 90 ml of distilled water.
(2) Agitate diluted sample mechanically (magnetic stirrer). Carefully immerse electrodes of pH meter into the solution and titrate with $0.1N$ NaOH from a 50-ml burette to a pH value of 8.1. This pH value is the pH at which phenolphthalein will turn from colorless to pink.
(3) Record the amount of sodium hydroxide used in the titration. Express total acids in the sample as percent of acid.

Calculation

Total acids are usually expressed as the dominant acid contained in the fruit. In fruits such as strawberries, raspberries, tomatoes and citrus fruits, this is taken as citric acid; in plums, cherries, peaches and apples as malic acid; and in grapes as tartaric acid. Total acid may be expressed on three different bases as explained in the text.

BIBLIOGRAPHY

ANON. 1948. Definition of pH and standards for pH measurements. U.S. Dept. Com., Nat. Bur. Standards, Letter Circ. *LC 933*.
ANON. Not Dated. The meaning, application and measurement of pH. Allis-Chalmers Manufacturing Co., Water Conditioning Dept. Inform. Bull. *51*.
GOULD, W. A. 1957. Changes in pH values after time, temperature of cook. Food Packer *38*, No. 9, 22, 38; No. 10, 16–17, 20.
GOULD, W. A. 1974. Tomato Production, Processing and Quality Evaluation. Avi Publishing Company, Westport, Conn.
NOLING, A. W., and O'BRIEN, J. H. 1957. Just how reliable is pH as a guide to food acidity. Food Eng. *29*, No. 2, 88–90, 93.

CHAPTER 30

Mold-Counting Methods and Principles

Mold growth in significant quantities is not found on sound raw products. The growth of mold on raw products breaks down the tissue, producing a rot (Continental Can Co. 1968). Therefore, the presence of large numbers of mold filaments in processed products indicates the use of unfit moldy raw material or contamination of the product by unclean, moldy equipment (Cruess 1958). Federal and State food law enforcement authorities consider the amount of mold in canned products to be an index of the care used by the packer—particularly in the sorting and trimming operations—to keep rot out of canned products (Continental Can Co. 1968).

As increasing proportions of visible rot are allowed to remain on the raw product going to processing, the mold count on the finished product increases, but the relationship is not an exact one because of the nature of the rot that is present. The character of the rotten portion varies greatly. Rots caused by some species of molds are soft and will be broken up during processing. Other types of rot are hard and tough, and will be largely discharged with processing waste without contributing much mold to the finished product (Continental Can Co. 1968).

HOWARD MOLD COUNT METHOD

In order to determine the degree of mold contamination and thus check the efficiency of the operations, the Howard Mold Count Method is used by food technologists (Eisenberg 1952). The Howard Mold Count Method was developed in 1910 by B. J. Howard of the Bureau of Chemistry and Soils, U. S. Department of Agriculture; the federal agency then charged with the enforcement of the Food and Drug Act of 1906. Since the decay of tomatoes is largely caused by molds, Howard conceived the idea of using the occurrence of mold filaments in the comminuted finished product as a means of ascertaining the presence of decayed tomato materials. Investigational studies were started in the fall of 1908 as a result of a conference between Howard and Dr. W. D. Bigelow, the chief of the Food Division of the Bureau of Chemistry. Studies were conducted by Howard in canning factories, and the official procedure was first described in U.S. Circular 68 (Howard 1911).

The Howard Method was designed for two purposes: (1) to give the manufacturers a method for checking their product by determining if their sorters and trimmers were doing good work; and (2) to enable the Federal food law-enforcement agencies to prevent shipment in interstate commerce of tomato pulp or other strained tomato products made from moldy or decomposed material.

Howard stated that the only equipment necessary for mold-counting was a good compound microscope giving magnifications of approximately 90X and ordinary slides (Howard 1911). His instructions were to place a drop of the product to be examined on a microscope slide, place a cover glass over it, and press down until a film of the product about 0.1 mm thick was obtained. A film thicker than this was too dense to be examined, and a thinner one gave a very uneven preparation.

When a satisfactory slide was obtained, it was to be examined at about 90X magnification. Each field of view was to cover approximately 1.5 sq mm, and approximately 50 fields were to be examined. The presence or absence of mold filaments in each field was noted, and then the slide was moved to a completely new field. If a field had more than one filament or clump of mold, it was still to be counted as one positive field. The percentage of the 50 fields showing mold present was then calculated. Results with this method showed that homemade ketchup had practically no positive fields, some manufactured ketchup had from 2-5%, and in some factories every field counted showed the presence of mold. Howard stated that with reasonable care, a mold count of 25% positive fields was possible (Howard 1911).

Methods for estimation of yeasts, spores, and bacteria were also published at this time, but counts on these organisms are done infrequently now.

Howard and Stephenson (1917) published slight modifications of this method, which had been developed during further study. By this time, a Howard mold-counting cell, which was a modification of the Thoma-Zeiss blood-counting cell, had been developed. The cell had an unruled center disk about 19 mm in diameter, and was constructed to give a sample thickness of 0.1 mm when properly prepared. Further instructions given were to clean the Howard cell so that Newton's rings were produced between the slide and cover glass. There were precise directions for preparing the slide. Using the knife blade or scalpel, a drop of thoroughly mixed sample was to be spread evenly over the central disk, so that molds and tomato fiber would not be concentrated in the center. No excess liquid was allowed to be drawn into the moat of the slide.

MOLD—COUNTING METHODS AND PRINCIPLES 413

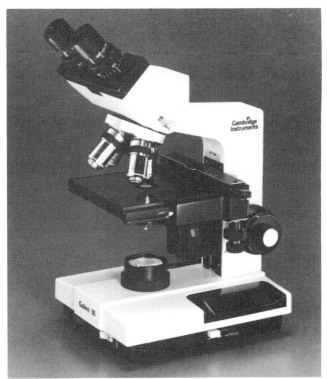

FIGURE 30.1 — Cambridge Instruments Galen III
Compound Microscope
Courtesy of Leica

It was again reported that each field of view was to be 1.5 sq mm, but it was added that this size could be obtained by adjusting the diameter of the microscope field to 1.382 mm. Again 50 fields prepared from two or more slides were to be examined, and it was further stated that no field could be considered positive unless the aggregate length of the mold filaments exceeded ¹/6th of the diameter of the field. These changes tended to make mold-counting more defined and less variable when done by different counters. This method was adopted as a tentative method of the AOAC for the microanalysis of tomato products.

A modified method for slide preparation was presented in 1956 by Kopetz *et al.* (1957). This modification was basically the inclusion of careful manipulation of the cover glass instead of the official prespreading technique. The cover glass could either be lowered at a slant from one edge or rapidly lowered from a parallel position so that the sample spread evenly over the entire surface. Kopetz stated that there were several advantages to this modification, e.g., elimination of scratching the glass, and that there were no significant differences in distribution of mold filaments. However, the modified technique was not accepted in the official final action.

A few techniques to make mold-counting easier without changing the official method have been developed. Beach reported that a 25-square grid drawn on the microscope stage and used with a rectangular Howard cell was beneficial in training mold counters faster (Beach 1951). Pitman (1943) reported a method to lessen the eye fatigue of the counter. The lower end of a camera cable release was attached to the microscope, and a piece of paper which moved with the slide was attached to the mechanical stage. Each time a positive field was counted, the cable was used to punch a hole in the paper, and thus the counter did not have to look up from the microscope.

The Howard method has been criticized as not being truly scientific, and as admitting wide variations in results among competent counters. While discrepancies have been encountered at times among analysts, the method has been used by the tomato industry for over 50 years, during which time no superior one has been advocated. It is now judicially recognized as the official method of determining the presence of mold in tomato products, and, on the basis of the counts obtained, as an index of whether decayed tomatoes were included in the product.

Characteristics of Mold

To the naked eye, the vegetative form of a mold resembles a bit of cotton and is called the mycelium. Each thread or filament of this mycelium is called a mycelial thread or hyphae. As with most fruits and vegetables, such growth on tomatoes results in (and/or occurs with) partial to entire rot and decomposition of the fruit. Hence, the Howard mold count to determine the amount of mold hyphae present is actually an index of the extent of decomposition of the tomatoes used in the product.

Mold filaments have been classified in a number of ways. Perhaps the classification that is of most significance to the person making a Howard mold count is the one which divides them on the basis of function into two types: vegetative and fertile. The vegetative filaments (hyphae) are those which grow underneath the surface of substances, and the fertile filaments (hyphae) are those which bear the fruiting bodies or reproductive spores and grow above the surface. The combination of these two form the entire mold plant; the submerged vegetative hyphae serve to anchor the mold more solidly like the roots of a plant, and the fertile aerial hyphae will be removed as the tomatoes pass through the soak tank, followed by a vigorous spray wash. The submerged vegetative hyphae, on the other hand, can be removed only by careful trimming and sorting, and if not removed they go into the finished product. It is therefore these vegetative hyphae that the counter must recognize and distinguish from other filaments with which they are closely associated. Rot itself consists usually of fruit tissue and mold.

Mold hyphae in all cases are tubular although they may appear to be flat under the microscope. Although different molds show great differences in diameter, in most instances the diameters of the tubes of any one filament are uniform and the cell walls appear under the microscope as parallel lines. Two notable exceptions are molds of the *Mucor* type and *Oospora,* which are often tapering. Because of these two important general characteristics, parallel walls of even intensity must be emphasized.

Mold filaments are also somewhat brittle and will break instead of forming sharp bends, like the sharp curves and oblique ends characteristic of broken spiral tubes from the fibrovascular bundles. Molds also break to form blunt ends and, except for growing ends which are slightly rounded, the ends of such filaments will always be blunt and not sharp or jagged.

Growing molds have living protoplasm within the tubular structure and may present a granular or stippled appearance. This characteristic may persist after the mold is killed in processing and is prominent in the *Mucor* and *Rhizopus* molds.

Many molds found in food products contain cross walls, which separate the mold filament into short sections; such molds are referred to as septate. The presence of cross walls assists in identifying positively many otherwise doubtful filaments. However, cross walls are generally absent in *Mucor* mold.

Most molds show an abundance of branching, and branches are frequently an aid in the positive identification of mold. Here again, however, many fragments do not show branching in a comminuted food product.

One condition frequently observed in mold filaments found in canned tomato products is the apparent breaks which appear to be present in a given filament. Such a filament is usually referred to as "broken" mold. Sometimes close observation of the filament under proper lighting and proper magnification will reveal fine parallel lines connecting the granulated or other visible parts of the filament. This condition apparently develops within the transparent cell wall or sheath where the protoplasm of the mold separates, possibly at certain of the original cell walls, to form small-to-large spaces between the granular area of the filament.

To overcome uncertainty in the identification of questionable filaments, as well as to check the official count of government agencies, all counters must be familiar with the following set of rules published by the National Canners Association (Pitman 1943).

Only filaments which have at least one of the following characteristics shall be classed as mold hyphae.
   (1) parallel walls of even intensity with both ends definitely blunt
   (2) parallel walls of even intensity with characteristic branching
   (3) parallel walls of even intensity with characteristic granulation
   (4) parallel walls of even intensity with definite septation
   (5) occasionally encountered, parallel walls of even intensity with one end blunt and the other end rounded
   (6) occasionally encountered, slowly tapering walls of even intensity with characteristic granulation or septation

As mentioned before, fertile or aerial hyphae are almost never seen in commercial products. Also, since they would be confused with other filaments, they are not counted as mold except when there are spores attached making identification positive.

TEST 30.1 — The Microscope

Proficiency in the microscopic examination of tomato products for mold requires skill in the use and care of the microscope (Continental Can Co. 1968). For mold-counting work, the microscope should be a basic compound microscope, either monocular or binocular. The latter is most desirable (USDA 1967).

It should be equipped with 10X Huygenian ocular (eyepiece) containing Howard Micrometer disc. Field diameter for the 10X objective—10X ocular combination must be 1.382 mm. (In the purchase of a new microscope, this diameter should be specified; a used or existing microscope should be modified by a microscope service specialist to give the correct field diameter. Some models of older microscopes have adjustable draw tubes. These can be manipulated to give the proper field diameter; frequent checking of the diameter is necessary to ensure that the body tube length has not been inadvertently changed.) It should also be equipped with an Abbe condenser and achromatic objectives of 10X and 20X. The microscope should contain a mechanical stage and substage lamp with daylight glass.

Construction of the Microscope

The basic compound microscope consists of two sets of optics mounted in a holder called the body tube. The set nearer the specimen, called the objective, magnifies the specimen a definite amount. The second lens system, the ocular, further magnifies the image formed by the objective, so that the image seen by the eye has a magnification equal to the product of the magnifications of the two systems. Each objective and ocular has its magnification factor engraved on it. Microscopes are generally supplied with several objectives, each having a different power. For mold-counting, two objectives are required: the 10X (16 mm) for counting, and the 20X (8 mm) for confirming identity of questionable filaments. The Howard micrometer disk is located within the ocular.

The body tube (containing the ocular and objective) and a stage to support the specimen are mounted on a stand. Depending upon the model of microscope, either the tube or the stage is equipped with a rack-and-pinion focusing mechanism. A mechanical stage is provided to enable the analysts to move the Howard slide with controlled longitudinal and lateral movements. With regard to moving the slide, an apparent movement of the slide to the left in the microscope field is actually movement to the right on the stage, since the image formed by a compound microscope is seen inverted and reversed.

A condenser and mirror are located beneath the stage. The function of the mirror is to focus light rays into the condenser. The mirror has two surfaces, one plane and one concave. Choice of surface depends, for the most part, on the type of illumination and the objective being used. With the low-powered objectives and with the microscope lamps in common use, the concave mirror generally gives more uniform illumination. Experience with the equipment will dictate whether the concave or plane mirror is preferable.

The condenser contains a series of lenses which assist in illuminating the field by providing a cone of light of sufficient angle to fill the aperture or opening of the objective. Some microscopes are equipped with condensers that are divisible. For low-power observations, such as at 100X magnifications, the top element of the condenser can be removed to achieve more uniform illumination. A rack-and-pinion device, permitting vertical travel of the condenser, facilitates proper light adjustment. The condenser is provided with an iris diaphragm which controls the amount of light entering the condenser and the angle of the emitted cone. Proper adjustment of the condenser is of extreme importance because it enables the analyst to control the light to achieve maximum visibility of microscopic particles in the specimen. If insufficient illumination reaches the specimen, visibility will be impaired; excess illumination will render invisible the smallest objects in the specimen (Continental Can Co. 1968).

Proper Use of the Microscope

(1) Place the slide containing the specimen onto the mechanical stage and move it to a point where the specimen is approximately centered over the opening in the stage.

(2) By means of the coarse adjustment, bring the objective to a point where it almost touches the cover glass. Do this only while observing the movement from the side of the microscope with your eyes near stage level. Do not look into the eyepiece while making this adjustment, because by doing so, you might damage the objective or break the cover slip. Illuminate the specimen by plugging the attached cord into a 110- to 120-volt electric outlet and turning on the light switch.

(3) While looking into the eyepiece, adjust the mirror to direct the light from the substage lamp, through the condenser and specimen, into the objective. Then adjust the condenser height and the iris diaphragm opening as necessary to give even but not brilliant illumination over the entire field.

(4) Look through the eyepiece and slowly move the objective away from the slide, by means of the coarse adjustment, until the image is fairly distinct. This movement must be slow so that the specimen image will not be missed. If there is too much light, so that the specimen cannot be seen easily against the glare, turn the disk diaphragm under the microscope stage until a smaller opening (aperture) is under the specimen.

(5) Focus with the fine adjustment until the specimen is clearly defined.

The specimen under examination may be moved as desired by means of the mechanical stage. During examination of each field of a Howard slide, the fine adjustment must be in almost continuous use in order to examine the total volume of sample within that field.

Use of binocular microscope differs from that of monocular microscope in that two additional adjustments must be made: (1) an adjustment of the separation of the two eyepieces to adapt it to the interpupillary distance of the analyst's eyes; and (2) an adjustment of the focusable eyepiece. The latter is done by focusing on the specimen through the

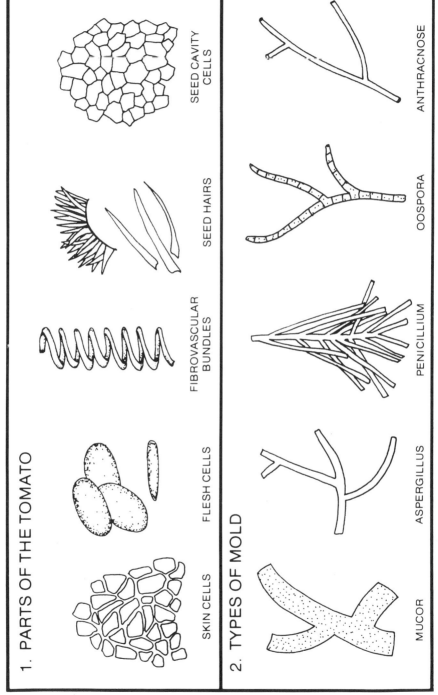

FIGURE 30.2 — Filiments Found In Tomato Products

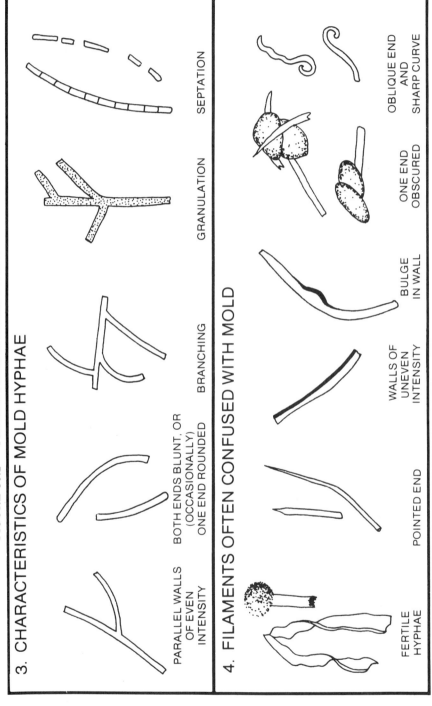

FIGURE 30.2 — Filiments Found In Tomato Products – continued

fixed eyepiece (use fine adjustment); then adjust the focusable eyepiece until the specimen is clearly defined. A single sharp image will now be seen through both eyepieces (Continental Can Co. 1968).

Daylight may be used for illuminating the microscope by turning the mirror so that light from a window is reflected into the microscope. If the concave side of the mirror is used, the light is more intense, because the concave mirror acts like a lens to condense the light from the whole area of the mirror onto that covered by the objective.

Daylight is of variable quality and not always available. Consequently, artificial light is generally used for microscopy. Sunlight, except for special applications, should never be allowed to fall on the mirror or into the lenses of the microscope. It is not good for the microscope and is likely to injure the eye of the observer.

Two kinds of microscope illuminators are in common use. One is a simple lamp which has a uniformly illuminated surface of ground or opal glass. The other type has a focusable condensing lens to project an image of the lamp filament into the microscope. A small light source of the illuminated surface type may be placed in front of the microscope, or the mirror may be removed and the lamp set underneath the microscope condenser.

The lamp should be moved towards or away from the mirror until the field examined is evenly filled with light. When a concave mirror is used, this distance may easily be determined by placing a pencil on the lamp and then moving the lamp until the pencil point is in focus with the specimen. To make sure that the illumination is good, remove the eyepiece from the microscope and look at the back lens of the objective. This lens should be evenly and completely filled with light. If not so filled, the lamp and mirror should be readjusted until the best illumination is obtained (Continental Can Co. 1968).

Care of the Microscope

Successful microscopy requires skill and the proper care of the instrument. The microscope is a precision instrument made from valuable materials by expert workmen; with reasonable care, it will give many years of reliable service.

The microscope should be carried by its arm in an upright position and, when not in use, should be placed in its case or properly covered to protect it from dust. When the microscope is brought from a cold to a warm room it should be allowed to warm up gradually before being used (Richards 1958).

The lenses must be kept meticulously clean. Dust should be loosened and brushed off with a camel's hair brush or blown off with an aspirator, such as an all-rubber ear syringe (available from a drug store) (Richards 1958; Continental Can Co. 1968). Lens paper (no substitute) may be used to remove smudges, such as those caused by eyelashes touching the upper lens of the eyepiece (Continental Can Co. 1968).

Dust on the eyepiece lenses is seen as specks which rotate when the eyepiece is turned while looking through it. Dirt on the objective prevents clear vision and the object appears as if it were in a fog. If a wet preparation touches the objective lens, the lens will have to be cleaned before one can see clearly through it. An eyepiece should always be kept in the tube to prevent dust from collecting on the back lens of the objective or on the prisms.

Should dust settle on the prisms or on the glass protection plates of the binocular body, blow it off with air from an aspirator. Compressed air from laboratory pipes may contain traces of moisture or of oil from the compressor, and should not be used unless an absorbent cotton filter is placed on the discharge tube. If the field does not appear clear, examine the lower surface of the objective with a magnifying glass. Any dirt or damage to the lens may then be seen easily.

Objective lenses are carefully adjusted at the factory, and should not be taken apart except where they have been made to separate. If they must be taken apart, this should be done at the factory, where facilities are available for testing reassembly.

The dry objectives, the condenser, and the eyepiece may be cleaned with distilled water when a liquid is necessary and the lens should be wiped dry with fresh lens paper immediately after cleaning.

In tropical regions, mold may grow on dirty lens surfaces when the relative humidity and the temperature exceed 80°F. When the optical surfaces are kept clean, mold growth can be avoided, or minimized, by keeping the optics in a desiccator, or the microscope in a warmed cabinet. The surface of the microscope is finished with enamel or metal plating and requires little more care than keeping it clean and free from dirt. These finishes resist most laboratory chemicals; ordinarily a little mild soap and water is all that is necessary for cleaning.

Careless handling or dropping may disturb the adjustment of the optical parts of the microscope. If the instrument does not seem to perform properly and there is no dirt on the objective or the eyepiece, it may mean that some of the prisms have become shifted. Do not attempt to adjust any of the prism systems, but send the instrument to the factory, where tools and tests are available for adjustment and for making certain that the adjustment has been done properly (Richards 1958).

TEST 30.2 — Histology Of The Tomato

In addition to training in the use and care of the microscope, the analyst should also study the microscopic appearance of each part of a sound tomato. To accomplish this, he should cut very thin slices of each of the various parts of the tomato, place them on a clean glass slide in a drop or two of water, cover with a thin glass cover and examine under 100X magnification (Continental Can Co. 1968). Before taking sections of the tomato, it may be advisable to place the tomato in alcohol in order to harden the fruit (Eisenberg 1952).

With the razor blade cut a small section of the epicarp (skin) from the tomato. If this section contains flesh cells, they may be removed from the epicarp by soaking in a chloral hydrate solution (Eisenberg 1952). Epicarp cells are greenish-yellow in color and polygonal in shape. Although separated by the so-called middle lamella, the individual cells fit together closely and show definite cell-wall outlines. The cell walls are rather thick and amber in color. When examined at 100X magnification, a piece of tomato skin closely resembles the surface of hammered aluminum.

Mount, in a similar manner, a thin cross section of mesocarp, flesh cells, and examine under low power. Flesh cells are clear, thin walled, and considerably larger than skin cells. They vary in shape from oval to circular and because of their appearance are often referred to as "cellophane footballs." The cells lying just beneath the skin and next to the edge of the seed cavity are smaller and narrower. The cell contents are sparsely granular and appear to be made up of tapering lines which, in reality, are folds in the cell wall.

Fibrovascular bundles are the white, thread-like veins in the tomato flesh that carry moisture and nutrient throughout the fruit. Under the microscope they appear dark in color, and the vascular elements of the bundles resemble a series of tightly coiled springs. Rectangular, brick-shaped companion cells may frequently be seen attached to the coiled bundles. In comminuted tomato products, the bundles are broken up into many small pieces, and occasionally a coil-like spring (resembling the letter "s" in shape) will be found in the finished product. These broken vascular cells can usually be distinguished from mold by a small loop or enlargement on the end, or an end that is pointed or slightly frayed.

Cut a section of the seed-cavity lining cells with a razor blade and place in the chloral hydrate solution. Soaking the section will remove some of the gelatinous material which interferes with the examination of these cells (Eisenberg 1952). The cells lining the seed cavity are smaller than flesh cells and are thin walled. They are irregular in shape and might be compared to the pieces of a jigsaw puzzle with interlocking edges. The cell contents are almost clear, except for occasional flecks of cellular material (Continental Can Co. 1968).

Remove some of the core cells and place a section on a slide, add water and a cover glass, and examine under low power (USDA 1967). Core cells are small and round and have rather thick walls. The contents of the larger cells are rather clear, while the smaller cells have a light amber color due to the more dense cellular materials. The large cells tend to occur in clumps surrounded by many small cells (Continental Can Co. 1968).

Lift out a portion of the gelatinous mass containing seeds and covering cells and place in the chloral hydrate solution. Soak and remove a seed from the solution. Hold the seed with the tweezers and cut the seed in half with the razor blade. Place sections on a slide, add a cover glass, and press gently to separate the cells. Examine under low power (Eisenberg 1952). The internal seed cells vary in shape from rectangular brick-like cells through cube-like to almost ovoid or egg-shaped cells. Also, considerable color variation exists, depending on the compactness of the cell structure and amount and size of the cell contents. The darker cells are more compact and appear to contain more cytoplasm.

Sometimes portions of stem or sepals are encountered in tomato products. Make thin tangential sections using a razor blade, clean in chloral hydrate solution, place on a slide, and study under the microscope. Stem cells vary in shape and size depending upon their location in the stem; however, they are usually rectangular in appearance. Compared to the flesh cells of the fruit, the stem cells are rather small. They are characterized mainly by the presence of chlorophyll, giving a green coloration to the cell. The epidermal hairs of the stem are long and tapering, and are usually divided into three segments.

TEST 30.3 — AOAC Mold Count Procedure

Preparation of Sample

For canned tomatoes and on products containing whole tomatoes, mold counts will be made on the drained liquid only, unless examination, visual observations, or other history indicate that mold counts should be made on pulped tomatoes also. If necessary to make mold counts on drained tomatoes, pass such tomatoes through laboratory cyclone pulper.

Tomato Juice and Tomato Sauce.—Use the juice or sauce as it comes from the container without dilution.

Tomato Paste, Tomato Puree, Concentrated Tomato Juice, and

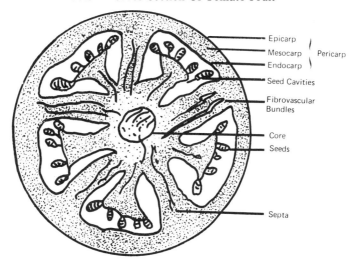

FIGURE 30.3 — Cross Section Of Tomato Fruit

Dehydrated Tomato Powders.—Add sufficient water to the puree, paste, or powder to make a mixture having a total solids content of 8.5–9.5% (refractive index 1.3440–1.3454, corrected to 25°C).

Chili Sauce, Pizza Sauce and Similar Products.—Pass the contents of the container through the laboratory pulper to remove seeds and large particles. If the container is a No. 10 can size (or larger) of chili or pizza sauce, pulp a 20-oz aliquot of a well-mixed sample from individual containers. If tomatoes are pulped, mix well the entire contents of individual container, and use a 5-lb aliquot.

Tomato Catsup.—Tomato catsup is diluted 1:1 with 0.5% sodium carboxymethylcellulose before analysis.

Equipment, Materials and Reagents

(1) Howard Mold Counting Chamber and the 33 × 33 cover glass
(2) Microscope: The microscope should be the basic compound microscope and can be either a monocular or binocular. The latter is most desirable. It should be equipped with an Abbe condenser and achromatic objectives of 10X and 20X. The 10X-16 mm objective must be calibrated with the eyepiece, 10X huygenian ocular with drop in micrometer disc cross ruled in sixths, to give a field diameter of 1.382 mm. This diameter is checked by comparing the field with the circle etched on the right-hand rail of the chamber. If the diameter is not 1.382 mm, the microscope must be standardized or erroneous results will be obtained. The microscope should contain a mechanical stage and a substage lamp with daylight glass.

Procedure

The following is quoted from the AOAC 11th Edition (Horowitz 1970):
Clean Howard cell, so the Newton's rings are produced between slide and cover glass. Remove cover and with scalpel (or knife blade) place portion of the well-mixed sample upon central disk; with same instrument, spread evenly over disk, and cover the glass so as to give uniform distribution. Use only enough sample to bring material to edge of disk. (It is of utmost importance that portion be taken from thoroughly mixed sample and spread evenly over slide disk. Otherwise, when cover slip is

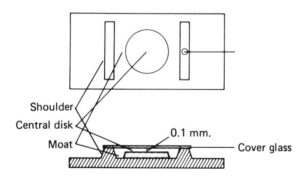

FIGURE 30.4 — Howard Mold Counting Chamber

put in place, insoluble material, and existing molds, may be more abundant at center of mount.) Discard any mount showing uneven distribution or absence of Newton's rings, or liquid that has been drawn across moat and between cover glass and shoulder.

Place slide under microscope and examine with such adjustment that each field of view covers 1.5 mm^2. (This area, which is essential, may frequently be obtained by so adjusting draw-tube of monocular scopes only—binocular scopes must be adjusted at the factory—that diameter to field becomes 1.382 mm. When such adjustment is not possible, make necessary drop in ocular diaphragm with aperture accurately cut to determined size. Diameter of area of field of view can be measured by use of stage micrometer, or the calibration generally etched on shoulders of Howard mold counting cell. When instrument is properly adjusted quantity of liquid examined per field is 0.15 cu mm.) Use magnification of 90–25X. In those instances where identifying characteristics of mold filaments are not clearly discernible in standard field, use magnification of about 200X (8 mm objective) to confirm identity of mold filaments previously observed in standard field.

The latest Howard Mold Counting Chamber has a rectangular chamber (15 × 20 mm) rather than a disk, as described above.

Modifications in Slide Preparation

Two alternate techniques described herein are acceptable. They are to overcome the uneven distribution of the insoluble solids and entrapment of air bubbles between the central disk and the cover glass. These techniques substitute careful manipulation of the cover glass for the prespreading technique. Both methods utilize the uniform spreading, equal distribution action resulting from lowering the cover glass into place.

Inclined Cover Glass Technique.—Using a spatulate instrument, transfer a portion of a well-mixed sample to an area on the central disk halfway between the center of the disk and the edge opposite the analyst (use of a dissecting knife is helpful). One edge of the cover glass is rested in a slanting position on the edges of the slide shoulders nearest the portion of test materials. The cover glass is lowered slightly until it is almost touching the material; then it is lowered rapidly but gently into place. This causes the material to spread evenly over the entire surface of the disk.

Parallel Cover Glass (or "Drop") Technique.—Using a spatulate instrument, place a portion of a well-mixed sample on the approximate center of the central disk. The cover glass is then held in a position parallel to the surface of the central disk and is lowered slowly until slight contact with the sample and cover glass is seen. While maintaining contact with the sample, the cover glass is lowered rapidly but gently until contact is made with the shoulders of the slide. This causes the sample to spread evenly over the entire surface of the disk.

In these techniques, as in the prespreading method, the cover glass should not be lowered too rapidly or a part of the sample may splash onto one or both shoulders, thereby ruining the count. Nor should it be lowered too slowly, as the insoluble material will not spread uniformly. With practice, one can soon learn to control this step in solids preparation and consistently prepare slides showing evenly distributed insoluble material.

Any mount showing uneven distribution of insoluble material, absence of Newton's rings, or liquid that has been drawn or splashed across the moat onto the shoulders must be discarded. A slide should not be counted unless it is properly prepared.

Counting Procedures

After the slide, microscope and light are properly prepared the sample is ready for counting. If the results are to be expressed as "percent of mold," proceed as follows.

Single Plan.—From each of two or more mounts examine at least 25 fields taken in such manner as to be representative of all sections of the mount. Observe each field carefully, noting presence or absence of mold filaments and recording results as positive when aggregate length of not more than 3 filaments present exceeds $1/6$ of the diameter of field. Calculate proportion of positive fields from results of examination of all observed fields and report as percent of fields containing mold filaments.

Although the "percent of positive fields of mold" is calculable from a minimum of 50 examined fields of the product, an examination of at least 100 fields is more desirable and reliable.

If the acceptability of the sample is to be determined by the multiple mold count procedure, the following criterion is to be used.

Multiple Plan.—Under this procedure, increments of 10 fields are counted according to the standard Howard Mold Count technique; the results are then applied to the appropriate sampling plan (see Table 30.1 to determine if the sample unit meets specifications, fails specifications, or whether additional fields must be counted.

These tables are designed to give the same producer-consumer protec-

TABLE 30.1 —
Multiple Sampling Plan For Mold Counts

	10%		20%		30%		40%		50%		60%		70%		80%	
Nc	$-c==r$		c	r	c	r	c	r	c	r	c	r	c	r	c	r
10	1[1]	4	1[1]	6	0	7	0	8	1	9	2	10	4	10	5	2[2]
20	1[1]	5	0	8	2	10	4	13	6	15	8	17	10	19	13	20
30	1	6	3	10	5	13	7	17	11	19	14	23	17	26	20	28
40	1	7	5	12	8	16	11	21	16	25	20	29	24	32	28	36
50	2	8	7	14	11	19	15	25	21	30	26	35	31	40	36	44
60	3	9	9	16	14	22	20	29	26	34	32	41	38	48	44	52
70	4	10	11	18	17	25	24	33	31	40	38	46	45	54	52	60
80	5	11	13	20	20	28	28	36	36	45	44	52	52	62	61	68
90	6	11	15	22	23	31	32	40	42	49	50	58	60	68	70	76
100	8	12	17	23	26	35	37	43	47	54	57	64	66	75	78	83
110	10	13	20	23	30	37	42	47	53	58	63	69	74	80	86	90
120	12	13	24	25	36	37	48	49	60	61	72	73	84	85	96	97

[1] The sample unit cannot be accepted at this level.
[2] The sample unit cannot be rejected at this level.
Nc The cumulative number of fields to count.
c   The maximum cumulative number of positive fields permitted to accept the sample unit for the appropriate percent mold level.
r   The minimum cumulative number of positive fields necessary to fail the sample unit for the appropriate percent mold level.

tion as a single plan consisting of 100 fields. The advantages of a multiple plan are listed.
(1) A decision can be made on very good sample units, or very bad sample units by counting considerable less than 100 fields.
(2) An objective guide is provided as to the number of fields to count in order to properly classify the sample units.
(3) It may be used as a systematic screening procedure for quality control during processing.

Application

(1) Referring to Table 30.1, select the plan which corresponds to the appropriate mold count acceptance level for the product being tested.
(2) Prepare the Howard mold count slide according to the official method and make counts in increments of 10 fields.
   (a) Accept the sample unit if the number of positive fields does not exceed the first value under the column c.
   (b) Fail the number unit if the number of positive fields equals or exceeds the first value under the column r.

(c) Continue counting an additional 10 fields if a decision to accept or fail cannot be reached on the first 10 fields, i.e., if the number of positive fields falls between c and r.
(4) If additional counting is required, continue in increments of 10 fields, comparing the cumulative results with the appropriate steps in Table 30.1 until a decision to accept or fail is reached. In some instances, it may be necessary to count 120 fields.
(5) The number of fields tested (Nc) and the number of positive fields found are cumulative. The results are recorded on a work sheet (Figure 30.5) as number of positive fields per number of fields counted and not as percent positive fields.

The plan is designed to count a maximum of 4 slides, 30 fields per slide. The rectangular type cell is best suited for 30 fields per slide however, the same plan is applicable to the round cell except that it may be necessary to prepare an extra slide if necessary to count 120 fields.

Lot Acceptance Criteria

(1) Sampling Plan Table for Lot Acceptance is as follows:

No. of mold counts	2	4	9	13
Ac^1	0	1	2	3

(For example, the Ac for one through 3 mold counts is 0, the Ac for 4 through 8 is 1 and so forth.) Whenever possible the number of mold counts made on lot in-plant inspection should correspond to the sampling plans above.

(2) By the use of multiple sampling plans in Table 30.1, classify each sample unit in one of the following categories.
   (a) Acceptable: meets the specific requirement.
   (b) A deviant: fails the specific requirement but meets the next higher level.
   (c) Worse than a deviant: fails the next higher level.
(3) Accept the lot as meeting the specific requirement for mold if:
   (a) none of the sample units is "worse than a deviant";
   (b) the number of "deviant" sample units in the sample does not exceed the acceptance number (Ac) in the sampling plan for Lot Acceptance.

The procedure for classifying a sample unit as a "deviant" or "worse than a deviant" is illustrated as follows, using the following portions of the multiple sampling plan from Table 30.1 at the 40% and 50% levels.

[1]Provided these deviants are acceptable at the next highest level and further provided rot counts are well within limits.

MOLD—COUNTING METHODS AND PRINCIPLES

FIGURE 30.5 — Mold Count Record

Slide No. _____ Positive Fields _____ Analyst _____ Date _____

Place the slide on the stage with number at the right.

The circles below correspond to the 25 fields on a slide. In each circle, sketch all of the mold filaments seen in the corresponding field, indicating the relative size and position of each filament. Below each circle put (+) or (−) according to your findings.

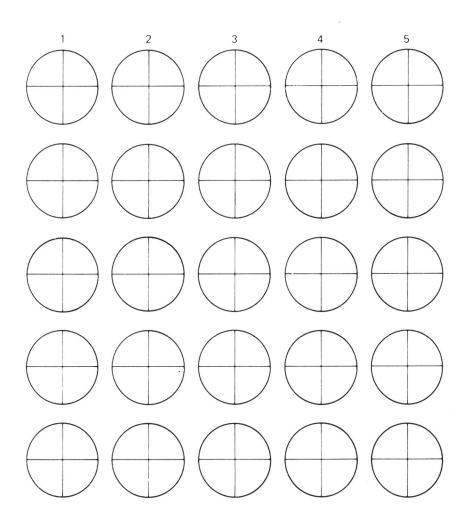

40% Level			50% Level		
Nc	c	r	Nc	c	r
10	0	8	10	1	9
20	4	13	20	6	15
30	7	17	30	11	19
40	11	21	40	16	25
50	15	25	50	21	30
60	20	29	60	26	34
70	24	33	70	31	40
80	28	36	80	26	45

The following counts were made first testing at the 40% level. Example:

Nc   Cumulative Number of Positive Fields Found

Nc	
10	4
20	8
30	11
40	16
50	21
60	26
70	30
80	36

After counting 80 fields, 36 positive fields were found. This fails the 40% level. These same results are not taken, 10 fields at a time, and compared with the 50% level. No decision can be reached until the third group of 10. The cumulative number of positive fields found after counting 30 fields is 11. The sample unit meets the 50% level, to the 40% requirement rather than "worse than a deviant."

Food and Drug Administration action levels for mold count for the various processed tomato products are shown as tolerances in Table 30.2. FDA recognizes the fact that special processing equipment such as mills and homogenizers increase the mold count. Under such conditions, allowances over the published tolerances may be made. In recognition of this fact, it is advisable to make mold counts both before and after the product is processed through any mill or comminuting equipment. Such information can be helpful in verifying the increase in mold counts resulting from the use of this equipment.

TEST 30.4 — AOAC Rot Fragments In Comminuted Tomato Products

General

One of the most difficult problems the processing industry faces today is to cope with insects on or in the products delivered to the plant for processing. Further, the requirements of the Food and Drug Administration make it mandatory for many products that the processors com-

TABLE 30.2 —

FDA Tolerance and Defect Action Level (DAL) For Mold Count

Product	Tolerance %	DAL's
Tomatoes (in juice)	15	16
Tomatoes (in puree)	25	26
Tomato juice	20	21
Tomato catsup	30	31
Tomato puree	40	41
Tomato paste	40	41
Tomato soup	40	41
Tomato sauce (undiluted)	40	41
Pizza sauce (6% tomato solids after pulping)	30	31
Other products	40	41

ply with the removal of all insects and insect parts or face the problem of seizures, court proceedings, etc. The methods of determining the insects and insect parts in or on foods are quite advanced when considering the processor's problem of removing same. Nevertheless, a quality-conscious packer can comply with most requirements if he uses care in preparing his product prior to processing, and knows how to determine whether or not he has removed the insects or insect parts from his product.

The following material is taken in part from the AOAC. The technologist making any microanalysis of foods should refer to the detailed methods, procedures and illustrations given in the AOAC, and those given in Food and Drug Circular 1 "Microanalysis of Foods and Drugs" (Anon. 1944) for specific methods on determination of rot in fruits, identification of molds in fruits and specific details and illustrations on identification and determination of insects in foods.

Equipment and Supplies

(1) Stabilizer solution: 0.5% sodium carboxymethyl cellulose (2.5 gm Cellulose gum blended with 500 ml boiling water in high-speed blender plus 10 ml chloroform). Alternative: 3–5% pectin or 1% algin—add 2 ml chloroform per 100 ml solution as preservative
(2) Crystal violet solution (10 gm dye/100 ml alcohol)
(3) 1 Wildman trap flask (1–2 liter)
(4) 1 Box of rapid solution filter paper with hard finish and 1–2 cm wider than the internal diameter of the funnel
(5) 1 5000-cc Separatory funnel

- (6) 1 2000-cc Separatory funnel
- (7) 1 Ring stand and clamp
- (8) 1 Suction flask—500 cc
- (9) 1 No. 8 sieve
- (10) 1 No. 10 sieve or 10 xx silk bolting cloth
- (11) Petri dishes
- (12) Wash bottle
- (13) Pipettes
- (14) Beakers
- (15) Graduate cylinders 100-cc and 500-cc capacity
- (16) Gasoline
- (17) Castor oil

Procedure

- (1) After determining the drained weight in the usual way and receiving the drained liquor, chop the greens into pieces of approximately 1–2 in. in length.
- (2) Weigh out 100 gm of the well mixed sample into a 1-liter beaker.
- (3) Add 10 drops of crystal violet solution, stir and allow to stand 3 min.
- (4) Add 25 gm of neutral lead acetate crystals.
- (5) Add 10 cc of glacial acetic acid.
- (6) Boil on a hot plate for 5–10 min, cool and transfer to a 2-liter Wildman trap flask.
- (7) Add 35 cc of gasoline and mix thoroughly so as to assure contact between the gasoline and all the leaves.
- (8) Fill the flask full with distilled or deaerated water.
- (9) Allow to settle for a few minutes. Most of the vegetable matter will sink to the bottom. Usually some tissues rise, probably held by entrapped globules of gasoline. To force this material to sink, pivot the lower end of the rod of the Wildman Trap on the bottom of the flask and rotate the upper part of the rod around the neck of the flask with a view to knocking the globules from vegetable tissues without at the same time breaking the interface and thus re-wetting the tissue with gasoline.
- (10) Add stabilizer solution.
- (11) Again allow to stand, trap off the gasoline layer and filter.
- (12) Determine the total aphids or other light filth in the entire liquor drained from the can by subjecting it to gasoline flotation in the

TABLE 30.3 —
FDA Tolerance and Defect Action Level (DAL) For Rot Fragment Count

Product	Tolarance/DAL count/gm
Product reguiring 41% mold count	200
Canned tomatoes with tomato juice	20
Drained Puree from canned tomatoes	100

These guides do not apply to homogenized (viscolized) products. However, they apply to tomato products or packing mediums which have been processed through a Fitzpatrick Comminutor (Fitzmill)

usual manner. (The liquor does not normally present any difficulty and the use of lead acetate is unnecessary.)

(13) Examine the filter paper with a Greenough microscope, 10X magnification. Rot fragments are particles of tomato cellular material with one or more mold filaments attached.

BIBLIOGRAPHY

AMERICAN CAN CO. 1954. The Howard Mold Count Method as Applied to Tomato Products. Am. Can Co., Maywood, Ill.

ANON. 1944. Microanalysis of foods and drugs. FDA Circ. *1*, U.S. Govt. Printing Office, Washington, D.C.

BEACH, P.L. 1951. Speed your mold count with this handy guide. Food Eng. *23*, No. 10, 151.

CONTINENTAL CAN CO. 1968. Mold Counting of Tomato Products. Continental Can Co. Tech. Center, Chicago, Ill.

CRUESS, W.V. 1958. Commercial Fruit and Vegetable Products, 4th Edition. McGraw Hill Book Co., New York.

EISENBERG, M.V. 1952. Observations and suggestions of factory control of rot and extraneous matter in tomato products. Natl. Canners Assoc., Conv. Issue *1371*.

GOLDBLITH, S.A., JOSLYN, M.A., and NICKERSON, J.T.R. 1961. An Introduction to the Thermal processing of Foods. Avi Publishing Co., Westport, Conn.

GOULD, W.A. 1974. Tomato Production, Processing and Quality Evaluation. Avi Publishing Co., Westport, Conn.

HOROWITZ, W. 1970. *In* Official Methods of Analysis, 11th Edition. Assoc. of Official Analytical Chemists, Washington, D.C.

HOWARD, B.J. 1911. Tomato ketchup under the microscope, with practical suggestions to insure a clean product. USDA Bur. Chem. Circ. *68*.

HOWARD, B.J., and STEPHENSON, C.H. 1917. Microscopical studies on tomato products. USDA Bull. *581*.

KOPETZ, A.A., TROY, V.S., and MCCALLUM, M.R. 1957. Modified slide preparation for the official Howard mold count method. J. Assoc. Offic. Agr. Chemists *40*, 905.

PITMAN, G.A. 1943. Mold counting recording device. J. Assoc. Offic. Agr. Chemists *26*, No. 4, 511.

RICHARDS, D.W. 1958. The Effective Use and Proper Care of the Microscope. American Optical Co., Buffalo, N.Y.

SMITH, R.H. 1954. Instructions in microanalytical methods. J. Assoc. Offic. Agr. Chemists *37*, No. 1, 170–177.

USDA. 1967. Methods of Analysis for Tomato Products, Mold Count. *1967*, U.S. Govt. Printing Office, Washington, D.C.

WEISER, H.H., MOUNTNEY, G.J., and GOULD, W.A. 1971. Practical Food Microbiology and Technology, 2nd Edition. Avi Publishing Co., Westport, Conn.

CHAPTER 31

Water Activity

The following is taken from the Food Processors Institute Manual "Canned Foods-Principles of Thermal Process Control, Acidification and Container Closure Evaluation" 1982.

"As late as 1940, food technologists thought that the percent water in a food product controlled microbial growth but gradually they learned that it is the availability of the water that is the most important factor influencing growth.

"A measure of the availability of water in a food can be made by determining the water activity, which is designated a_W. Water activity is the water vapor pressure of the solution divided by the vapor pressure of pure water at the same temperature. Mathematically it is expressed as:

$$a_W = \frac{P_X}{P_W} \text{ or } \frac{\text{Vapor pressure of the food system}}{\text{Vapor pressure of pure water under identical conditions}}$$

"When substances are dissolved there is substantial reaction between the substance and the water. A number of the molecules of the water are bound by the molecules of the substances dissolved. All of the substances dissolved in the water reduce the number of unattached water molecules and this way reduce the activity of the water. The extent to which the water activity is lowered depends primarily on the total concentration of all dissolved substances. Thus, if some ingredient such as sugar, salt, raisins, dried fruits, etc., is added to food it competes with the bacteria for available water. The water binding capacity of a particular ingredient influences the amount of water left for the growth of the bacteria."

Methods for Determining a_W.

"There are several methods of determining the water activity of a given food but the one commonly used is an electric hygrometer with a sensor to measure equilibrium relative humidity (ERH). The instrument was actually devised by weathermen and the sensors are the same as those which they use to measure the relative humidity of the air.

"The equilibrium humidity above the food in a closed container is a measure of the moisture available or the water activity. A single measurement of ERH on a food provides information as to which specific spoilage danger it is most susceptible and how close it is to the safety limits.

"Most foods have a water activity above 0.95, and most bacteria, yeasts, and molds will grow above this point. Clostridium botulinum are generally inhibited at an a_W of about 0.93 or less. Thus, if the amount of water available to spores is decreased to a point where they are inhibited and mild heat is applied to destroy the vegetative cells, we have a method of preservation of food products whose quality is sensitive to heat. Examples of such foods are brick or jar cheeses, peanut butter, honey, syrups, jams and jellies, canned breads, and confectionary preparation such as toppings."

"Under the FDA Regulation Part 113, if reduced water activity is used an adjunct to the process, the maximum water activity must be specified. Under the regulation, canned food with a water activity greater than 0.85 and a pH greater than 4.6 is considered a low-acid food and its minimum heat process will have to be filed by the individual packer."

"If the water activity of any food is adjusted to 0.85 or less, it requires no thermal processes and is not covered, regardless of pH, by the Acidified Food-Part 114 or Low-Acid Food Regulations-Part 113. If the pH is 4.6 or less and the a_W is $> .85$, it is covered by acidified food regulation-Part 114 and requires only enough heat to destroy vegetative bacterial cells.

"It is apparent from the table that as far as *C. botulinum* is concerned a water activity of 0.85 provides a large maring of safety. Studies with this organism show that an accurate water activity of 0.93 plus pasteurization will give commercial sterility, however at this stage in time, there is some question about the precision or accuracy of the instruments and methods used to determine water activity and about some factors which control water activity. Therefore, if water activity plus pasteurization is used to

TABLE 31.1 — Water Activity Of Some Microorganisms

Microorganism	Minimum a_W for Growth
Molds (e.g., *Aspergillus*)	0.75
Yeasts	0.88
C. botulinum	0.93
Staphylococcus aureus*	0.85
Salmonella*	0.93

*No-spore forming food poisoning bacteria readily destroyed by heat. Symptoms severe but death is rare.

control commercial sterility, data must be obtained and records kept to show that the process yields commercial sterility."

The critical factors in the control of water activity as an adjuvant in preservation are the ingredients in the final product and their effect on water binding capacity which is measured by the ERH (water activity, a_W).

In determining the ERH (a_W) several hours are required for the water vapor (relative humidity) to reach equilibrium in the headspace above the food in the closed container. Therefore, the formulation of the product to give the required a_W must be predetermined and very accurately compounded at the time of packing. The critical points for supervision are the product preparation and the achievement of the required center temperature in the final product.

Samples of the final product should be checked as frequently as necessary to ensure that a water activity of 0.93 or less is being achieved.

Salt and a_W

Another method of preservation is the use of salt. This is particularly applicable to salt-cured meats and fish. In meat, salt is usually supplemented with ingredients which aid in spoilage prevention. In all cases the salt is relied on to inhibit the growth of spore-forming bacteria such as *C. botulinum* and only enough heat is applied to kill the non-heat resistant types. Strains of *C. botulinum* are known which will grow in a suitable food containing 7 percent salt, but their growth is inhibited at a concentration of 10 percent, which is equivalent to a water activity of approximately 0.93. Although growth can occur at 7 percent, no toxin has yet been demonstrated in this concentration of salt.

TABLE 31.2 — Water Activity Of Some Common Foods

Food	a_W
Cheese Spread	0.947
Soy Sauce	0.803
Fudge Sauce	0.832
Soft moist Pet Food	0.83
Peanut Butter 15% total moisture	0.70
Dry Milk 8% total moisture	0.70
Liverwurst	0.96
Salami	0.82

APPENDIX I

Temperature Conversion Tables

Temperature			Temperature		
Celsius	°C or F	Fahr.	Celsius	°C or F	Fahr.
-40.0	-40	-40.0	-15.0	+ 5	+41.0
-39.4	-39	-38.2	-14.4	+ 6	+42.8
-38.9	-38	-36.4	-13.9	+ 7	+44.6
-38.3	-37	-34.6	-13.3	+ 8	+46.4
-37.8	-36	-32.8	-12.8	+ 9	+48.2
-37.2	-35	-31.0	-12.2	+10	+50.0
-36.7	-34	-29.2	-11.7	+11	+51.8
-36.1	-33	-27.4	-11.1	+12	+53.6
-35.6	-32	-25.6	-10.6	+13	+55.4
-35.0	-31	-23.8	-10.0	+14	+57.2
-34.4	-30	-22.0	- 9.4	+15	+59.0
-33.9	-29	-20.2	- 8.9	+16	+60.8
-33.3	-28	-18.4	- 8.3	+17	+62.6
-32.8	-27	-16.6	- 7.8	+18	+64.4
-32.2	-26	-14.8	- 7.2	+19	+66.2
-31.7	-25	-13.0	- 6.7	+20	+68.0
-31.1	-24	-11.2	- 6.1	+21	+69.8
-30.6	-23	- 9.4	- 5.5	+22	+71.6
-30.0	-22	- 7.6	- 5.0	+23	+73.4
-29.4	-21	- 5.8	- 4.4	+24	+75.2
-28.9	-20	- 4.0	- 3.9	+25	+77.0
-28.3	-19	- 2.2	- 3.3	+26	+78.8
-27.8	-18	- 0.4	- 2.8	+27	+80.6
-27.2	-17	+ 1.4	- 2.2	+28	+82.4
-26.7	-16	+ 3.2	- 1.7	+29	+84.2
-26.1	-15	+ 5.0	- 1.1	+30	+86.0
-25.6	-14	+ 6.8	- 0.6	+31	+87.8
-25.0	-13	+ 8.6	.0	+32	+89.6
-24.4	-12	+10.4	+ 0.6	+33	+91.4
-23.9	-11	+12.2	+ 1.1	+34	+93.2
-23.3	-10	+14.0	+ 1.7	+35	+95.0
-22.8	- 9	+15.8	+ 2.2	+36	+96.8
-22.2	- 8	+17.6	+ 2.8	+37	+98.6
-21.7	- 7	+19.4	+ 3.3	+38	+100.4
-21.1	- 6	+21.2	+ 3.9	+39	+102.2
-20.6	- 5	+23.0	+ 4.4	+40	+104.0
-20.0	- 4	+24.8	+ 5.0	+41	+105.8
-19.4	- 3	+26.6	+ 5.5	+42	+107.6
-18.9	- 2	+28.4	+ 6.1	+43	+109.4
-18.3	- 1	+30.2	+ 6.7	+44	+111.2
-17.8	0	+32.0	+ 7.2	+ 45	+113.0
-17.2	+ 1	+33.8	+ 7.8	+ 46	+114.8
-16.7	+ 2	+35.6	+ 8.3	+ 47	+116.6
-16.1	+ 3	+37.4	+ 8.9	+ 48	+118.4
-15.6	+ 4	+39.2	+ 9.4	+ 49	+120.2

APPENDIX I — TEMPERATURE CONVERSION TABLES

Temperature			Temperature		
Celsius	°C or F	Fahr.	Celsius	°C or F	Fahr.
+10.0	+50	+122.0	+35.0	+95	+203.0
+10.6	+51	+123.8	+35.6	+96	+204.8
+11.1	+52	+125.6	+36.1	+97	+206.6
+11.7	+53	+127.4	+36.7	+98	+208.4
+12.2	+54	+129.2	+37.2	+99	+210.2
+12.8	+55	+131.0	+37.8	+100	+212.0
+13.3	+56	+132.8	+38.3	+101	+213.8
+13.9	+57	+134.6	+38.9	+102	+215.6
+14.4	+58	+136.4	+39.4	+103	+217.4
+15.0	+59	+138.2	+40.0	+104	+219.2
+15.6	+60	+140.0	+40.6	+105	+221.0
+16.1	+61	+141.8	+41.1	+106	+222.8
+16.7	+62	+143.6	+41.7	+107	+224.6
+17.2	+63	+145.4	+42.2	+108	+226.4
+17.8	+64	+147.2	+42.8	+109	+228.2
+18.3	+65	+149.0	+43.3	+110	+230.0
+18.9	+66	+150.8	+43.9	+111	+231.8
+19.4	+67	+152.6	+44.4	+112	+233.6
+20.0	+68	+154.4	+45.0	+113	+235.4
+20.6	+69	+156.2	+45.6	+114	+237.2
+21.1	+70	+158.0	+46.1	+115	+239.0
+21.7	+71	+159.8	+46.7	+116	+240.8
+22.2	+72	+161.6	+47.2	+117	+242.6
+22.8	+73	+163.4	+47.8	+118	+244.4
+23.3	+74	+165.2	+48.3	+119	+246.2
+23.9	+75	+167.0	+48.9	+120	+248.0
+24.4	+76	+168.8	+49.4	+121	+249.8
+25.0	+77	+170.6	+50.0	+122	+251.6
+25.6	+78	+172.4	+50.6	+123	+253.4
+26.1	+79	+174.2	+51.1	+124	+255.2
+26.7	+80	+176.0	+51.7	+125	+257.0
+27.2	+81	+177.8	+52.2	+126	+258.8
+27.8	+82	+179.6	+52.8	+127	+260.6
+28.3	+83	+181.4	+53.3	+128	+262.4
+28.9	+84	+183.2	+53.9	+129	+264.2
+29.4	+85	+185.0	+54.4	+130	+266.0
+30.0	+86	+186.8	+55.0	+131	+267.8
+30.6	+87	+188.6	+55.6	+132	+269.6
+31.1	+88	+190.4	+56.1	+133	+271.4
+31.7	+89	+192.2	+56.7	+134	+273.2
+32.2	+90	+194.0	+57.2	+ 135	+275.0
+32.8	+91	+195.8	+57.8	+ 136	+276.8
+33.3	+92	+197.6	+58.3	+ 137	+278.6
+33.9	+93	+199.4	+58.9	+ 138	+280.4
+34.4	+94	+201.2	+59.4	+ 139	+282.2

Temperature			Temperature		
Celsius	°C or F	Fahr.	Celsius	°C or F	Fahr.
+60.0	+140	+284.0	+85.0	+185	+365.0
+60.6	+141	+285.8	+85.6	+186	+366.8
+61.1	+142	+287.6	+86.1	+187	+368.6
+61.7	+143	+289.4	+86.7	+188	+370.4
+62.2	+144	+291.2	+87.2	+189	+372.2
+62.8	+145	+293.0	+87.8	+190	+374.0
+63.3	+146	+294.8	+88.3	+191	+375.8
+63.9	+147	+296.6	+88.9	+192	+377.6
+64.4	+148	+298.4	+89.4	+193	+379.4
+65.0	+149	+300.2	+90.0	+194	+381.2
+65.6	+150	+302.0	+90.6	+195	+383.0
+66.1	+151	+303.8	+91.1	+196	+384.8
+66.7	+152	+305.6	+91.7	+197	+386.6
+67.2	+153	+307.4	+92.2	+198	+388.4
+67.8	+154	+309.2	+92.8	+199	+390.2
+68.3	+155	+311.0	+93.3	+200	+392.0
+68.9	+156	+312.8	+93.9	+201	+393.8
+69.4	+157	+314.6	+94.4	+202	+395.6
+70.0	+158	+316.4	+95.0	+203	+397.4
+70.6	+159	+318.2	+95.6	+204	+399.2
+71.1	+160	+320.0	+96.1	+205	+401.0
+71.7	+161	+321.8	+96.7	+206	+402.8
+72.2	+162	+323.6	+97.2	+207	+404.6
+72.8	+163	+325.4	+97.8	+208	+406.4
+73.3	+164	+327.2	+98.3	+209	+408.2
+73.9	+165	+329.0	+98.9	+210	+410.0
+74.4	+166	+330.8	+99.4	+211	+411.8
+75.0	+167	+332.6	+100.0	+212	+413.6
+75.6	+168	+334.4	+100.6	+213	+415.4
+76.1	+169	+336.2	+101.1	+214	+417.2
+76.7	+170	+338.0	+101.7	+215	+419.0
+77.2	+171	+339.8	+102.2	+216	+420.8
+77.8	+172	+341.6	+102.8	+217	+422.6
+78.3	+173	+343.4	+103.3	+218	+424.4
+78.9	+174	+345.2	+103.9	+219	+426.2
+79.4	+175	+347.0	+104.4	+220	+428.0
+80.0	+176	+348.8	+105.6	+222	+431.6
+80.6	+177	+350.6	+106.7	+224	+435.2
+81.1	+178	+352.4	+107.8	+226	+438.8
+81.7	+179	+354.2	+108.9	+228	+442.4
+82.2	+180	+356.0	+110.0	+230	+446.0
+82.8	+181	+357.8	+111.1	+232	+449.6
+83.3	+182	+359.6	+112.2	+234	+453.2
+83.9	+183	+361.4	+113.3	+236	+456.8
+84.4	+184	+363.2	+114.4	+238	+460.4

APPENDIX II

Weights and Measures

WEIGHTS AND MEASURES

Domestic Weights

1 grain	=	.002286 ounces = .0001429 pounds
7,000 grains	=	1 pound
16 ounces	=	1 pound
2,000 pounds	=	1 short ton

Metric Weights

1,000 micrograms (4)	=	1 milligram (mg)
1,000 milligrams	=	1 gram (gm)
1,000 grams	=	1 kilogram (kg)
1,000 kilograms	=	1 metric ton

Domestic Equivalents of Metric Measures

1 gram	=	.035274 ounces
1 kilogram	=	2.204622 pounds
1 metric ton	=	2,204.622 pounds
1 liter	=	2.1134 pints, liquid measure
1 liter	=	1.05671 quarts, liquid measure
1 liter	=	.26418 gallons, liquid measure

Metric Equivalents of Domestic Measure

1 grain	=	64.799 milligrams
1 ounce, avoir.	=	28.3495 grams
1 pound, avoir.	=	453.5924 grams
1 short ton	=	907.185 kilograms
1 short ton	=	0.9072 metric tons
1 pint, liquid measure	=	.47317 liters
1 quart, liquid measure	=	.9463 liters
1 gallon, liquid measure	=	3.785 liters

Food Weights and Measures

One pinch or dash	=	1/16 teaspoon
60 drops	=	1 teaspoon
3 teaspoons	=	1 tablespoon = ½ oz. liquid
4 tablespoons	=	¼ cup = 2 oz. liquid
1 gill	=	½ cup = 4 oz. liquid
1 cup	=	16 tablespoons = 8 oz. liquid
2 cups	=	1 pint = 16 oz. liquid
2 pints	=	1 qt. = 32 oz. liquid
1 liter	=	1.05 quarts liquid
4 quarts	=	1 gallon = 128 oz. liquid
31½ gallons	=	1 barrel
2 barrels	=	1 hogshead
8 quarts	=	1 peck
4 pecks	=	1 bushel

Total Quality Assurance References

SUGGESTED BASIC REFERENCE TEXTS AND JOURNALS

AMERINE, M.A. et al. The Technology of Wine Making. Avi Publishing Company. 1980.
ANON. THE ALMANAC. Edward E. Judge & Sons, Inc. 1987.
ANON. Laboratory Manual for Food Canners and Processors, Volumes I and II. Avi Publishing Company. 1980.
CROSBY, Philip B. Quality is Free. McGraw Hill Book Co. 1984.
CROSBY, Philip B. Quality Without Tears. McGraw Hill Book Co. 1984.
CROSBY, Philip B. Running Things. McGraw Hill book Co. 1986.
FEIGENBAUM, A.V. Total Quality Control-Engineering and Management. McGraw Hill Book Company, Inc. 1961.
GOULD, Wilbur A. Tomato Production, Processing and Quality Evaluation. Avi Publishing Company. 1983.
HERSCHDOERFER, S.M. Quality Control in the Food Industry. Academic Press. Volumes I, II, and III. 1967.
JURAN, J.M., L.A. Seder and F.M. Gryna. Quality Control Handbook. McGraw Hill Book Company, 1962.
KRAMER A. and B.A. Twigg. Quality Control for the Food Industry, Volumes I and II. Avi Publishing Company. 1970.
LOPEZ, Anthony A Complete Course in Canning, Volumes I, II, and III. The Canning Trade, Inc. 1987.
LUH, B.S. and J.S. Woodruff. Commercial Vegetable Processing. Avi Publishing Company. 1975.
MOUNTNEY, George J. and Wilbur A. Gould. Practical Food Microbiology and Technology. Van Nostrand Reinhold Company. 1988.
PETERS, T.J. and R.H. Waterman, Jr. In Search of Excellence, Harper and Row Publishers. 1981.
POMERANZ, Y. and C.E. Meloan. Food Analysis: Theory and Practice. Avi Publishing Company, Inc. 1978.
STOK, Th. L. The Worker and Quality Control. The University of Michigan. 1965.
WOODRUFF, J.S. and B.S. Luh. Commercial Fruit Processing. Avi Publishing Company. 1975.

JOURNALS
Food Technology. Institute of Food Technology
Food Science. Institute of Food Technology
Quality Progress. American Society for Quality Control.

INDEX

Absolute pressure, 156
Acceptance Number, 48
Acid levels, 31, 194
Acidity, 408-410
Adulteration, 13
Agtron E4, 246-247, 262-263
  E5, 248-249, 264
  E10, 249
  F, 247-248, 263-264
  M30, 250-251, 265
  M400, 250, 265
Alcohol, 379
Alcohol Tables, 382-383
Alcohol Insoluble Solids, 299-300
  determination of, 302-303
Analysis of variance, 23-24, 77-80, 123, 197
ANSI Standards, 131
Anthoycyanin, 273
  beets, 273
  pigment table, 258
Appearance, 345
Apples, maturity, 295
  moisture, 309
  size, 279
  texture, 315, 323
Apricots, defects, 346
  mositure, 310
  size, 280
Aroma, 215-216
Asparagus, color, 276
  fiber, 315, 320
  maturity, 295
  moisture, 310
  size, 281
  texture, 295, 317
ASQC Standards, 131
Average, 61, 62, 71, 122

Bailey-Walker method, 387
Bananas, 281
Baume 161-165
Beckman pH meter, 404
Beef, 318
Beer-Lambert Law, 259
Beets, Anthocyanin, 273
  size, 283-284

Bitterness, 190, 194
  definition, 190
  threshold, 194
Bougher's law, 260
Brine, clearness, 168
  cloudiness, 169
  control factors, 169
  cylinders, 175
  evaluation, 149
  maturity by flotation, 297, 304
  measurement, 161-164
  off-colored, 168
  percent salt, 165
  quality, 168
  sediment, 170
  table, 168, 170
Brix, 163
  correction table, 168
  definition, 164
  determination, 164
  hydrometers, 161-164

Can code, 140
Can seam, definition, 141
  double seam dimensions, 145-146
  double seam requirements, 145
  evaluation, 142
  external inspection, 143
  measurements, 144
  reports, 147, 155
  terms, 145
  visual examinations, 142, 154
Can opener, 23
Carotenoids, and carotenes, 275
  pigment table, 258
Catalase, 370
Catechol, 377
Catsup, 339
CEDAC, 6, 35-39
Celery, 317
Character, 295
Characteristic, 53
  curve, 54-55
Cheese, 216
Chemical methods, 13, 15

INDEX

Chicken, 318
Chlorine, 171-172
Chlorophylls, 277
  a and b, 277
  pigment table, 256
Christel Texture Meter, 315
Chroma, 239-240
Citrus, 283
Closures, 146
Code, 140, 148, 193
  Color, 235
  Chemical evaluation, 256
  definition, 235
  instruments, 246
  systems, C.I.E., 243
  Macbeth-Munsell Colorimeter, 243-244
  Macbeth-Munsell Disk Colorimeter, 260
  Maerz and Paul Color Dictionary, 240
  Objective color evaluation, 246
  Rideway Charts, 240
  subjective color evaluation, 238
  USDA Standards, 240
Commercial enzymes, 276
Common Causes, 4
Consumer acceptance, 189
Consistency, see Viscosity
Consistometers, see Viscosimeters
Container, 137
  can, 141
  code, 140
  flexibles, 157
  foil, 157
  glass, 141
  size, 140, 145
  vacuum, 142
Control charts, 122
Control limits, 75
  formula, 76
Corn, alcohol insoluble solids, 302
  cream style consistency, 339
  cream style defects, 346
  maturity, 306
  moisture, 309
  pericarp, 319, 355
  worksheet, 355
Crinkle slices, 288
Crispness, 313
Cultivar, 31

Defect Action Levels, 360, 361, 433, 435
Defects,
  action levels, aphids and thrips, 360
    Drosophila eggs and larvae, 360
    inspect fragments, 361
    mold count, 433
    rot fragment count, 432
  Apricots, 345
  Beans, 347
  Pears, 348
  Pickles, 349
  Potato chips, 351-353
  Potatoes, 350
  Tomatoes, 354
  types, entomological, 345
    extraneous, 345
    genetic, 345
    mechanical, 345
    pathological, 345
    reasons for, 345
Dehydrated vegetables, 300
Descriptive terms, 192
Deviant, 48
DiChromat salt analyzer, 182-183
Dilatant, 326
Dilution test, 191
Drained weights, 151
Dried fruits, 301
Drosophila, characteristics, 357
  defect action levels, 359
  description, 358
  life cycle, 357-358
  methods of detection, 360
    AOAC, 363
    GOSUL, 361
    staining, 365
    Wildman trap flask, 367

Ebulliometer, 279
Environment, 35
Enzymes, activity, 369
  characteristics, 372
  classifications, 370
  commercial, 376
  control, 370
  determination, catalase, 370
    catechol, 377
    peroxidase, 371
  reactions, desirable, 371
    undesirable, 372

INDEX

F table, 81-82
Fats, 385
Fiber, green beans and asparagus, 317-320
  percentage, 320
Fill of container, drained weight vs water capacity, 147
  general method, 150
  percent total capacity, 152
  volumetric determination, 153
  water capacity methods, 151
Finish, 145
Flat slice, 289
Flavor, attributes and concentration, 194-195
  chlorine, effect of, 171
  defects, Crocker, 216
  difference method, 190
  definition, 209, 211-214
  descriptive methods, 191
  evaluation methods, 189
  importance, 189
  measurement, 190
    dilution test, 191
    numerical scoring, 191
    paired comparison, 190
    ranking test, 191
    triangle-scoring, 190
    triangle taste, 190
    threshold level, 190
Flavor difference method, 199
  evaluation, 199
  interpretation, 199-203
  score card, 202
  significance, 200
  summary sheet, 204
Fly egg method AOAC, 361
Food Technologist, 18
Free fatty acids, 391
Freeze dried vegetables, 301
FTC Texture Meter and recorder, 318, 319, 323

Gauges, countersink, 144, 145
  headspace, 25
  sizing, 26
  vacuum, 24
Glass-calomel system, 400
Glass containers, 141
  closures, 146-147
  composition, 146

Glass, containers, Continued —
  parts of, 146
  size, 145
Glucose, 175
GOSUT Texture Meter, 317
Grade standards, see Standards
Grading and grader, 48
  scale, 25
  screens, 25
  table, 22-23
  trays, 26
Grapefruit, 284, 295
GOSUL, fly eggs, 360

HACCP, 36
Hard water, 167
Headspace, evaluation of, 150
  gauge, 149
Histogram, 61, 122
Histology, 423
Howard Mold Count Method, see Mold Count
Hunter Color and Color-Difference Meter, 251, 266-268
  D25 Color and Color Difference Meter, 251, 268-269
  D25 D3A Digital Color Difference Meter, 264, 270-271
  USDA Hunterlab D6 Tomato Colorimeter, 255
Hue, 239-249
Hydrometer, Baume 168
  description, 161
  method, 175
  potato, 298, 307-308
  temperature, 168
  Brix scale, 163
  percentage, 163
  specific gravity, 163-164

Imperfections, 345
Infrared method, 302
Insect fragment, 366
Inspector, 48
Inspection, 49
Iodine value, 386
ISO Standards, 131

INDEX

Jams and jellies, sugar standards, 173-174
Judges, 192-196
Juliene strip, 290

Laboratory, equipment, 27
　functions, 20-23
　procedure, 28
　reports, 31
　taste-panel, 197-199
Least significant difference, 79, 197, 210
Light and lighting, 236
　color temperature, 239
Lima beans, maturity, 282, 315-318, 323
　moisture, 298
　size, 282
Lot, 49
Lycopene, 274-275

MacBeth Examolites, 237-240
MacBeth Munsell Disk Colorimeter, 243
Machines, 4, 35
Manager, 4, 35
Manpower, 4, 35
Magness-Taylor pressure tester, 315
Materials, 4, 35
Maturity, 32, 295
　definition, 295
　measurement, 304
Mean, 62
Median, 62
Metal containers, 137
Methods, 4, 35
Microscope, care, 421
　construction, 417
　slide preparation, 425-427
　use, 418
Milk and cheese, 216
Mode, 62
Moisture, 300, 310
　content, 310
　infrared, 302
　table of, 310
　toluene distillation, 301-302
　vacuum oven method, 300
Mold, characteristics, 414
　definition, 411
　filaments, 411-412
　hypae, 411-412, 416, 418
　types, 411

Mold counting, acceptance criteria 428, 430,
　AOAC method, 425-432
　application, 429
　Howard method, 411
　chamber, 427
　multiple sampling plan, 428-429
　procedures, 425-428
　record, 431
Money, 3
Munsell color percentages, 241
Mushrooms, 289

Net weight, 149
Newtonian, 322, 323
Normal curve, 63
Numerical scoring, 191
　interpretation, 204
　method, 197, 199
　score card, 201
Nutrition, sampling procedures, 53

Objective methods, 13, 213
　color, 244
　definition, 213
Odor, classification, 213, 216
　defects, Crocker, 219
　definition, 217
　directory, Crocker-Dillion, 218
　packaging materials, 218
　standards, Crocker-Dillion, 217
　　Crocker-Henderson, 217-219
　test methods, 232
　transitions, 216
　water, 219
Oils, 385
Olives, 284
Operating curve, 54-55, 60
Oranges, moisture, 310
　size, 283
Organoleptic, 213
Oval shred, 290
Overall significant difference, 208

Paired comparison test, 190
Panel, difference, 190
　judges, 192
Pareto chart, 123
Peach, maturity, 295-296
　moisture content, 310

INDEX

Peach, Continued —
  size, 284
  texture, 315
Pear, defects, 348
  moisture, 310
  size, 284,
Peas, alcohol insoluble solids, 303-305
  maturity, 305
  moisture, 309
  size, 286
  texture 315-316
People, 42-45
Pericarp content, 319-320
Peroxidase, 371
Peroxide number 370
Peroxide value, 394-396
pH, 397
  definition, 397
    colorimetric, 399
    electrodes, 400
  effect on processing, 402
  factors affecting, 402
  indicators, 401
  measurement, 401
  relationship to acid, 400
  relationship to alkalinity, 400
  scale, 399
  values, 399-400, 403
Pickles, defects, 288-289
  definition and measurement, 287, 289
  firmness rating, 315
  size, 285-286
Pigments, 258
Plantkeeping, 102
Potatoes, bruise, 350-351
  chips, 351-353
  moisture, 310
  size, 285, 289
  specific gravity, 308-309
  styles, 293
PONC, 123
Popcorn, 350
Productivity, 41
Pseudoplastic, 322
Put in-Cut out value, 173

Quality, 9
  audit, 9
  assurance 9, 12-16, 121

Quality, — Continued
  circle, 6-7, 43
  control, 9
  communicating, 121
  definition, 48
  equipment, 23-28
  factors, 270-271
  fundamentals, 17
  laboratory, 20-22
  measurements, 20
  notebook, 121
  organization, 17-19
  personnel, 153
  procedures, 28
  reports, 31-32, 122-123
  specifications, 127-128
  standards. See Standards and specifications,
    Quality work life, 41

Range chart, 61-62, 67, 73
Ranking test, 191
Raw materials, 87
Refractive index, definition, 166, 386
  determination 389
  method, 165
  table, 165
Refractometers, advantages, 165
  care, 164
  temperature correction table, 168
  types, 164-165
Research, 113
  and development 113-119
  new products, 118
  plans, 116-117
Rheology, see viscosity
Return on Investment, 122-123
Rot fragment count, 432-435
Run chart, 122

Salometer, 26, 163
  table, 185
Salt, brine table, 165
  determination, 181-183
  Mohr titration method, 182
  threshold levels, 194-195
  percent salt method, 180
  percent salt table, 165, 180

INDEX

Saltiness, 190
  definition, 197
  threshold level, 193-195
Samples and sampling, acceptance levels, 47
  accessibility, 50
  accept-reject, 49
  certificate of, 47
  class, 47
  condition, 47
  continuous, 48
  deviant, 47
  inspection, 48
  lot compliance, 48, 51
  nutrient, 51
  pack, 48
  plan, 51
  selection, 51-53
  shipping, 51
  USDA, 47
Salometer, 26, 163
Sanitation, 101
  evaluation, 110
  exterior, 102
  interior, 103
  inspection, 109
  program, 111
  responsibilities, 101
  sanitarian, 110
  suppliers, 110
  value, 101
Saponification number, 386
Schaal test, 386
Screen sieves, 292
Seasoning, 161, 186-187
Sensory, definition, 213
  terms, 209-214
Shape and symmetry, 279
Shrimp, 318
Sieves, Tyler Standard Screen Scale, 25
Sirup, See Brine; Sugar and Sirup.
Size, 279
  container, 137, 147
  gauges, 25
Snap beans, defects, 347
  fiber count, 320
  moisture, 310
  size, 282
  texture values, 318-319
Smoke point, 386, 394

Sodium chloride, 182-183
Soluble solids, 174-175
  determination, 174
Sour, definition, 213
  threshold levels, 193
Spaghetti, 318
Specification, 10, 127-132
Specific gravity, 163, 295
  measurement, 306
  maturity, 306-307
  potatoes, 163, 306
  total solids, 295
Spectrophotometer, 255-257, 272-273
Standard deviation, 63-64
Standards, grade, 11-13
  industry, 13
  legal, 13
Statistical analysis of data, 23, 77-78, 123
  frequency distribution, 64-67
  histogram, 61-62
  measures of central tendency, 62
  measures of dispersion, 62-63
Statistical Quality Control, advantages, 61, 82
  definition, 61
  records, 72
  terms, 61-63, 69-70
Statistical Process Control, 61
Strawberries, 296
Style, 280
  beets, 287
  mushrooms, 291
  white potatoes, 291
  frozen french fries, 293
Subjective Methods, color, 238
Succulence, 298-300, 309-311
  values, 300
Sugar and sirup, composition of sucrose, 164
  sugar, 161
  purpose, 172
  Put in-Cut out values, 173
  Standards of identity for jams, jellies, fruit butters, 173
  types, 172
Supervisor, 7, 43, 44
Sweet, definition, 213
  threshold levels, 194

Taste, definition, 213

INDEX

Taste panel, consumer acceptance, 189
  judges, 192
  panel difference method, 190
  preparation, 193
  product differences, 190
  table, 193-195
Temperature, correction of, 166
Tenderness, definition, 313
Texture, 313
  apples, 314
  asparagus, 317
  definition, 313
  Magness Taylor Texture Test, 313
  peaches, 314
  pickles, 314
Texturometers, FTC Texture Press and Recorder, 318
  GOSUT, 317
  Magness Taylor Pressure Test, 315
  Wm. F. Christel Co., Texture Meter, 315-316
Thermometer, 27
Threshold level, 193-197
  determination, 193
  flavor quality control form, 196
  interpretation, 197
Toluene distillation method, 300
Tomato, catsup consistency, 332, 338
  characteristics of, 419-420
  color, grades, 339
    measurement, 266
  color disk specifications, 241
  defects, 354
  histology, 423-425
  Hunter-Munsell Chromaticity Diagram, 269
  juice color, 240, 260
    viscosity, 342
  pigment 274-276
Total acidity, 397
  common acids, 398
  measures, 408-410
Total quality assurance, 3
Total solids, measurement, 295, 309
  maturity, 296
Triangle taste panel, 197
Triangular scoring flavor panel, 190, 197
  method, 197
  significance, 199

Vacuum, definition, 142, 147
  evaluation, 142
  gauge, 142
  reasons for, 142
Vacuum oven method, AOAC, 300
Variable control chart, 73
Variable, 89
  Machines, 91
  Manpower, 95
  Methods, 93
  Raw Materials & Ingredients, 87
V-cut shred, 290
V-cut slice, 289
Viscometers, Adams consistometer, 332, 339, 340
  Bostwick consistometer, 332, 338, 339
  Brabender Recording, 336
  Brookfield, 336
  Cannon & Fenske, 327
  constant force types, 334
  constant speed devices, 335
  Drage, 336
  Dudley, 328
  Engler & Scott, 328
  Fenske, 327
  Feranti-Shirley, 337
  Fisher Electroviscometer, 330
  Fisher-Gardner, 337
  Fisher-Irany, 327
  FMC, 337
  Ford, 328
  Gardner Mobilometer, 330
  GOSUC Efflux tube, 328, 342
  Hagan, 337
  Hercules, 337˙
  Hoeppler, 330
  MacMichael, 335
  Orifice Constriction, 327
  Ostwald, 327
  Saybolt, 328
  Sil Koehler Mobilometers, 330
  Stormer, 334, 340
  Ubelohde, 327
  Universal, 331
  Zahn, 328
Viscosity, classification of instruments, 327
  definition, 325
  instruments, cone penetrometer, 331
    constant force type, 334

INDEX

Viscosity, Instruments, — Continued
   constant speed devices, 335
   falling ball, 330
   falling plate devices, 330
   flow rate, 327
   flow under pressure, 328
   gravity flow, 327
   limit of flow devices, 331
   orifice constriction, 327
   rotation, 334
   spreading mass, 315
  Newtonian Liquid, 325
  Non-Newtonian Liquid, 326
   dilatant materials, 327
   Non-Newtonians, 327
   plastic materials, 325
   pseudoplastic materials, 326
   thixotropic, 326

Volumetric determination, 153

Water, activity, 437-440
  chlorine, 168-169
  hardness, 167
  minerals, 167-168
  vapor transmission rate, 157
Wax beans, 347
Wines, 379

YSI analyzer, 175-176, 179

QC AND EVALUATION TESTS

Test 2.1 - Testing the Significance Between Two Sample Means	85
Test 15.1 - Container Size, Vacuum, Fill & Drained Weight	152
Test 15.2 - Can Seam Evaluation	154
Test 15.3 - Permeability of A Barrier Film	156
Test 15.4 - Water Vapor Transmission Rate	157
Test 15.5 - Burst Test	157
Test 16.1 - Determination of Soluble Solids - Refractometric Method	174
Test 16.2 - Hydrometer Method	175
Test 16.3 - Glucose Determination by the YSI Analyzer	175
Test 16.4 - Sowokinos Testing for Sucrose Rating	177
Test 16.5 - Measurement of Sucrose by the Yellow Spring Instrument	179
Test 16.6 - Titrimetric Determination of Salt Content in Food Products	181
Test 16.7 - Determination of Salt-Mohr Titration Method	182
Test 16.8 - DiCromat Salt Analyzer Method	182
Test 16.9 - Sodium Electrode Method	184
Test 16.10 - Determination of Percent Seasoning in Snack Foods	186
Test 17.1 - Determination of Threshold Levels of Panel Members	193
Test 17.2 - Triangular Taste Evaluation	197
Test 17.3 - Numerical Scoring Taste Evaluation	197
Test 17.4 - Flavor Difference Taste Evaluation	199
Test 18.1 - Evaluating Undesirable Odors in Packaging Materials	218
Test 19.1 - Tomato Juice Color Evaluation Using MacBeth-Munsell Disk Colorimeter	260
Test 19.2 - Fresh Tomato Color Measurement with Model E Agtron	262
Test 19.3 - Color Evaluation with Agtron Model F	263
Test 19.4 - Color Measurement with Agtron E5	264
Test 19.5 - Color Evaluation Using Agtron M-400 and M-30	265
Test 19.6 - Color Evaluation Using Hunter Color & Color Difference Meter	266
Test 19.7 - Color Evaluation Using Hunter Lab D25 Color & Color Difference Meter	268
Test 19.8 - Color Evaluation Using Hunter Lab Digital Color Difference Meter, D25D3A	270
Test 19.9 - Evaluation of Product Color, Using G.E. Recording Spectrophotometer	272
Test 19.10 - Determination of Red Anthocyanin Color in Red Beets	273
Test 19.11 - Qualitative Determination of Lycopene Pigment in Tomatoes	274
Test 19.12 - Quantitative Determination of Tomato Pigments	275
Test 19.13 - Qualitative Green Color of Raw and Frozen Asparagus	276
Test 19.14 - Spectrometric Method for Total Chlorophyll and the A and B Components	277
Test 21.1 - Vacuum Oven Method - Moisture	300
Test 21.2 - Toluene Distillation - Moisture	301
Test 21.3 - Infrared Moisture	302
Test 21.4 - Determination of Alcohol Insoluble Solids Content of Peas or Corn	302
Test 21.5 - Flotation in Brine Solution	304
Test 21.6 - Specific Gravity Determination	306
Test 21.7 - Specific Gravity Evaluation of W.K. Corn Weight Method	307
Test 21.8 - Succulence - Moisture Content	309
Test 22.1 - Pericarp Content of Sweet Corn	319
Test 22.2 - Determination of Fiber Content in Snap Beans and Asparagus	320
Test 22.3 - FTC Texture Press and Recorder Using Apple Slices	323
Test 23.1 - Consistence of Food Products with the Bostwick Consistometer	338
Test 23.2 - Consistence of Food Products with the Adams Consistometer	339

QC AND EVALUATION TESTS

Test 23.3 - Viscosity with the Stormer Viscosimeter	340
Test 23.4 - GOSUC Procedure for Determining Consistency	342
Test 24.1 - Procedure for Evaluation of Quality of Canned Cream Style Corn for Defects	346
Test 24.2 - Determining Kernel Count & Percent Defects of Raw Popcorn	350
Test 24.3 - Bruise Detection in Raw Potatoes	350
Test 24.4 - Testing for Freedom from Defects by Visual Examination	351
Test 25.1 - GOSUL Method	361
Test 25.1 - AOAC Method	363
Test 25.3 - Staining Method	365
Test 25.4 - Insect Fragment Determination	366
Test 26.1 - Enzyme Determination	370
Test 27.1 - Determining the Alcoholic Content of Beverages by Use of Ebulliometer	379
Test 28.1 - Testing for Percent of Oil in Foods	387
Test 28.2 - Testing for Free Fatty Acid Content	391
Test 28.3 - Testing for Peroxide Value of Fats and Oils	394
Test 29.1 - pH Measurement of Fruits and Vegetables	404
Test 29.2 - Determination of Acidity - Titration	408
Test 29.3 - Electrometric Method for Highly and Slightly Colored Solutions	410
Test 30.1 - The Microscope	416
Test 30.2 - Histology of the Tomato	423
Test 30.3 - AOAC Mold Count Procedure	425
Test 30.4 - AOAC Rot Fragments in Comminuted Tomato Products	432

TABLES

Table 2.1 - Common Physical Tests of Food Products	14
Table 2.2 - Common Chemical Tests of Food Products	15
Table 2.3 - Organization for Quality Assurance	19
Table 5.1 - Canned Sampling Plans and Acceptance Levels	56
Table 5.2 - Frozen Sampling Plans and Acceptance Levels	57
Table 5.3 - Canned, Frozen Sampling Plan of a Comminuted Fluid	58
Table 5.4 - Dehydrated Sampling Plan and Acceptance Levels	59
Table 6.1 - Definition of Terms Used in Statistical Quality Control	69
Table 6.2 - Factors for Computing Variable Control Limits	75
Table 6.3 - 5% Points for the Distribution of "F"	81
Table 6.4 - 1% Points for the Distribution of "F"	82
Table 6.5 - Allowance Score Points for Drained Wgt. Indices and Min. Wgts. of Tomatoes	83
Table 6.6 - Squares, Square Roots and Reciprocals	84
Table 14.1 - FDA Food Standards by CFR Part No.	133
Table 15.1 - Container Size Conversion - Tin & Glass	141
Table 15.2 - Double Seam Dimensions of Can Sizes	145
Table 15.3 - Physical Properties of Some Films	158
Table 16.1 - Composition of Sucrose Sugar Solutions at 20°C	164

TABLES

Table 16.2 - Baume, Specific Gravity, Percent Salt	165
Table 16.3 - Correction Table for Determining Brix or % Solids	168
Table 16.4 - Brine Table for Vegetables for Canning	170
Table 16.5 - Effect of Chlorine Treatment on Flavor of Canned Foods	171
Table 16.6 - Relationship of Put-in Syrup Versus Cut-out Syrup	173
Table 16.7 - Percent of Salt	180
Table 17.1 - Selected Code Numbers Assigned to Each Treatment in Each Replicate	193
Table 17.2 - Basic Flavor Attributes & Concentration for Training Panel Members	194
Table 17.3 - Deciding Whether Panel Has Ability to Differentiate Between Samples in Traingular Test	200
Table 17.4 - Rule for Deciding Whether Panel Has Ability to Classify Correctly in Triangular Test	201
Table 17.5 - Flavor Difference Summary Sheet, Tomato Juice	204
Table 17.6 - Flavor Difference Multipliers in Computing Judge's Performance	207
Table 17.7 - Flavor Difference Multipliers in Computing Least Significant Differences	210
Table 17.8 - Flavor Difference Minimum Percentage of Acceptable Judgements	211
Table 18.1 - Standards for Odor Analysis	218
Table 18.2 - American Perfumer and Essential Oil Odor Directory	220
Table 19.1 - Munsell Color Percentages for Scoring Tomato Juice	241
Table 19.2 - Mathematical Relationships Between Color Scales	245
Table 19.3 - Effect of Cooking Conditions Upon Vegetable Pigments	258
Table 19.4 - Tomato Color Disk Specifications and Percentages by Grade for Products	261
Table 19.5 - Comparison of Transmittance and PPM Chlorophyll	277
Table 20.1 - Size of Apples	280
Table 20.2 - Size of Apricots	280
Table 20.3 - Size of Asparagus (Raw)	281
Table 20.4 - Size of Asparagus (Canned)	281
Table 20.5 - Size of Bananas	281
Table 20.6 - Size of Lima Beans	282
Table 20.7 - Size of Snap Beans	282
Table 20.8 - Size of Whole Beets	283
Table 20.9 - Size of Oranges	283
Table 20.10 - Size of Grapefruit	284
Table 20.11 - Size of Olives	284
Table 20.12 - Size of Peach & Pear Halves	284
Table 20.13 - Size of Peas	286
Table 20.14 - Size of Pickles	286
Table 20.15 - Tyler Standard Screen Scale Sieves	292
Table 20.16 - Relationship of Tuber Size to Various Physical Constants For Tubers of Different Diameters	293
Table 21.1 - Relative Succulometer, Moisture and Insoluble Solids Values for Corn	300
Table 21.2 - Percent AIS Determination for Peas	303
Table 21.3 - Percent AIS Determination for Corn	304
Table 21.4 - Scoring Canned Peas for Tenderness and Maturity Using Brine Flotation	305
Table 21.5 - Scoring Frozen Peas for Tenderness and Maturity Using Brine Flotation	306
Table 21.6 - Relationship of Specific Gravity, Water Content, Dry Matter & Starch in Potatoes	308
Table 21.7 - Specific Gravity Values for Fresh & Processed Corn by USDA	309
Table 21.8 - Moisture Content in Fresh Fruits & Vegetables	310

TABLES

Table 22.1 - Relationship Between Magness-Taylor Press Test and Degree of Ripeness for Apples	314
Table 22.2 - Range of Firmness, Maturity and Ripeness of Eastern Grown Peaches at Harvest	314
Table 22.3 - Firmness Rating of Processed Pickles Using Magness-Taylor Pressure Tester	315
Table 22.4 - Relationship Between GOSUT Texture Values and Sieve Size for Snap Beans	318
Table 22.5 - Relationship Between Canned Whole Kernel Corn and % Pericarp	320
Table 22.6 - Relationship Between Canned Snap Beans and % Fiber	322
Table 22.7 - Maturity Scores Based on Shear-Press Values for Lima Beans	323
Table 23.1 - Tomato Catsup Consistency Score Points with Bostwick	339
Table 23.2 - Cream-Style Corn Consistency Score Points	340
Table 24.1 - Apricots - Tolerances for Defects	346
Table 24.2 - Green or Wax Beans Maximum Allowances for Defects	347
Table 24.3 - Pears-Maximum Allowance for Defects	348
Table 24.4 - Pickles-Maximum Allowances for Defects	349
Table 24.5 - Potato Chips - Defect Evaluation	353
Table 24.6 - Maximum Defects for Tomatoes	354
Table 25.1 - FDA Defect Action Level for Drosphila Eggs and Larvae	360
Table 25.2 - FDA Defect Action for Aphids of Thrips	360
Table 25.3 - FDA Defect Action Level for Insect Fragments in Vegetables	361
Table 26.1 - Characteristics of Enzymes	372
Table 26.2 - Promotion of Desirable Enzyme Reaction	374
Table 26.3 - Prevention on Undesirable Enzyme Changes	375
Table 26.4 - Available Commercial Enzymes & Their Applications	376
Table 27.1 - Simplified Alcohol Tables	382
Table 28.1 - Relationship of Free Fatty Acid with Smoke Point of Frying Oils	394
Table 29.1 - Common Acids in Foods & Their Weights and Factors	399
Table 29.2 - Relationship of pH Value to Concentration of Acid	400
Table 29.3 - Indicators Showing Both pH Ranges and Color Changes	401
Table 29.4 - Fruits and Vegetables Classed According to Acidity	403
Table 29.5 - pH Values of Some Commercially Canned Foods	406
Table 30.1 - Multiple Sampling Plan for Mold Counts	429
Table 30.2 - FDA Action Level for Mold Count	433
Table 30.3 - FDA Defect Action Level for Rot Fragment Count	435
Table 31.1 - Water Activity of Some Microorganisms	439
Table 31.2 - Water Activity of Some Common Foods	440

FIGURES

Figure 1.1 - Organizational Plan for a Food Firm	3
Figure 2.1 - Quality Assurance Dept.	12
Figure 2.2 - Floor Plan for a Quality Evaluation & Analysis Laboratory	23
Figure 2.3 - Grading Table Arrangements with Lighting	24
Figure 2.4 - Determination of Drained Wgt. Using Scale and Screens	25
Figure 3.1 - Cause and Effects Diagram	35
Figure 3.2 - Example of CEDAC for Potato Chip Manufacture	37

FIGURES

Figure 3.3 - CEDAC Showing Cards	38
Figure 3.4 - CEDAC with Control Chart	38
Figure 3.5 - Possible Factors Affecting Flavor of Potato Chips	39
Figure 5.5 - Operating Curve	
Figure 6.1 - Histogram for Vacuums in Canned Soup	62
Figure 6.2 - Approximate Areas for a Typical Curve	63
Figure 6.3 - Expected Wgt. Distribution of Product in Control and Underweight	64
Figure 6.4 - Frequency Distribution for Vacuum in Canned Soup	65
Figure 6.5 - Ascending Cumulative Frequency Distribution of Vacuums for Canned Soup	67
Figure 6.6 - Descending Cumulative Frequency Distribution of Vacuums for Canned Soup	68
Figure 12.1 - R & D Organization Plan	115
Figure 12.2 - R & D Organization by Discipline	115
Figure 12.3 - R & D Organization by Product	116
Figure 12.4 - R & D Organization by Process	116
Figure 12.5 - Ideal Organization Plan	117
Figure 15.1 - Standard Terms Identifying Basic Parts of a Can	143
Figure 15.2 - Required Measurements for Evaluating Quality of Double Seams	144
Figure 15.3 - Basic Parts of a Glass Container	146
Figure 15.4 - Determination of Net Weight	149
Figure 15.5 - Determination of Headspace	150
Figure 15.6 - Determination of Drained Weight	152
Figure 16.1 - Brix Hydrometer	162
Figure 16.2 - Relationship Between Total Solids and Specific Gravity of Potatoes	163
Figure 16.3 - Abbe Refractometer for Determining Refractive Values	166
Figure 19.1 - Color Temperature of Various Light Sources	236
Figure 19.2 - MacBeth Executive Light	238
Figure 19.3 - MacBeth-Munsell Disc Colorimeter	244
Figure 19.4 - Autocal Agtron M-35 & M-45	251
Figure 19.5 - Three Dimensional Coordinate System for Colors	254
Figure 19.6 - USDA - Hunter Tomato Colorimeter - D6	255
Figure 19.7 - Hunterlab Digital Color Difference Meter	257
Figure 19.8 - Hunter-Munsell Chromaticity Diagram	269
Figure 20.1 - Curved Pickle	288
Figure 20.2 - Crooked Pickle	288
Figure 20.3 - Mishapen Pickle	288
Figure 20.4 - Crinkle Slice	289
Figure 20.5 - V-Cut Slice	289
Figure 20.6 - Oval Shred	290
Figure 20.7 - V-Cut Shred	290
Figure 21.1 - Infrared Moisture Determination	297
Figure 21.3 - Corn Succulometer	298
Figure 21.4 - Alcohol Insoluble Solids	299
Figure 22.1 - Magness-Taylor Pressure Tester	315
Figure 22.2 - Christel Texturometer	316
Figure 22.3 - GOSUT Texturometer	317
Figure 22.4 - FTC Model TMS-90 Texture Measurement System	319
Figure 22.5 - Determination of Percent Fiber	321
Figure 23.1 - GOSUC Consistometer	329
Figure 23.2 - Bostwick Consistometer	332

FIGURES

Figure 23.3 - Adams Consistometer	334
Figure 25.1 - Life Cycle of *Drosophila Melanogaster*	358
Figure 25.2 - Wildman Trap Flask	367
Figure 26.1 - Apparatus for Catalase Determination	377
Figure 28.1 - Standard Curve	391
Figure 28.2 - Free Fatty Acid Testing Apparatus	392
Figure 29.1 - pH and Total Acid Evaluation	405
Figure 30.1 - Compound Microscope with Howard Mold Counting Chamber	413
Figure 30.2 - Filiments Found in Tomato Products	419
Figure 30.3 - Cross Section of Tomato Fruit	425
Figure 30.4 - Howard Mold Counting Chamber	426
Figure 30.5 - Mold Count Record	431

QC DATA FORMS

Form 2.1 - Examination of Processed Foods	30
Form 6.1 - SQC Frequency Distribution for Measured Variables	66
Form 6.2 - SQC Record and Data Form	72
Form 6.3 - SQC Statistical Analysis	76
Form 6.4 - SQC Testing the Significance Between Two Sample Means	77
Form 6.5 - SQC Two-Way Analysis of Variance	78
Form 6.6 - SQC Two-Way Analysis of Variance for Yield of Green Beans	79
Form 15.1 - Daily Double Seam Report	155
Form 17.1 - Basic Flavor Attribute Quality Control Form	196
Form 17.2 - Flavor Difference Evaluation Judge's Score Card	198
Form 17.4 - Flavor-Numerical Scoring Quality Control Form	202
Form 17.5 - Score Sheet Used by Panel Members for Evaluating Flavor	205
Form 24.1 - Work Sheet for Evaluating Quality of Canned Whole Kernel Corn	355